Recent Advances in Geological Oceanography II

Recent Advances in Geological Oceanography II

Editors

George Kontakiotis
Assimina Antonarakou
Dmitry A. Ruban

MDPI

Basel • Beijing • Wuhan • Barcelona • Belgrade • Novi Sad • Cluj • Manchester

Editors

George Kontakiotis
National and Kapodistrian University of Athens
Athens
Greece

Assimina Antonarakou
National and Kapodistrian University of Athens
Athens
Greece

Dmitry A. Ruban
Southern Federal University
Rostov-on-Don
Russia

Editorial Office
MDPI AG
Grosspeteranlage 5
4052 Basel, Switzerland

This is a reprint of articles from the Special Issue published online in the open access journal *Journal of Marine Science and Engineering* (ISSN 2077-1312) (available at: https://www.mdpi.com/journal/jmse/special_issues/16D321F9EU).

For citation purposes, cite each article independently as indicated on the article page online and as indicated below:

Lastname, A.A.; Lastname, B.B. Article Title. *Journal Name* **Year**, *Volume Number*, Page Range.

ISBN 978-3-7258-2557-8 (Hbk)
ISBN 978-3-7258-2558-5 (PDF)
doi.org/10.3390/books978-3-7258-2558-5

Contents

About the Editors

George Kontakiotis

George Kontakiotis achieved his Ph.D. in Paleoceanography in 2012 at the University of Athens, Greece, where he later worked as Laboratory and Teaching Staff in the fields of Marine Geology and Sedimentology and recently as an Associate Professor in Sedimentology–Paleoceanography–Chemostratigraphy. His major research contributions include developing novel approaches to the distribution and pathways of diagenesis in Mg/Ca paleothermometry. He has further worked on sedimentological and paleoceanographic reconstructions at different time scales by means of marine cores and land sections. His main research topics are summarized as follows: environmental sedimentology; marine petroleum systems; exploitation of natural energy resources; integrated bio-cyclo-tephro-stratigraphy; carbonate reservoirs; calibration–validation–application of geochemical proxies for sea surface temperature (SST) and salinity (SSS); applied environmental micropaleontology as a bio-monitoring tool; sea-level variations; ocean/climate changes. He has published more than 100 peer-reviewed articles in international journals.

Assimina Antonarakou

Assimina Antonarakou is a Professor of Marine Geology–Micropaleontology–Didactics on Geosciences and Head of the Faculty of Geology and Geoenvironment at the University of Athens, Greece. Her PhD thesis dealt with Miocene cyclic sedimentary successions of the eastern Mediterranean in terms of orbital periodicities and paleoclimatic variations based on planktonic foraminiferal assemblages. Her main research topics are summarized as follows: planktonic foraminiferal eco-biostratigraphy, geobiology, and paleoceanography; astronomical frequencies in paleoclimates; extreme geological events; marine environmental monitoring; ocean dynamics and sea-level changes; natural and human environmental stressors; foraminiferal trace metals and stable isotopes. She has participated in several national and international projects focused on multiproxy ecosystem responses to past/present environmental events, and she is the co-author of more than 90 peer-reviewed publications in international journals.

Dmitry A. Ruban

Dmitry A. Ruban is an Associate Professor at the Southern Federal University, Russia. He received his C.Sci. (Geology and Mineralogy) degree from the Rostov State University (Russia) in 2004 and his Ph.D. (Geology) from the University of Pretoria (South Africa) in 2009, followed by an M.Sci. (State and Municipal Governance) degree from the Southern Federal University in 2021. His research interests are linked to Earth and environmental sciences (stratigraphy and sedimentology, palaeobiology, tectonics, physical geography and geomorphology, and environmental issues), tourism (also geotourism), and economics and management. In particular, some of his research projects have focused on Phanerozoic mass extinction, megaclast nomenclature, and geoheritage assessment. He has published more than 200 peer-reviewed articles in international journals, and he is a member of the editorial board of several international and national journals.

Preface

In the current Special Issue, we present a collection of articles that delve into the marine environment, including depositional and diagenetic controls, geochemistry, and the economic potentiality of marine deposits worldwide. The published papers included in this Special Issue also fill in some of the knowledge gaps on geological oceanography, such as the initial oceanization to seafloor hydrothermal activity and from very ancient tsunamis to plastic cycling in modern seashore environments. As we reflect on the wealth of insights presented in this Special Issue, it is evident that the study of the marine environment remains a dynamic and multifaceted field ripe for future exploration. By continuing to unravel the complexities of such environments and addressing key knowledge gaps, we can better understand past environmental changes, decipher the geological evolution, and inform future research endeavors.

George Kontakiotis, Assimina Antonarakou, and Dmitry A. Ruban
Editors

Journal of *Marine Science and Engineering*

MDPI

Editorial

Recent Advances in Geological Oceanography II

George Kontakiotis [1,*], Assimina Antonarakou [1] and Dmitry A. Ruban [2]

1 Department of Historical Geology and Palaeontology, Faculty of Geology and Geo-Environment, National and Kapodistrian University of Athens, Panepistimiopolis, 15784 Athens, Greece; aantonar@geol.uoa.gr
2 Department of Organization and Technologies of Service Activities, Institute of Tourism, Service and Creative Industries, Southern Federal University, 23-ja Linija Street 43, Rostov-on-Don 344019, Russia; ruban-d@mail.ru
* Correspondence: gkontak@geol.uoa.gr

1. Introduction

Marine geology is a well-known [1,2] and still-developing field of research that deals with geological bodies and processes below the sea level and sometimes refers to ancient marine environments. Moreover, it is evident that the geological peculiarities of seas and oceans can influence their physical and biological states and overall dynamics [3]. Such influences are partly moderated by the geological-scale activities of humans. It is very reasonable to talk about geological oceanography as a more general discipline focusing on both elementary and highly complex phenomena, including the interactions of geological and non-geological objects and processes.

Significant achievements have already been made in the field of geological oceanography in the 2020s. Huang et al. [4] examined the diversity of sedimentation rates across the World Ocean in the Quaternary and registered several intriguing patterns. Martin Erin et al. [5] reconsidered ocean evolution via a general tectonic concept known as Wilson's cycle. Pohl et al. [6] explained how major plate tectonic processes affected oxygen concentrations in seawater during the Phanerozoic. According to Klausen et al. [7], it was sea-level changes that drove the early evolution of the dinosaurs. Gao et al. [8] established that the construction of artificial reefs is related to the deep involvement of plastics in marine sedimentation. These examples demonstrate the outstanding width of the agenda of geological oceanography, as well as the high complexity of the problems it seeks to solve.

Despite numerous achievements, much is yet to be understood about the geological context of our seas and oceans. For instance, it is very possible that the diversity of models describing the birth and tectonic evolution of our oceans under different conditions is incompletely understood [9–11]. Peculiarities of depositional and diagenetic processes in ancient seas can be documented only via multiple studies in starkly different regions and geological time slices [12,13]. Our knowledge of geophysical processes, including those related to submarine seismicity and hydrothermal growth, must be further enhanced [14–16]. Such phenomena as pre-Quaternary tsunamis [17,18] and ocean palaeoproductivity changes [19–22] are known, but our knowledge is still fragmentary. Geochemical changes on the shores and at bottom of the seas and oceans related to increasing anthropogenic pressure and the input of new, non-natural materials (such as plastics) require monitoring, and the our information on them requires regular updating [23,24]. Finally, the geological resources of on our ocean floors remain a subject demanding further exploration [25–28]. These noted gaps in our knowledge underpin the urgency of research in the field of geological oceanography, as well as the breadth of the relevant topics and directions.

The present Special Issue, entitled "Recent Advances in Geological Oceanography II", aims to contribute to the international growth of the field of geological oceanography. It contains a collection of research and review articles by specialists from different countries and regions, address a number of the principal problems remaining in this field. In particular, they contribute to filling in some of the knowledge gaps mentioned above, from

Citation: Kontakiotis, G.; Antonarakou, A.; Ruban, D.A. Recent Advances in Geological Oceanography II. *J. Mar. Sci. Eng.* **2024**, *12*, 1887. https://doi.org/10.3390/jmse12101887

Received: 19 September 2024
Accepted: 17 October 2024
Published: 21 October 2024

the initial oceanization to seafloor hydrothermal activity and from very ancient tsunamis to plastic cycling in modern seashore environments.

2. Published Papers

Liu et al. (Contribution 1) investigated and addressed the increasing complexity of the depositional conditions and their effects on organic matter accumulation and preservation derived from a series of Lower Paleozoic shale samples at the northeastern margin of the Yangtze platform in south China. The analyses incorporate TOC content, mineral composition, and major, trace, and rare earth elements into the investigation. This enables the analysis of how different factors such as paleoredox state, paleoproductivity, preservation, paleoclimate, and terrestrial influx intensity impact organic matter enrichment in this setting.

A comprehensive review by Mikhailenko and Ruban (Contribution 2) provides valuable insights into the global geographical distribution of plastics and five associated heavy metals (cadmium, chromium, mercury, nickel, and zinc) on sea beaches, highlighting both real and potential risks to the environment. The results indicate that the geographical extent of our knowledge of Hg-bearing plastics is highly limited. Overall, the findings of this literary study widen the focus on Anthropocene marine geochemistry.

Sang et al. (Contribution 3) present an analytical model for the punctiform breakup and initial oceanization of the central Red Sea rift. The integrated model incorporates new reflection seismic profiles and gravity modeling results and focuses on the density structure, tectonic evolution, breakup mechanism, and future evolution of this continuously spreading setting.

Yutsis et al. (Contribution 4) provide new insights into the seamount structure of the northern part of the Ninetyeast Ridge in the Indian Ocean. The investigative geophysical methods used in this study include bathymetric, seismostratigraphic, and magnetic data. By integrating these aspects, the study provides a comprehensive approach, enabling comparisons with similar settings in a global context.

Wang et al. (Contribution 5) represent a numerical simulation-based analysis of the seafloor hydrothermal plumes in Carlsberg Ridge in the northwestern Indian Ocean. The proposed model is based on the topography of the region and long-term current monitoring data and allows for the reconstruction of the present hydrothermal plume in terms of its structure, velocity field, and temperature field. The findings of this study provide useful information for tracing the hydrothermal vents, prospecting the submarine polymetallic sulfide resources, and designing long-term observation networks, and provide a foundation for future studies on element cycling and the energy budget.

Qamar et al. (Contribution 6) offer an in-depth exploration of the sedimentology, diagenesis, and sequence stratigraphy of the middle Jurassic carbonate deposits of north Pakistan, employing a multi-proxy approach which involves field observations, paleontological analysis, and sedimentological microfacies characterization. This study provides valuable insights into the complex diagenetic history and sequence stratigraphy of the Middle Jurassic Samana Suk Formation in the Hazara basin, shedding light on the depositional stages and highlighting the reservoir characteristics of these shallow-to-marginal marine carbonates for possible hydrocarbon exploration in the future.

Gao et al. (Contribution 7) examine the provenance of the Lower Jurassic Badaowan and Sangonghe Formations in the Junggar Basin and the constraint it poses on the Karamaili Ocean. The interpretation of the sandstone-derived petrological and geochronological results allow for a better understanding of the evolution of the Paleo-Karamaili orogenic belt as part of the Paleo-Asian Ocean, supported by three distinct evolutionary stages.

Through a systematic review, Ruban and Yashalova (Contribution 8) present a synthesis of the available information related to Ordovician tsunamis. In particular, the authors describe a summary of potential hypotheses based on 24 events in different localities around the world, with further implications for geoheritage resources. In this regard, the outcomes of this study contribute to a better awareness of the world's geoheritage resources during the Phanerozoic period, and they also, through comparisons with similar events throughout the Cenozoic era, form a basis for further study.

Shi et al. (Contribution 9) represent a new biomonitoring approach with which to determine the biotoxic effects of Ag nanoparticles on coasts. The experiment was conducted on *Skeletonema costatum*, the most typical phytoplankton species in coastal settings. The results of this environmental study shed light on the biological toxicity of nanometals and their possible toxicity mechanism.

The study of Krylov et al. (Contribution 10) is devoted to the features of seismological observations in the Arctic seas, which are complicated by harsh climatic conditions, the presence of ice cover, stamukhi and icebergs, and limited navigation. This study could be considered a reference study, referring, in great detail, to the features and difficulties of seismological operation systems in such "noisy" environments.

Vakalas and Zananiri (Contribution 11) present a sedimentological study of the North Aegean Sea (Greece) based on their investigation of 323 surficial marine sediments. The sediment transport patterns were analyzed based on the grain size parameters (mean, sorting, and skewness). The dominant factors affecting sediment transport are river discharge and longshore drift near the shoreline, while open sea water circulation controls sediment distribution patterns at the open shelf. Moreover, the strong heterogeneity of the sediment textural parameters across the study area suggests that seafloor sediments are further reworked in areas where water masses are highly energetic.

3. Perspectives

The articles in this Special Issue fill many particular gaps in our knowledge of the geological context of seas and oceans. Taken together, they amply demonstrate that the field of geological oceanography is wide, complex, and highly diverse. This is an arena ripe for inter- and multidisciplinary research where "pure" geologists and oceanographers can cooperate with geophysicists, geochemists, environmental scientists, marine geologists, sedimentologists, and geomorphologists. In this field, classical and advanced research projects can co-exist (and benefit one another), and field-based, modeling, and conceptual studies are also possible. Indeed, one single Special Issue cannot (and should not) represent the entire research field (especially one so vast as geological oceanography), but it can try to demonstrate its central ideas and potential.

In solving some important questions, the contributors to the present Special Issue raise new ones, including (but not limited to) the following:

(1) Did the Early Paleozoic organic matter enrichment in the Yangtze Platform reflect the actions of any planetary-scale mechanisms?
(2) What is the true diversity of seamount structures in the World Ocean?
(3) Do the Middle Jurassic marine carbonates of North Pakistan record any global-scale events?
(4) Were there differences and irregularities in the preservation of tsunami records in the time slices of the Phanerozoic?
(5) Is it possible to develop a single protocol for seismological investigations at high latitudes?

These are specific questions for specialized projects. More generally, this Special Issue highlights two important perspectives for future research. The first perspective is linked to identifying and subsequently examining those features that mark the diversity of the geological environments of our seas and oceans. The second perspective refers to the geological (and geological-scale) activities of humans in marine environments that alter, disrupt, or modify natural geological cycles and trends.

Author Contributions: Conceptualization, G.K.; investigation, G.K., D.A.R. and A.A.; writing—original draft preparation, G.K. and D.A.R.; writing—review and editing, G.K., D.A.R. and A.A. All authors have read and agreed to the published version of the manuscript.

Conflicts of Interest: The authors declare no conflicts of interest.

List of Contributions

1. Liu, P.; Liu, C.; Guo, R. Depositional Environment and Organic Matter Enrichment in the Lower Paleozoic Shale from the Northeastern Margin of the Yangtze Platform, South China. *J. Mar. Sci. Eng.* **2023**, *11*, 501. https://doi.org/10.3390/jmse11030501
2. Mikhailenko, A.V.; Ruban, D.A. Plastics and Five Heavy Metals from Sea Beaches: A Geographical Synthesis of the Literary Information. *J. Mar. Sci. Eng.* **2023**, *11*, 626. https://doi.org/10.3390/jmse11030626
3. Sang, Y.-D.; Adam, B.M.T.; Li, C.-F.; Huang, L.; Wen, Y.-L.; Zhang, J.-L.; Liu, Y.-T. Punctiform Breakup and Initial Oceanization in the Central Red Sea Rift. *J. Mar. Sci. Eng.* **2023**, *11*, 808. https://doi.org/10.3390/jmse11040808
4. Yutsis, V.; Levchenko, O.; Ivanenko, A.; Veklich, I.; Turko, N.; Marinova, Y. New Insights into the Seamount Structure of the Northern Part of the Ninetyeast Ridge (Indian Ocean) through the Integrated Analysis of Geophysical Data. *J. Mar. Sci. Eng.* **2023**, *11*, 924. https://doi.org/10.3390/jmse11050924
5. Wang, K.; Han, X.; Wang, Y.; Cai, Y.; Qiu, Z.; Zheng, X. Numerical Simulation-Based Analysis of Seafloor Hydrothermal Plumes: A Case Study of the Wocan-1 Hydrothermal Field, Carlsberg Ridge, Northwest Indian Ocean. *J. Mar. Sci. Eng.* **2023**, *11*, 1070. https://doi.org/10.3390/jmse11051070
6. Qamar, S.; Shah, M.M.; Janjuhah, H.T.; Kontakiotis, G.; Shahzad, A.; Besiou, E. Sedimentological, Diagenetic, and Sequence Stratigraphic Controls on the Shallow to Marginal Marine Carbonates of the Middle Jurassic Samana Suk Formation, North Pakistan. *J. Mar. Sci. Eng.* **2023**, *11*, 1230. https://doi.org/10.3390/jmse11061230
7. Gao, Y.; Zhang, G.; Li, S.; Guo, R.; Zeng, Z.; Cheng, S.; Xue, Z.; Li, L.; Zhou, H.; Liu, S.; et al. Provenance of the Lower Jurassic Badaowan and Sangonghe Formations in Dongdaohaizi Depression, Junggar Basin, and Its Constraint on the Karamaili Ocean. *J. Mar. Sci. Eng.* **2023**, *11*, 1375. https://doi.org/10.3390/jmse11071375
8. Ruban, D.A.; Yashalova, N.N. Ordovician Tsunamis: Summary of Hypotheses and Implications for Geoheritage Resources. *J. Mar. Sci. Eng.* **2023**, *11*, 1764. https://doi.org/10.3390/jmse11091764
9. Shi, K.; Yao, Y.; Xue, J.; Cheng, D.; Wang, B. The Biotoxic Effects of Ag Nanoparticles (AgNPs) on Skeletonema costatum, a Typical Bloom Alga Species in Coastal Areas. *J. Mar. Sci. Eng.* **2023**, *11*, 1941. https://doi.org/10.3390/jmse11101941
10. Krylov, A.A.; Novikov, M.A.; Kovachev, S.A.; Roginskiy, K.A.; Ilinsky, D.A.; Ganzha, O.Y.; Ivanov, V.N.; Timashkevich, G.K.; Samylina, O.S.; Lobkovsky, L.I.; et al. Features of Seismological Observations in the Arctic Seas. *J. Mar. Sci. Eng.* **2023**, *11*, 2221. https://doi.org/10.3390/jmse11122221
11. Vakalas, I.; Zananiri, I. Net Transport Patterns of Surficial Marine Sediments in the North Aegean Sea, Greece. *J. Mar. Sci. Eng.* **2024**, *12*, 512. https://doi.org/10.3390/jmse12030512

References

1. Seibold, E.; Berger, W.H. *The Sea Floor: An Introduction to Marine Geology*, 3rd ed.; Springer: Berlin/Heidelberg, Germany, 2017.
2. Wu, X.; Zhang, T.; Song, S.; Wang, J.; Niu, Y. Development trend of international marine geology research. *Geol. Bull. China* **2021**, *40*, 233–242.
3. Agiadi, K.; Hohmann, N.; Gliozzi, E.; Thivaiou, D.; Bosellini, F.R.; Taviani, M.; Bianucci, G.; Collareta, A.; Londeix, L.; Faranda, C.; et al. The marine biodiversity impact of the Late Miocene Mediterranean salinity crisis. *Science* **2024**, *385*, 986–991. [CrossRef] [PubMed]
4. Huang, T.; Ma, C.; Jin, S.; Yang, Y.; Hu, X.; Hou, M. Quaternary sedimentation rate revealed by semi-quantitative analysis in global ocean. *Mar. Pet. Geol.* **2024**, *166*, 106900. [CrossRef]
5. Martin Erin, L.; Cawood Peter, A.; Murphy, J.B.; Nance, R.D.; Heron Phillip, J. The tectonics of introversion and extroversion: Redefining interior and exterior oceans in the supercontinent cycle. *Geol. Soc. Lond. Spec. Publ.* **2024**, *542*, 15–29. [CrossRef]
6. Pohl, A.; Ridgwell, A.; Stockey, R.G.; Thomazo, C.; Keane, A.; Vennin, E.; Scotese, C.R. Continental configuration controls ocean oxygenation during the Phanerozoic. *Nature* **2022**, *608*, 523–527. [CrossRef]
7. Klausen, T.G.; Paterson, N.W.; Benton, M.J. Geological control on dinosaurs' rise to dominance: Late Triassic ecosystem stress by relative sea level change. *Terra Nova* **2020**, *32*, 434–441. [CrossRef]
8. Gao, C.; Liang, B.; Zhang, S. Accumulation characteristics and ecological risk evaluation of microplastics in sediment cores from the artificial reef area and surrounding seas of Haizhou Bay, north China. *Sci. Total Environ.* **2024**, *925*, 171789. [CrossRef]
9. Zhang, B.-C.; Fan, J.-J.; Luo, A.-B.; Zeng, X.-W.; Duan, M.-L.; Sun, S.-L. Tectonic evolution of the Meso-Tethys Ocean in the Jurassic: Birth, growth, and demise of a missing intra-oceanic arc. *Gondwana Res.* **2023**, *122*, 41–59. [CrossRef]
10. Capponi, G.; Festa, A.; Rebay, G. Birth and death of oceanic basins: Geodynamic processes from rifting to continental collision in Mediterranean and circum-Mediterranean orogens: GEOPROB Project. *Geol. Mag.* **2018**, *155*, 229–232. [CrossRef]

11. Seton, M.; Müller, R.D.; Zahirovic, S.; Gaina, C.; Torsvik, T.; Shephard, G.; Talsma, A.; Gurnis, M.; Turner, M.; Maus, S.; et al. Global continental and ocean basin reconstructions since 200 Ma. *Earth-Sci. Rev.* **2012**, *113*, 212–270. [CrossRef]
12. Bilal, A.; Yang, R.; Mughal, M.S.; Janjuhah, H.T.; Zaheer, M.; Kontakiotis, G. Sedimentology and Diagenesis of the Early–Middle Eocene Carbonate Deposits of the Ceno-Tethys Ocean. *J. Mar. Sci. Eng.* **2022**, *10*, 1794. [CrossRef]
13. Bilal, A.; Yang, R.; Janjuhah, H.T.; Mughal, M.S.; Li, Y.; Kontakiotis, G.; Lenhardt, N. Microfacies analysis of the Palaeocene Lockhart limestone on the eastern margin of the Upper Indus Basin (Pakistan): Implications for the depositional environment and reservoir characteristics. *Depos. Rec.* **2023**, *9*, 152–173. [CrossRef]
14. Varnavas, S.P.; Papavasiliou, C. Submarine hydrothermal mineralization processes and insular mineralization in the Hellenic Volcanic Arc system: A review. *Ore Geol. Rev.* **2020**, *124*, 103541. [CrossRef]
15. Chen, J.; Tong, S.; Han, T.; Song, H.; Pinheiro, L.; Xu, H.; Azevedo, L.; Duan, M.; Liu, B. Modelling and detection of submarine bubble plumes using seismic oceanography. *J. Mar. Syst.* **2020**, *209*, 103375. [CrossRef]
16. Hudson, T.S.; Kendall, J.M.; Blundy, J.D.; Pritchard, M.E.; MacQueen, P.; Wei, S.S.; Gottsmann, J.H.; Lapins, S. Hydrothermal Fluids and Where to Find Them: Using Seismic Attenuation and Anisotropy to Map Fluids Beneath Uturuncu Volcano, Bolivia. *Geophys. Res. Lett.* **2023**, *50*, e2022GL100974. [CrossRef]
17. Dawson, A.G.; Stewart, I. Tsunami deposits in the geological record. *Sediment. Geol.* **2007**, *200*, 166–183. [CrossRef]
18. Ramírez-Herrera, M.T.; Coca, O. A global database of tsunami deposits. *Geosci. Data J.* **2024**, *11*, 974–994. [CrossRef]
19. Liu, S.; Zhang, H.; Cao, P.; Liu, M.; Ye, W.; Chen, M.-T.; Li, J.; Pan, H.-J.; Khokiattiwong, S.; Kornkanitnan, N.; et al. Paleoproductivity evolution in the northeastern Indian Ocean since the last glacial maximum: Evidence from biogenic silica variations. *Deep Sea Res. Part I Oceanogr. Res. Pap.* **2021**, *175*, 103591. [CrossRef]
20. Paytan, A. Ocean Paleoproductivity. In *Encyclopedia of Paleoclimatology and Ancient Environments*; Gornitz, V., Ed.; Springer: Dordrecht, The Netherlands, 2009; pp. 644–651.
21. Lyu, J.; Auer, G.; Bialik, O.M.; Christensen, B.; Yamaoka, R.; De Vleeschouwer, D. Astronomically-Paced Changes in Paleoproductivity, Winnowing, and Mineral Flux Over Broken Ridge (Indian Ocean) Since the Early Miocene. *Paleoceanogr. Paleoclimatol.* **2023**, *38*, e2023PA004761. [CrossRef]
22. Giamali, C.; Kontakiotis, G.; Antonarakou, A.; Koskeridou, E. Ecological Constraints of Plankton Bio-Indicators for Water Column Stratification and Productivity: A Case Study of the Holocene North Aegean Sedimentary Record. *J. Mar. Sci. Eng.* **2021**, *9*, 1249. [CrossRef]
23. Coyle, R.; Hardiman, G.; Driscoll, K.O. Microplastics in the marine environment: A review of their sources, distribution processes, uptake and exchange in ecosystems. *Case Stud. Chem. Environ. Eng.* **2020**, *2*, 100010. [CrossRef]
24. Jambeck, J.R.; Geyer, R.; Wilcox, C.; Siegler, T.R.; Perryman, M.; Andrady, A.; Narayan, R.; Law, K.L. Plastic waste inputs from land into the ocean. *Science* **2015**, *347*, 768–771. [CrossRef] [PubMed]
25. Seijmonsbergen, A.C.; Valentijn, S.; Westerhof, L.; Rijsdijk, K.F. Exploring Ocean Floor Geodiversity in Relation to Mineral Resources in the Southwest Pacific Ocean. *Resources* **2022**, *11*, 60. [CrossRef]
26. Guo, X.; Fan, N.; Liu, Y.; Liu, X.; Wang, Z.; Xie, X.; Jia, Y. Deep seabed mining: Frontiers in engineering geology and environment. *Int. J. Coal Sci. Technol.* **2023**, *10*, 23. [CrossRef]
27. Seibold, E.; Berger, W.H. Resources from the Ocean Floor. In *The Sea Floor: An Introduction to Marine Geology*; Seibold, E., Berger, W.H., Eds.; Springer: Berlin/Heidelberg, Germany, 1996; pp. 277–302.
28. Wen, Z.; Wang, J.; Wang, Z.; He, Z.; Song, C.; Liu, X.; Zhang, N.; Ji, T. Analysis of the world deepwater oil and gas exploration situation. *Pet. Explor. Dev.* **2023**, *50*, 1060–1076. [CrossRef]

Journal of
Marine Science and Engineering

MDPI

Article

Depositional Environment and Organic Matter Enrichment in the Lower Paleozoic Shale from the Northeastern Margin of the Yangtze Platform, South China

Peng Liu [1], Changjie Liu [2,*] and Ruiliang Guo [3,*]

1 College of Safety Science and Engineering, Xi'an University of Science and Technology, Xi'an 710054, China
2 Northwest Institute of Eco-Environment and Resources, Chinese Academy of Sciences, Key Laboratory of Petroleum Resources, Lanzhou 730000, China
3 School of Earth Sciences and Engineering, Xi'an Shiyou University, Xi'an 710065, China
* Correspondence: liuchangjie@nieer.ac.cn (C.L.); 210106@xsyu.edu.cn (R.G.)

Abstract: In this study, twenty-six core shale samples were collected from the marine Lower Paleozoic shale in a well in the northeastern margin of the Yangtze Platform. Analyses of TOC content, mineral composition, major elements, along with trace and rare earth elements were conducted on the samples. The results were used to investigate the depositional conditions and their effects on organic matter accumulation and preservation. Generally, the sedimentation period of Niutitang Formation shale was in a cold and arid climate with anoxic marine environments, while the shale from Wufeng-Longmaxi Formation was formed in a warm and humid climate with oxic marine environments. In addition, the Wufeng-Longmaxi and Niutitang formations are characterized by low paleo-productivity. The organic matter enrichment for shale in this study could be simultaneously controlled by paleo-redox state and paleo-productivity. Organic matter enrichment of the Niutitang shale is mainly driven by preservation rather than productivity, while the dominant driving factor is the opposite for the Wufeng-Longmaxi shale. Additionally, palaeoclimate and terrestrial influx intensity were found to significantly impact the organic matter enrichment in the Wufeng-Longmaxi shale. The findings have implications for the understanding of the sedimentary processes, organic matter enrichment and preservation and shale gas potential of the study area.

Keywords: Yangtze Platform; Lower Paleozoic; black shale; depositional condition; organic matter

Citation: Liu, P.; Liu, C.; Guo, R. Depositional Environment and Organic Matter Enrichment in the Lower Paleozoic Shale from the Northeastern Margin of the Yangtze Platform, South China. *J. Mar. Sci. Eng.* **2023**, *11*, 501. https://doi.org/10.3390/jmse11030501

Academic Editors: George Kontakiotis, Assimina Antonarakou and Dmitry A. Ruban

Received: 22 January 2023
Revised: 17 February 2023
Accepted: 20 February 2023
Published: 25 February 2023

1. Introduction

Shale gas is an important unconventional natural gas resource that has gained worldwide attention, and the marine shale gas in the Sichuan Basin of China has been the subject of significant breakthroughs in recent years [1–3]. Additionally, the widely developed and distributed Lower Paleozoic black shale near the Micangshan-Hannan Uplift in the northeastern margin of the Yangtze Platform, which is adjacent to the north Sichuan Basin, is considered to have shale gas potential due to its high TOC content, thickness, and maturity level [4]. However, studies of the Lower Paleozoic shale in the northeastern margin of the Yangtze Platform are still limited compared to that in the Sichuan Basin.

Unlike natural gas from conventional (e.g., sandstone) reservoirs, shale gas is generated and preserved within the source rock [5]. The organic matter in shale can contribute to the generation of hydrocarbons via thermal decomposition processes as the precursors of petroleum and natural gas [6]. Studies have shown that the adsorption capacity of marine shale is positively correlated with its total organic carbon (TOC) content due to the organic pores in the shale matrix [7–9]. As a result, the organic matter in marine shale could have a considerable impact on hydrocarbon generation as well as the occurrence of gas hydrocarbons in shale. Depositional conditions, such as paleo-productivity, preservation conditions (e.g., redox conditions, paleo-water depth), terrestrial influx, and paleo-salinity, can all impact the accumulation and preservation of organic matter in sediments [10–13]. Depositional conditions can undoubtedly affect the accumulation and dispersion of elements in sediments [14]. Hence, inorganic geochemical proxies, including specific element abundances and ratios, have been applied to infer paleo depositional conditions for sediments and indicate the enrichment and preservation of organic matter [15–17]. However, specific paleo-environmental geochemical proxies show weather-related alteration in outcrop samples [18], suggesting that drilling core samples could better record the paleo-environmental characteristics. In a previous study, Liu et al., (2021) studied the organic matter accumulation of the Niutitang Formation in the SND-1 Well in the northeastern margin of the Yangtze Platform [19]. However, the depositional environment and organic matter enrichment of the shale in the Lower Paleozoic strata of the northeastern margin of the Yangtze Platform, mainly including the Wufeng-Longmaxi shale and Niutitang shale, have yet to be extensively studied through core samples.

Hence, core shale samples were collected from a well in the northeastern margin of the Yangtze Platform in this study, consisting of Wufeng-Longmaxi shale samples and Niutitang shale samples. The TOC content, mineral composition, major elements, trace elements and rare earth elements analyses were conducted to investigate the depositional conditions and their impacts on the accumulation and preservation of organic matter. The results of this study contribute to an improved understanding of the depositional conditions and their effects on the enrichment of organic matter in the Lower Paleozoic shale in the northwestern margin of the Yangtze Platform.

2. Geological Setting

The Micangshan-Hannan Uplift is located in the northwestern margin of the Yangtze Platform and acted as the coupling unit of the South Qinling Dabie orogenic belt and Sichuan Basin [20–22]. The Micangshan-Hannan Uplift is adjacent to the Mianlue suture zone to the north, the Sichuan Basin to the south, the Bikou block to the northwest, the Longmenshan Tectnoic Zone to the southeast, the Dabashan Folding Belt to the east (Figure 1) [23].

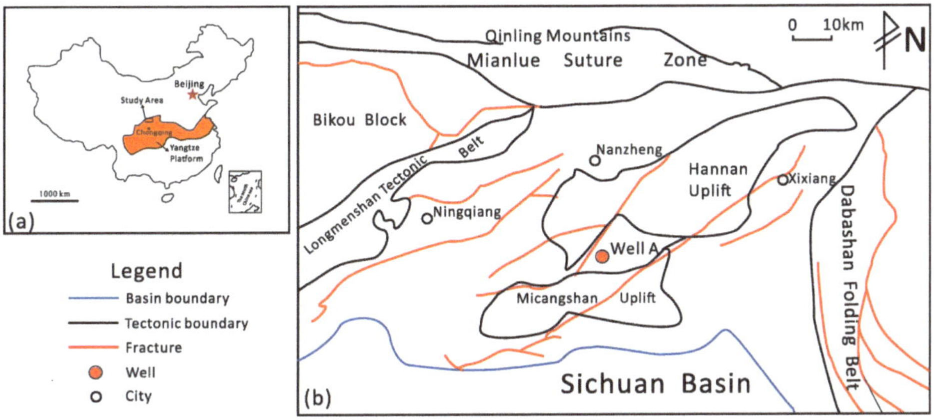

Figure 1. Location of the study area in the Yangtze Platform (**a**) and the geological map of the study area in the northern part of the Yangtze Platform and well location (**b**). The geological map of the study area is modified from Chen et al., (2019) [21].

The Yangtze Platform was developed on the Precambrian basement [24]. During the Jinning movement, the Micangshan area evolved from a continental margin basin to a back-arc basin. The ancient Qinling Ocean was formed in the early Jinning period and the ancient Qinling Oceanic block was subducted under the Yangtze Platform, forming the arc-basin system [25]. In the Early Sinian, the west and north margin of the Yangtze

Platform was a stretched state and eventually formed the Longmenshan-Micangshan continental rift belt [25]. The Micangshan-Hannan Uplift was the result of continuous collision and subduction of the Yangtze Plate and the North China Plate, and it was uplifted from the Late Triassic [26]. The Micangshan area formed the present geological structural characteristics in the Late Himalayan period. With the increasing sea level in the Early Cambrian, the Niutitang shale was widely distributed in the Sichuan Basin and its adjacent areas, including Micangshan area [27,28]. The shale is of relatively high maturity level and high TOC contents [29]. The Silurian Wufeng-Longmaxi formations were also widely deposited and distributed in the Sichuan basin and its adjacent areas [30]. However, a regional uplift occurred in the north margin area of the Yangtze Platform and the termination period of the uplift varied in different areas, which led to different depositional characteristics of the Ordovician-Silurian strata in Micangshan area [31].

3. Samples and Analyses

3.1. Samples

The stratigraphic sequences of the Wufeng-Longmaxi and the Niutitang Formations in Well A are shown in Figure 2. The location of the well in the Huijunba syncline is shown in Figure 1b. 26 core shale samples were collected from this well. Due to the non-continuous drilling and core sampling of this well, only upper and lower samples of the Longmaxi Formation are available. Nine of them were collected from the Wufeng-Longmaxi Formations with depths ranging from 1366.90 m to 1428.21 m. Samples 1 to 4 were collected from the upper part of the Longmaxi Formation, samples 5 to 7 were collected from the lower part of the Longmaxi Formation, and samples 8 and 9 were collected from the Wufeng Formation. Seventeen shale core samples were collected from the Niutitang Formation and the depths range from 2168.93 m to 2384.77 m. The lithology of the collected core samples was shale.

3.2. Analyses

Total organic carbon (TOC), mineral compositions, major elements, and trace and rare earth elements analyses were performed on the core shale samples.

(1) **TOC analysis**: The shale samples were crushed and ground to greater than 200 mesh. The powdered sample was subjected to hydrochloric acid treatment ($V_{\text{analytically pure HCl}}$: V_{water} = 1:7) and kept at temperatures between 60 and 80 °C for two hours to thoroughly dissolve the carbonate minerals. The remaining material was dried at 100 °C after being rinsed with distilled water to a neutral pH. Finally, samples were analyzed on the CS-230 Carbon-Sulfur analyzer (LECO). The TOC contents are reported as wt%, after taking into account the material lost by acid treatment. The analytical uncertainty is less than 0.5%.

(2) **Mineral compositions analysis:** A Bruker D8 Advance X-ray diffractometer with a Cu tube was used to analyze the mineral compositions after the shale had been crushed to more than 200 mesh. The Tube voltage and electric current were ≤40 kV and ≤40 mV, respectively. The scan ranges from 0 to 140° with a rate of 2°/min and step size of 0.02°. The analytical uncertainty is less than 5%.

(3) **Major elements analysis:** The powdered samples (greater than 200 mesh) were dried at 105 °C and compressed into a specimen (32 mm i.d.) under the pressure of 30 tons using boric acid to rim the substrate. Then, the compressed specimen was measured by an Axios Panalytical X-ray fluorescence(XRF) spectrometry to obtain the concentration of the major elements. The analytical uncertainty is less than 3%.

(4) **Trace and rare earth elements analysis:** Trace and rare earth elements concentration analysis was carried out by inductively coupled plasma mass spectrometry (ICP-MS) (Nu Attom, UK). The powder samples with >200 mesh were dried at 55 °C for 12 h. After that, the powder sample was digested by acid solution (HNO_3 + HF). Then, the solutions were transferred and diluted for ICP-MS analysis to access the trace and rare earth elements concentration. Generally, the analytical uncertainty for most elements is less than 2%.

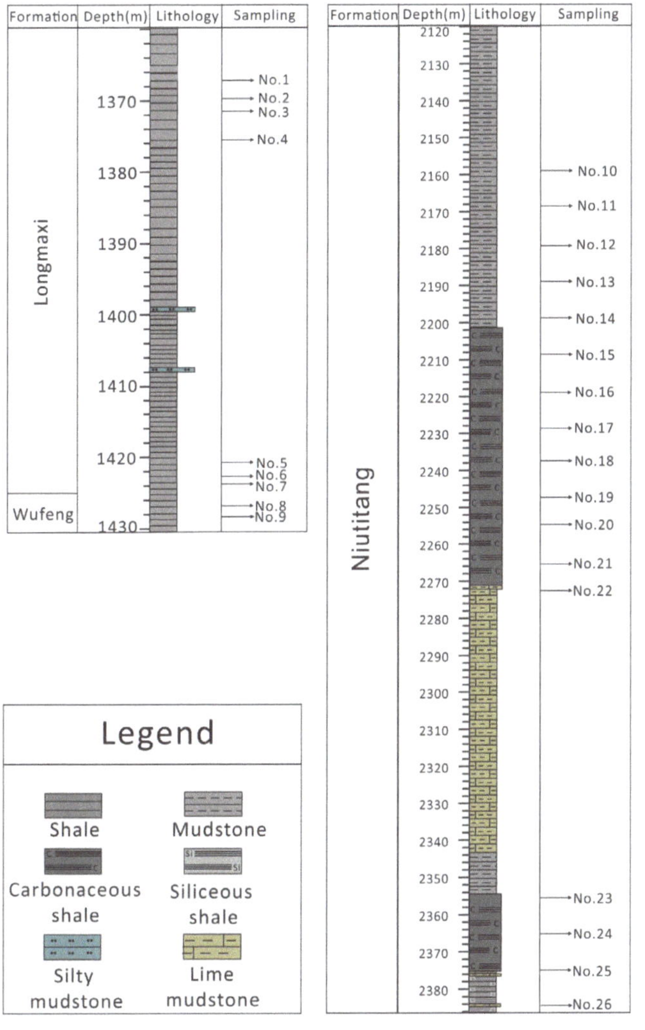

Figure 2. Stratigraphic sequences and the sampling depths of the Wufeng-Longmaxi Formation and the Niutitang Formation.

4. Results

4.1. TOC Content and Mineral Composition

TOC content and mineral composition for shale samples are given in Table 1. The TOC contents of the shale from the Upper part of the Longmaxi Formation, Lower part of the Longmaxi Formation, Wufeng-Longmaxi shale and Niutitang shale range from 0.06% to 0.26%, 1.15% to 1.63%, 1.42% to 2.86% and 0.15% to 2.56%, respectively. The mineral composition characteristics of the Wufeng-Longmaxi shale and Niutitang shale are different. In general, the shale is dominated by quartz, feldspar, and clay but of varying average proportions in different formations. The Wufeng-Longmaxi shale has average proportions of quartz, feldspar, and clay of 38.3%, 12.2%, and 27.9%, respectively. The average proportions for the same major minerals in the Niutitang shale are 29.2%, 31.4%, and 20.8%, respectively.

Table 1. TOC contents and mineral compositions of Wufeng-Longmaxi and Niutitang samples.

Formation	Sample Number	Depths (m)	TOC	Mineral Compositions									
				Quartz	Feldspar	Berlinite	Muscovite	Pyrite	Clay	Calcite	Dolomite	Total
			%	%	%	%	%	%	%	%	%	%
Wufeng-Longmaxi	1	1366.90	0.08	39.6	9.3	/	11.0	/	40.1	/	/	100.0
	2	1369.90	0.06	39.6	21.4	6.4	15.6	/	17.0	/	/	100.0
	3	1371.86	0.09	44.5	13.7	8.4	8.3	/	25.1	/	/	100.0
	4	1375.60	0.26	29.6	26.7	8.7	15.7	/	19.3	/	/	100.0
	5	1420.90	1.15	39.8	6.5	/	23.9	/	29.8	/	/	100.0
	6	1422.75	1.63	33.4	8.6	12.4	18.0	/	27.6	/	/	100.0
	7	1423.75	1.40	35.3	4.8	8.8	20.5	/	30.6	/	/	100.0
	8	1426.50	2.86	40.4	6.7	6.5	12.5	/	32.5	/	1.4	100.0
	9	1428.21	1.42	42.5	12.4	7.5	8.2	/	29.4	/	/	100.0
Niutitang	10	2158.70	0.97	27.7	24.9	9.1	5.5	1.6	28.3	2.1	0.8	100.0
	11	2168.93	0.71	35.4	14.4	7.6	18.2	/	22.0	1.2	1.2	100.0
	12	2179.25	0.66	31.2	22.4	4.6	17.7	/	18.9	4.4	0.8	100.0
	13	2188.60	0.79	29.3	24.7	4.6	13.9	/	22.9	2.8	1.8	100.0
	14	2198.00	0.98	32.1	14.9	/	12.5	1.4	34.3	3.3	1.5	100.0
	15	2208.00	1.97	26.6	36.6	6.0	13.7	/	15.8	1.3	/	100.0
	16	2218.50	1.58	35.7	22.7	7.8	16.0	/	17.8	/	/	100.0
	17	2228.84	1.22	24.4	32.0	7.4	10.9	0.8	24.5	/	/	100.0
	18	2237.38	1.28	36.3	33.2	5.4	6.6	/	18.5	/	/	100.0
	19	2246.86	0.90	37.2	30.9	7.4	6.0	/	18.5	/	/	100.0
	20	2254.60	1.10	24.9	27.4	9.4	12.9	1.1	24.3	/	/	100.0
	21	2265.80	1.18	24.5	39.4	9.5	4.6	1.1	20.9	/	/	100.0
	22	2273.20	0.15	21.9	49.7	/	3.7	1.0	23.0	/	0.7	100.0
	23	2355.80	2.56	29.1	48.1	5.4	4.3	/	13.1	/	/	100.0
	24	2365.66	2.54	25.9	46.9	4.0	11.6	/	11.6	/	/	100.0
	25	2375.95	1.96	24.4	35.3	8.6	11.6	1.7	18.4	/	/	100.0
	26	2384.77	1.68	31.3	29.6	10.5	6.2	1.8	20.6	/	/	100.0

4.2. Major Elements

Major element concentrations of the shale samples are presented in Table 2. No significant difference in major element oxide distribution can be observed between the Wufeng-Longmaxi and the Niutitang samples. The dominant major element oxides are SiO_2, Al_2O_3, and TFe_2O_3. The Wufeng-Longmaxi shale has an average SiO_2 content of 63.96%, with a range of 62.22% to 67.94%, while the SiO_2 abundance in the Niutitang shale ranges from 53.69% to 65.74% with an average of 60.41%. The average Al_2O_3 and TFe_2O_3 contents in the Wufeng-Longmaxi shale are 16.14% and 6.32%, respectively, while the Niutitang shale has average Al_2O_3 and TFe_2O_3 contents of 13.95% and 5.37%, respectively. Other major element oxides, including CaO, MgO, K_2O, Na_2O and P_2O_5, are present in relatively low amounts (less than 5%).

Table 2. Major elements data of Wufeng-Longmaxi and Niutitang samples.

Formation	Sample Number	Depths (m)	SiO_2 %	Al_2O_3 %	TFe_2O_3 %	CaO %	MgO %	K_2O %	Na_2O %	P_2O_5 %	TiO_2 %	MnO %	Other %	Total %
Wufeng - Longmaxi	1	1366.90	63.24	16.68	6.79	0.32	2.30	4.09	1.22	0.11	0.74	0.08	4.12	99.71
	2	1369.90	67.94	14.18	5.54	0.65	2.04	3.14	1.33	0.13	0.73	0.10	3.97	99.75
	3	1371.86	66.18	14.92	6.21	0.44	2.15	3.35	1.31	0.13	0.74	0.08	4.25	99.74
	4	1375.60	67.08	15.38	6.44	0.36	2.21	3.46	1.28	0.12	0.72	0.08	2.59	99.72
	5	1420.90	62.41	17.29	6.43	0.71	2.38	4.50	1.08	0.10	0.70	0.07	3.74	99.39
	6	1422.75	62.22	16.90	6.61	1.13	2.42	4.40	0.98	0.09	0.72	0.07	3.64	99.17
	7	1423.75	65.65	15.88	5.64	0.67	2.21	4.20	1.08	0.10	0.65	0.06	3.22	99.35
	8	1426.50	58.57	16.16	7.83	1.29	2.24	4.09	1.05	0.10	0.79	0.07	5.67	97.84
	9	1428.21	62.38	17.87	5.44	1.02	1.79	4.67	1.00	0.09	0.49	0.04	3.64	98.43
	average		63.96	16.14	6.32	0.73	2.19	3.99	1.15	0.11	0.46	0.26	0.26	99.23
Niutitang	10	2158.70	56.22	14.72	6.66	4.79	3.17	3.53	1.24	0.19	0.72	0.09	7.48	98.81
	11	2168.93	55.08	14.56	6.48	5.82	3.27	3.40	1.20	0.21	0.76	0.12	8.01	98.89
	12	2179.25	56.57	14.82	6.07	5.26	3.21	3.49	1.26	0.23	0.78	0.09	7.27	99.05
	13	2188.60	53.69	14.65	6.14	6.82	3.53	3.52	1.16	0.24	0.75	0.12	8.50	99.10
	14	2198.00	54.42	14.61	5.93	6.76	3.41	3.51	1.18	0.24	0.75	0.11	8.21	99.13
	15	2208.00	58.60	13.80	5.58	4.68	2.70	3.13	1.62	0.17	0.75	0.07	7.68	98.79
	16	2218.50	61.13	14.21	5.36	3.02	2.62	3.30	1.61	0.19	0.73	0.06	6.75	98.96
	17	2228.84	63.43	12.52	4.66	3.19	2.24	2.76	1.94	0.21	0.74	0.06	7.37	99.11
	18	2237.38	62.71	14.15	5.07	2.65	2.76	3.21	1.65	0.18	0.75	0.06	6.00	99.19
	19	2246.86	61.92	13.77	5.32	2.73	2.70	3.20	1.62	0.19	0.75	0.06	6.90	99.15
	20	2254.60	62.84	13.92	5.41	2.26	2.61	3.18	1.72	0.19	0.75	0.05	6.12	99.06
	21	2265.80	65.74	12.22	4.77	3.00	2.01	2.64	2.20	0.24	0.71	0.05	5.26	98.85
	22	2273.20	62.88	12.22	4.00	3.74	2.21	2.52	3.03	0.47	0.73	0.05	7.48	99.32
	23	2355.80	62.65	12.79	4.88	1.93	2.10	3.12	1.84	0.22	0.80	0.04	8.44	98.80
	24	2365.66	63.53	12.98	4.74	2.85	2.09	3.00	2.10	0.24	0.81	0.06	6.45	98.85
	25	2375.95	63.03	15.45	5.31	1.53	2.37	3.61	1.93	0.21	0.83	0.04	4.04	98.35
	26	2384.77	62.49	15.83	4.94	2.06	1.92	3.60	1.59	0.25	0.82	0.04	4.58	98.10
	average		60.41	13.95	5.37	3.71	2.64	3.22	1.70	0.23	0.76	0.07	6.86	98.91

4.3. Trace Elements

The abundances of selected trace elements of the Wufeng-Longmaxi and Niutitang samples in the study area are shown in Table 3. The UCC-normalized trace elements of shale from the Wufeng-Longmaxi Formation and the Niutitang Formation are illustrated in Figure 3. The distribution characteristics of trace elements in the Wufeng-Longmaxi samples are different, as demonstrated by the findings in Figure 3a. Cu, Sr, Mo, and Pb are depleted in the shale samples from 1366.90 m to 1375.60 m (Nos. 1–4). However, Cu, Mo, Pb and U are obviously enriched in samples from 1420.90 m to 1428.21 m (Nos. 5–9). The UCC-normalized trace element distributions of the Niutitang samples show a similar distribution pattern compared to the Wufeng-Longmaxi shale with depths ranging from 1420.90 m to 1428.21 m (Figure 3b). In general, the majority of the samples from the Niutitang samples are enriched in Mo, Cd and U.

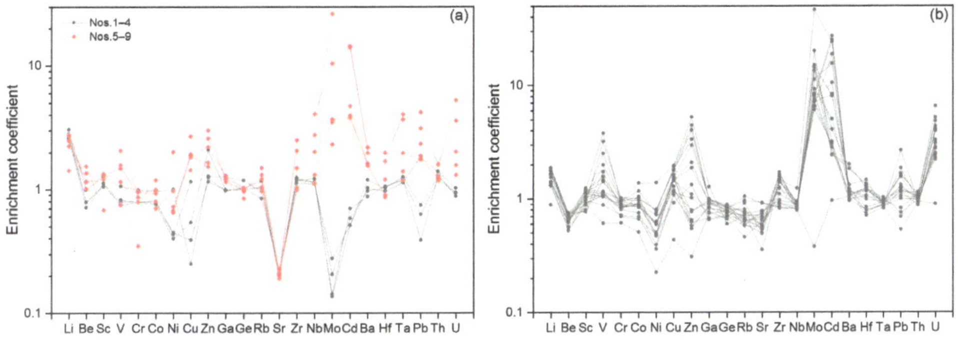

Figure 3. UCC-normalized trace element distributions for the Wufeng-Longmaxi shale (**a**) and the Niutitang shale (**b**). The enrichment coefficient can be expressed as X_{sample}/X_{UCC}, where X represents the weight concentrations of elements X.

4.4. Rare Earth Elements

Rare earth element concentrations of the samples are presented in Table 4 and Figure 4. The average ΣREE contents of the Wufeng-Longmaxi and the Niutitang samples are 209.93 μg/g and 166.90 μg/g, respectively. The Niutitang and the Wufeng-Longmaxi samples have average light rare earth elements (ΣLREEs) of 187.36 μg/g and 146.51 μg/g, respectively. The average heavy rare earth elements (ΣHREEs) in the Wufeng-Longmaxi and the Niutitang samples are 22.57 μg/g and 20.38 μg/g, respectively.

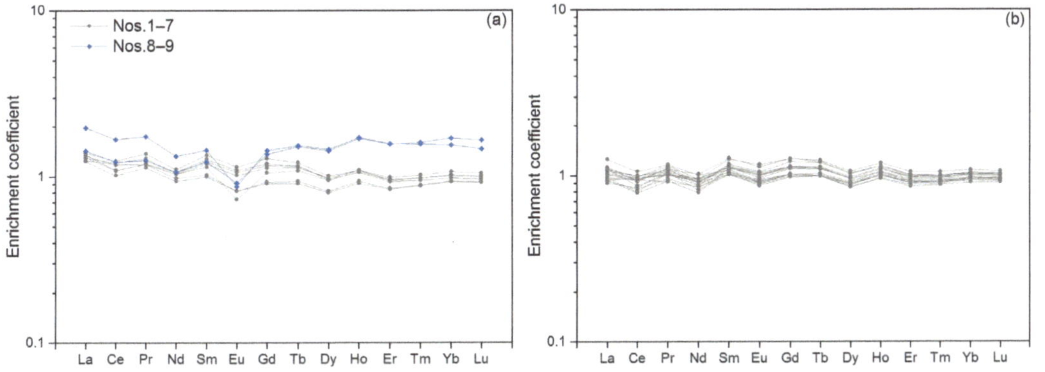

Figure 4. NASC —normalized REEs distribution for the samples from the Wufeng-Longmaxi Formations (**a**) and the Niutitang Formation (**b**). NASC = The North American shale composite from McLennan (1989) [32]. The enrichment coefficient can be expressed as X_{sample}/X_{NASC}, where X represents the weight concentrations of elements X.

Table 3. Trace elements data of Wufeng-Longmaxi and Niutitang samples.

Formation	Sample Number	Depths	Li	Be	Sc	V	Cr	Co	Ni	Cu	Zn	Ga	Ge	Rb	Sr	Zr	Nb	Mo	Cd	Ba	Hf	Ta	Pb	Th	U
		(m)												μg/g											
	1	1366.90	61.40	2.98	17.45	113.39	84.64	16.43	44.29	6.28	91.14	21.51	1.89	109.01	72.91	226.31	14.56	0.20	0.05	655.30	5.95	1.26	12.61	13.44	2.63
	2	1369.90	51.17	2.14	15.27	79.88	67.41	12.92	33.98	13.49	148.59	16.73	1.60	116.26	71.51	237.03	13.60	0.21	0.07	480.08	6.16	1.17	10.67	12.76	2.59
	3	1371.86	53.16	2.38	14.37	84.94	67.81	13.75	36.68	28.86	88.59	16.56	1.64	94.73	76.76	228.66	13.63	0.31	0.06	532.89	6.07	1.20	6.55	13.81	2.46
Wufeng-	4	1375.60	56.71	2.36	15.19	88.15	67.53	13.49	37.91	9.75	82.15	16.93	1.66	130.53	71.36	213.07	13.33	0.41	0.05	563.13	5.61	1.13	30.96	14.87	2.86
	5	1420.90	54.87	3.04	17.81	223.38	81.40	15.87	57.71	48.92	109.31	19.63	1.60	147.25	72.75	188.34	13.41	3.49	0.39	852.31	5.01	1.18	31.68	12.51	3.67
Longmaxi	6	1422.75	55.91	3.48	18.04	158.22	82.48	17.04	62.26	45.76	118.28	21.11	1.64	136.39	74.56	196.92	15.84	5.57	0.37	884.21	5.20	1.39	39.97	13.67	5.64
	7	1423.75	50.17	3.50	16.46	169.05	73.82	13.64	55.76	47.26	157.69	22.05	1.65	117.23	67.15	283.33	24.47	5.29	0.46	894.24	6.97	1.96	29.60	13.03	4.39
	8	1426.50	45.48	4.07	18.29	123.72	67.38	20.34	173.12	67.93	216.24	21.12	1.54	168.28	79.86	392.94	33.33	39.84	1.38	1089.65	9.79	4.06	52.88	16.99	14.74
	9	1428.21	28.70	4.66	9.32	81.46	29.63	12.03	83.17	36.09	186.41	20.76	1.35	108.56	70.68	478.97	49.12	15.65	1.41	1210.13	11.46	3.70	71.91	16.91	10.04
	10	2158.70	37.45	2.24	13.85	211.89	81.58	19.90	64.61	47.53	314.88	16.49	1.38	72.27	231.66	170.23	9.70	12.42	2.65	614.29	4.56	0.87	44.95	9.80	8.59
	11	2168.93	37.56	2.11	13.89	269.18	81.63	17.47	67.41	44.35	287.45	16.42	1.35	65.58	271.11	173.93	10.16	11.93	2.44	539.22	4.63	0.88	31.38	9.85	7.93
	12	2179.25	37.27	2.22	12.29	342.91	86.53	17.03	68.38	45.01	238.01	16.90	1.41	82.91	274.90	183.65	10.77	9.78	1.83	668.00	4.86	0.94	26.27	9.77	7.69
	13	2188.60	33.02	2.24	10.52	155.66	71.24	14.74	50.17	36.64	148.43	15.58	1.32	117.19	318.98	160.62	9.84	11.13	1.03	667.63	4.39	0.89	22.39	10.19	7.72
	14	2198.00	35.15	2.08	10.90	184.15	73.09	15.24	61.53	38.61	211.33	16.92	1.33	107.88	322.79	156.54	9.68	16.81	1.51	631.54	4.11	0.85	19.43	9.35	9.14
	15	2208.00	32.65	1.95	14.08	122.95	68.34	17.82	40.56	38.32	46.10	14.24	1.21	79.39	253.23	233.45	10.07	22.47	0.31	545.17	6.10	0.90	28.14	11.10	11.04
	16	2218.50	33.55	2.11	13.40	119.10	71.29	15.29	39.72	35.98	78.17	13.92	1.26	79.24	204.75	209.37	10.39	13.91	0.50	653.60	5.72	0.94	20.73	11.02	9.17
	17	2228.84	30.83	1.71	13.09	94.50	61.08	11.78	30.66	27.01	44.36	11.49	1.14	70.68	203.72	266.21	10.27	9.18	0.30	610.31	7.00	0.91	17.91	10.90	6.16
Niutitang	18	2237.38	35.18	2.03	12.49	116.63	68.64	12.76	33.34	28.38	72.05	12.67	1.25	65.85	174.23	264.79	10.60	12.56	0.27	526.23	7.09	1.02	12.75	10.58	7.01
	19	2246.86	37.23	1.99	16.11	166.77	76.01	14.33	53.32	36.64	132.33	15.81	1.35	86.13	203.01	236.76	11.16	10.71	0.81	626.99	6.13	0.97	11.72	12.15	6.44
	20	2254.60	35.81	1.91	15.67	151.43	80.99	15.06	42.65	40.45	54.40	16.76	1.35	74.83	186.09	272.48	11.34	10.84	0.27	603.84	6.75	0.96	19.94	11.20	6.82
	21	2265.80	26.05	1.58	12.60	114.38	59.39	11.31	41.41	23.05	94.09	11.15	1.12	64.35	196.47	324.33	9.73	22.46	0.79	727.90	8.46	0.86	16.94	10.51	11.46
	22	2273.20	26.06	1.67	12.03	65.48	52.30	8.57	19.29	10.83	21.88	12.89	1.07	51.24	224.14	281.72	9.49	0.56	0.09	574.39	7.32	0.82	9.03	10.15	2.50
	23	2355.80	27.76	1.86	15.18	111.23	76.39	15.35	50.48	32.03	38.50	15.17	1.29	90.41	205.50	296.12	10.51	29.87	0.24	1004.28	7.57	0.92	19.20	12.20	13.49
	24	2365.66	28.29	1.88	15.78	119.70	72.30	15.08	51.04	29.57	56.47	15.65	1.26	71.96	244.04	308.91	10.87	21.12	0.40	1006.17	7.76	0.96	14.02	11.97	14.56
	25	2375.95	26.69	2.25	16.20	124.07	83.55	16.43	50.78	38.34	40.94	14.24	1.11	102.50	172.50	270.43	11.17	20.08	0.24	1101.38	7.06	0.97	21.13	10.91	12.17
	26	2384.77	17.83	2.05	16.82	407.74	80.86	23.34	117.94	49.02	372.50	21.61	0.96	79.66	124.23	280.09	14.71	69.21	2.36	578.59	7.11	0.92	26.99	11.80	18.26
	UCC *		20.00	3	13.6	107	85	17	44	25	71	17	1.6	112	350	190	12	1.5	0.098	550	5.8	1	17	10.7	2.8

* The UCC data was from Taylor and McLennan (1985) [33] and McLennan et al. (1995) [34].

Table 4. Rare earth elements abundances of Wufeng-Longmaxi and Niutitang samples.

Formations	Sample Number	Depth (m)	La	Ce	Pr	Nd	Sm	Eu	Gd	Tb	Dy	Ho	Er	Tm	Yb	Lu	ΣREE
			μg/g														
Wufeng-Longmaxi	1	1366.90	40.41	86.92	9.41	34.30	6.82	1.27	5.86	0.97	5.69	1.15	3.37	0.51	3.30	0.50	200.48
	2	1369.90	42.05	79.59	9.49	35.15	7.29	1.36	6.22	0.98	5.59	1.10	3.16	0.48	3.04	0.46	195.96
	3	1371.86	41.71	88.73	10.16	34.98	7.08	1.32	6.11	0.98	5.70	1.13	3.29	0.49	3.13	0.48	205.30
	4	1375.60	44.56	91.18	10.87	36.57	7.72	1.42	6.68	1.04	5.84	1.14	3.25	0.48	3.05	0.46	214.25
	5	1420.90	39.70	74.75	9.03	31.05	5.70	1.02	4.71	0.77	4.63	0.95	2.85	0.44	2.87	0.44	178.93
	6	1422.75	43.42	80.58	9.37	32.46	5.87	1.02	4.85	0.80	4.77	0.97	2.91	0.45	2.91	0.44	190.82
	7	1423.75	42.73	86.04	9.41	34.24	6.54	0.90	5.49	0.93	5.49	1.11	3.23	0.49	3.18	0.49	200.25
	8	1426.50	45.92	90.08	9.96	35.45	7.02	1.15	7.13	1.29	8.31	1.75	5.32	0.81	5.27	0.80	220.27
	9	1428.21	62.97	122.46	13.81	43.84	8.28	1.08	7.47	1.31	8.45	1.79	5.37	0.79	4.79	0.71	283.11
NIutitang	10	2158.70	31.27	69.06	8.21	30.15	6.61	1.26	5.88	0.94	5.58	1.13	3.32	0.50	3.19	0.49	167.59
	11	2168.93	30.01	71.13	7.87	31.30	7.20	1.45	6.60	1.05	6.16	1.24	3.57	0.53	3.34	0.51	171.97
	12	2179.25	32.48	71.16	8.00	30.23	6.37	1.29	5.93	0.96	5.69	1.16	3.38	0.50	3.21	0.49	170.82
	13	2188.60	33.84	73.22	8.51	30.15	6.37	1.24	5.80	0.94	5.57	1.13	3.32	0.49	3.10	0.48	174.14
	14	2198.00	30.70	69.19	7.30	28.34	5.83	1.18	5.31	0.86	5.14	1.05	3.06	0.46	2.92	0.45	161.80
	15	2208.00	35.40	71.33	9.12	31.43	5.93	1.11	5.24	0.85	5.11	1.04	3.08	0.46	2.97	0.46	173.54
	16	2218.50	32.64	68.22	8.73	28.96	5.97	1.14	5.28	0.86	5.22	1.07	3.16	0.47	3.02	0.46	165.22
	17	2228.84	30.01	57.51	7.36	26.21	6.34	1.24	5.95	0.95	5.51	1.09	3.14	0.46	2.93	0.45	149.15
	18	2237.38	28.78	61.94	8.19	27.18	5.74	1.10	5.14	0.84	5.09	1.04	3.09	0.46	2.97	0.46	152.02
	19	2246.86	34.72	72.77	8.29	31.22	6.43	1.25	5.81	0.95	5.59	1.13	3.29	0.49	3.13	0.48	175.56
	20	2254.60	32.90	70.59	8.05	30.39	6.21	1.15	5.30	0.88	5.31	1.09	3.24	0.49	3.17	0.49	169.27
	21	2265.80	31.18	58.75	8.11	28.80	6.74	1.30	6.34	1.02	5.94	1.18	3.42	0.50	3.22	0.50	157.00
	22	2273.20	29.89	60.36	7.58	28.05	5.83	1.13	5.26	0.84	4.95	1.00	2.91	0.44	2.83	0.44	151.51
	23	2355.80	40.19	77.66	9.26	33.38	6.13	1.13	5.26	0.84	4.99	1.01	2.99	0.45	2.91	0.45	186.63
	24	2365.66	35.17	68.46	8.60	33.68	7.27	1.40	6.59	1.04	5.95	1.18	3.41	0.50	3.21	0.49	176.94
	25	2375.95	36.04	66.93	8.86	29.98	5.80	1.08	5.10	0.84	5.08	1.05	3.10	0.46	3.00	0.47	167.79
	26	2384.77	34.87	63.39	8.09	31.57	6.37	1.20	5.70	0.93	5.55	1.13	3.33	0.50	3.17	0.49	166.29
NASC *			32	73	7.9	33	5.7	1.24	5.2	0.85	5.8	1.04	3.4	0.5	3.1	0.48	/

* The North American shale composite (NASC) is from McLennan (1989) [32].

The NASC normalized REE distributions of the Wufeng-Longmaxi samples suggest a relative enrichment of LREE. Samples 8 and 9 from the Wufeng Formation have a considerably enriched HREE content. The shale samples from the Niutitang Formation do not show any notable enrichment or depletion of REE compared to NASC (Figure 4).

5. Discussion

5.1. Chemical Weathering and Palaeoclimate

The palaeoclimate during sedimentation can be reflected by the chemical index of alteration (CIA), which indicates the degree of weathering [35]. The CIA is calculated as follows:

$$CIA(\%) = 100 \times Al_2O_3/(Al_2O_3 + CaO^* + Na_2O + K_2O) \quad (1)$$

CaO^* stands for calcium that is presented as the silicate. In this study, the CaO^* was calibrated and the CIA (%) was estimated using the method proposed by McLennan (1993) [36]. Chemical weathering is beginning to occur, as indicated by the CIA (%) value range of 50 to 60. CIA (%) values of 60–80 and >80 indicate moderate chemical weathering and strong chemical weathering, respectively [35,37]. CIA (%) in the Wufeng-Longmaxi samples range from 67.27 to 70.37 with an average of 68.62, while the average CIA (%) in the Niutitang samples is 60.25 (Table 5). Chemical alterations become weaker under arid conditions [38]. Therefore, the Wufeng-Longmaxi shale was deposited in warmer and more humid conditions compared to Niutitang shale. However, with decreasing depth, the CIA (%) for the Niutitang shale rises from about 60 to 65, indicating a change in the palaeoclimate conditions.

5.2. Terrestrial Influx Intensity, Provenance, and Tectonic Setting

Influx of terrestrial materials can dilute the organic matter and thereby affect its accumulation. The Mn/Ca ratio serves as a proxy for the terrestrial influx intensity [39,40]. Figure 5a demonstrates a cross-plot of Mn/Ca (ppm/%) and Ca (%), which shows variations in terrestrial influx intensity among the samples from different formations. Generally, the Niutitang samples have the lowest terrestrial influx intensity among the shale, while the highest intensity is observed in the shale from the upper of the Longmaxi Formation. The

TiO_2/Zr ratio can be used to classify the origin of the clastic input to sediment [41], and the results of plotting TiO_2 and Zr suggest a provenance of felsic igneous rock (Figure 5b). Plotting La/Yb versus ΣREE can also be applied to classify the provenance of the sedimentary rocks as REEs inherit the characteristics of parental rocks [32,42,43]. The results of plotting La/Yb versus ΣREE are consistent with the interpretation of the TiO2/Zr crossplot and indicate a mixed provenance of granite, sedimentary rock, and alkalic basalts (Figure 5c). The shale from various formations reflects distinct tectonic settings. The majority of the Wufeng-Longmaxi samples are located within the passive margin, whereas the Niutitang samples are mostly located within the active continental margin (Figure 5d).

Table 5. Geochemical proxies of the Wufeng-Longmaxi and Niutitang samples.

Formations	Sample Number	Depths	Palaeclimate	Terrestrial Influx	Paleoredox Conditions				Paleo-Productivity
		(m)	CIA	Mn/Ca (ppm/%)	Th/U	δU	Mo_{EF}	U_{EF}	P/Ti
Wufeng-Longmaxi	1	1366.90	70.37	2650.03	5.12	0.74	0.23	0.96	0.11
	2	1369.90	67.70	1668.85	4.92	0.76	0.29	1.11	0.13
	3	1371.86	69.37	1920.55	5.62	0.70	0.39	1.00	0.13
	4	1375.60	70.28	2278.78	5.21	0.73	0.51	1.13	0.12
	5	1420.90	68.57	1118.61	3.41	0.94	3.82	1.29	0.10
	6	1422.75	67.93	716.73	2.42	1.11	6.23	2.04	0.09
	7	1423.75	67.85	985.68	2.97	1.01	6.30	1.68	0.11
	8	1426.50	67.27	573.60	1.15	1.44	46.58	5.56	0.10
	9	1428.21	68.20	465.01	1.68	1.28	16.55	3.43	0.13
Niutitang	10	2158.70	65.07	207.62	1.14	1.45	15.94	3.56	0.19
	11	2168.93	65.65	231.71	1.24	1.41	15.49	3.32	0.20
	12	2179.25	65.15	191.60	1.27	1.41	12.47	3.16	0.21
	13	2188.60	65.79	193.14	1.32	1.39	14.36	3.21	0.23
	14	2198.00	65.54	170.65	1.02	1.49	21.75	3.81	0.23
	15	2208.00	61.30	169.88	1.01	1.50	30.77	4.88	0.17
	16	2218.50	61.59	206.73	1.20	1.43	18.50	3.94	0.19
	17	2228.84	57.23	194.78	1.77	1.26	13.86	3.00	0.21
	18	2237.38	61.39	256.56	1.51	1.33	16.79	3.02	0.17
	19	2246.86	61.07	243.33	1.89	1.23	14.70	2.85	0.18
	20	2254.60	60.42	255.42	1.64	1.29	14.73	2.99	0.18
	21	2265.80	54.72	189.00	0.92	1.53	34.76	5.72	0.24
	22	2273.20	49.06	152.62	4.06	0.85	0.87	1.25	0.47
	23	2355.80	56.38	230.11	0.90	1.54	44.14	6.43	0.20
	24	2365.66	52.19	217.88	0.82	1.57	30.77	6.84	0.21
	25	2375.95	61.04	314.68	0.90	1.54	24.57	4.80	0.19
	26	2384.77	60.72	207.92	0.65	1.65	82.62	7.03	0.22

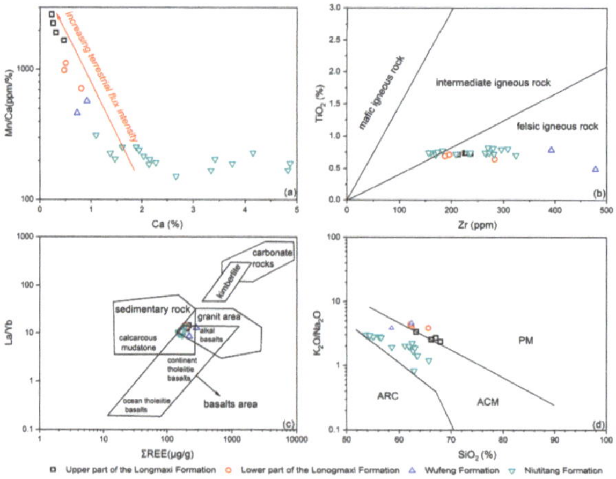

Figure 5. (**a**) Mn/Ca (ppm/%) versus Ca (%) for samples from different formations. (**b**) TiO_2 versus Zr for samples from different formations. The interpretation lines are from Hayashi et al., (1997) [41]. (**c**) La/Yb versus ΣREE for samples from different formations. The interpretation lines are from Zhao (2016) [44]. (**d**) K_2O/Na_2O versus SiO_2 for samples from different formations. The interpretation lines are from Roser and Korsch (1986) [45]. ARC: oceanic island arc margin; ACM: active continental margin; PM: passive margin.

5.3. Paleoredox Conditions and Water Mass Restriction

Specific ratios for redox-sensitive elements, such as V, Cr, Ni, Co, Mo and U, were employed as redox indicators to evaluate the paleo-redox conditions during sediment deposition [46–49]. In an oxidizing seawater environment, uranium (U) is presented as dissolved U (VI); however, it can be reduced to solid U (IV) and accumulate in sediments under anoxic conditions [14]. In contrast, thorium (Th) is a very stable element that may remain in sediments as undissolved thorium (IV) [47]. Thus, redox conditions may be revealed using the Th/U ratio. Th/U ratios of 2–7 typically indicate oxic marine environments, whereas a ratio of <2 reflects anoxic environments [50,51]. In addition, the Th/U ratio may exceed 7 in a terrestrial oxic environment [51]. Additionally, the paleo-redox conditions can be categorized using the δU value (2U/(U + 1/3Th)) [49]. A δU value of >1 suggests anoxic conditions, whereas a value of <1 denotes an oxic environment.

Table 5 presents the Th/U and δU ratios for the samples. The data for the Niutitang samples suggest a mainly anoxic environment, as indicated by Th/U ratios that vary from 0.64 to 4.06, with an average of 1.37. Shale from the Wufeng Formation have a Th/U ratio that varies from 1.15 to 1.68 with an average of 1.42. Samples 5–7 from the Lower Longmaxi Formation have Th/U ratios that vary from 2.42 to 3.41 with an average of 2.93, while the Longmaxi samples 1–4 are characterized by high Th/U with an average of 5.21 (Table 5). These results suggest that the upper Longmaxi samples were deposited in more oxic conditions than the samples from the lower Longmaxi Formation, while the conditions for the Wufeng samples were more anoxic, and similar to the Niutitang Formation during the Wufeng and Longmaxi sedimentary periods. The redox condition interpreted by δU is consistent with Th/U ratios. Most Niutitang, Wufeng, and Longmaxi samples (Nos. 5–7) display δU values greater than 1, indicating anoxic water conditions. The samples 1–4 from the Longmaxi Formation have low δU with an average of 0.73, suggesting oxic water conditions.

Molybdenum (Mo) can exist in sea water as a stable and soluble molybdate oxyanion (MoO_4^{2-}), while the MoO_4^{2-} can be converted to thiomolybdates in euxinic conditions. It has been previously shown that thiomolybdates can be adsorbed onto humic substances and captured by Mn-Fe oxyhydroxides at the sediment surface [14,52,53]. Hence, the Mo-U covariance can be an indication of the paleo-redox conditions [48,52,54]. The enrichment factor (EF) for Mo and U were calculated using the equation EF = [(X/Al)sample/(X/Al)PAAS], where X and Al represent the weight concentrations of elements X and Al, respectively. A plot of Mo_{EF} versus U_{EF} is presented in Table 5 and Figure 6. The plot shows that Niutitang samples were deposited in sulfidic/euxinic waters with higher EFs in Mo (EF > 10) compared to U, while the Mo_{EF} versus U_{EF} plot also indicates that paleo-redox conditions went from anoxic in the Wufeng Formation to oxic in the upper Longmaxi Formation.

Basinal water mass restriction is correlated to paleo-redox condition [55], and it is mainly controlled by changes in sea level [56]. In a restricted sea, the Mo (ppm)/TOC (%) ratio in sediments decreases as the restriction increases, due to the low resupply rate of Mo in an aqueous medium [14], In contrast, the ratio is higher in open marine sediments. This makes the Mo (ppm) to TOC ratio a useful tool in classifying paleo-hydrographic conditions. As shown in Figure 6b, the samples from the Niutitang and Wufeng formations were deposited in a moderately restricted basin. The ratio in the Longmaxi samples is lower than the Black Sea (~4.5), suggesting a strongly restricted condition with reducing sea level during this geological period.

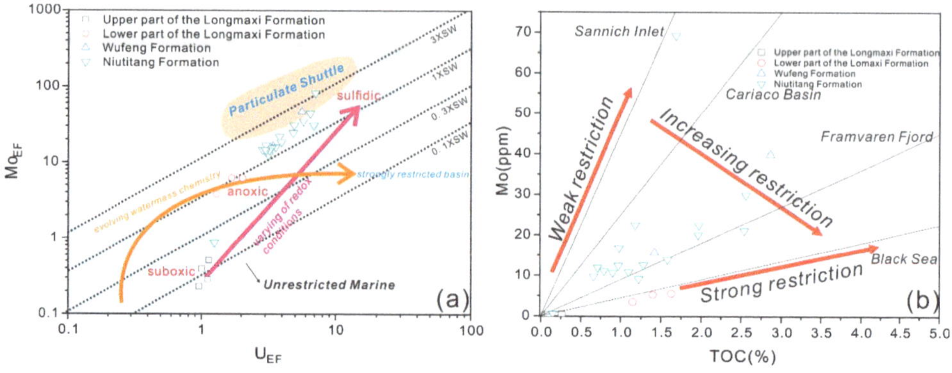

Figure 6. (**a**) Mo_{EF} versus U_{EF} in samples from the Wufeng-Longmaxi and the Niutitang Formations. The post-Archean average shale (PAAS) composition data used in this study are from Taylor and McLennan (1985) [33]. The interpretation lines are from Algeo and Tribovillard (2009) [52] and Tribovillard et al., (2012) [55]; (**b**) Mo (ppm) versus TOC (%) in samples from the Wufeng-Longmaxi and the Niutitang formations. The interpretation lines are from Algeo and Lyons (2006) [57] and Tribovillard et al. (2012) [55].

5.4. Paleoproductivity

Primary paleo-productivity presents the total amount of organic matter produced per unit time and area [58], and hence good source rocks were deposited in areas of basins with elevated primary paleo-productivity [59,60]. Phosphorus (P) is a crucial nutritional element that is found in the skeleton [14,61]. P can exist as both dissolved and precipitated phases in seawater [62]. Sources of P can be divided into detrital and organic. Since the concentration of detrital P is usually very low, total P may serve as a proxy to estimate paleo-productivity [63]. P_2O_5 (%) and Ti (ppm) show no positive correlation in the samples of this study (Figure 7), indicating that P is of organic origin and can be used as a paleo-productivity proxy. A P/Ti ratio less than 0.34 reveals low paleo-productivity, and $0.34 < P/Ti < 0.79$ and $P/Ti > 0.79$ indicate medium and high paleo-productivities, respectively [64]. The P/Ti ratio in Niutitang shale ranges from 0.17 to 0.47 with an average of 0.22, while the average P/Ti ratio in Wufeng-Longmaxi shale is 0.11. Hence, the Wufeng-Longmaxi and Niutitang formations are characterized by low paleo-productivity. In addition, Wufeng-Longmaxi shale possess a lower paleo-productivity compared to the Niutitang shale in the study area (Table 5).

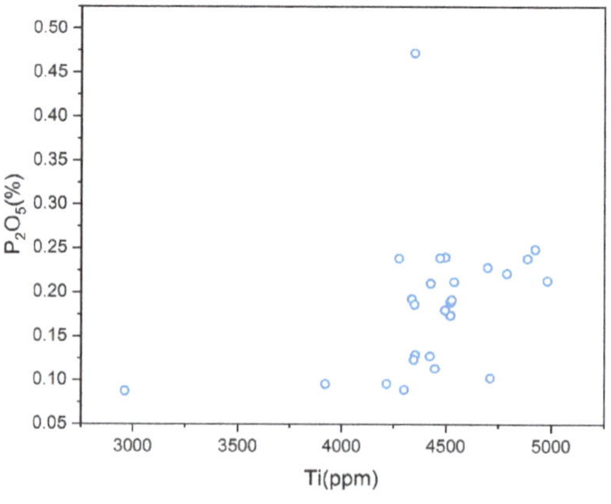

Figure 7. Relationship between P_2O_5 (%) and Ti (ppm) in the samples.

5.5. Influence of Deposition Conditions on Organic Matter Enrichment

Total organic carbon (TOC) contents represent the amount of organic carbon, and it could be utilized to evaluate the richness of organic carbon and reveal hydrocarbon generation potential for hydrocarbon source rock [65]. The enrichment of organic matter in sediment is a complex process that is controlled by the input and preservation of organic matter [49], as well as its dilution by clastic input. Trends of TOC content and paleodeposition-related inorganic geochemical parameters with depth, in the samples, are shown in Figure 8. The relationship between different inorganic geochemical indicators and TOC contents are presented in Figure 9.

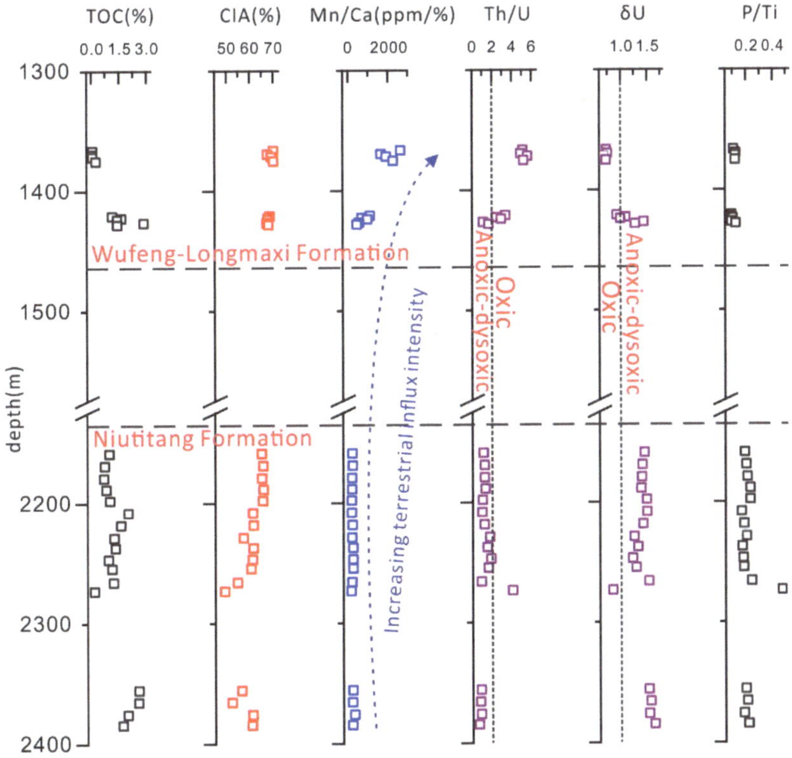

Figure 8. Trends of TOC content and paleodeposition-related inorganic geochemical parameters with depth.

A fair negative correlation can be observed in CIA and TOC content in the Wufeng-Longmaxi shale (R^2 = 0.49), whereas it shows a poor correlation in the Niutitang samples (Figure 9a). Mn/Ca ratios are negatively related to TOC content for the Wufeng-Longmaxi samples (R^2 = 0.72), while no clear relationship can be observed in the Niutitang samples (Figure 9b), indicating that the high terrestrial influx intensity in the Wufeng-Longmaxi samples diluted the TOC content and resulted in a negative correlation between Mn/Ca ratio and TOC content. In contrast, the terrestrial influx intensity for the Niutitang samples is low and had a minor effect on diluting TOC content. The regression analyses of Th/U, δU, and Mo_{EF} with TOC contents are presented in Figure 9c–e. The results reveal a clear correlation between the redox parameters with TOC contents, suggesting anoxic conditions which helped with preservation of the organic matters. The study also found a positive relationship between P/Al and TOC content in the Wufeng-Longmaxi shale (Figure 9f). In addition, the correlation of P/Ti and TOC content is less (R^2 = 0.24) in the Nititang samples compared to the Wufeng-Longmaxi samples (R^2 = 0.46).

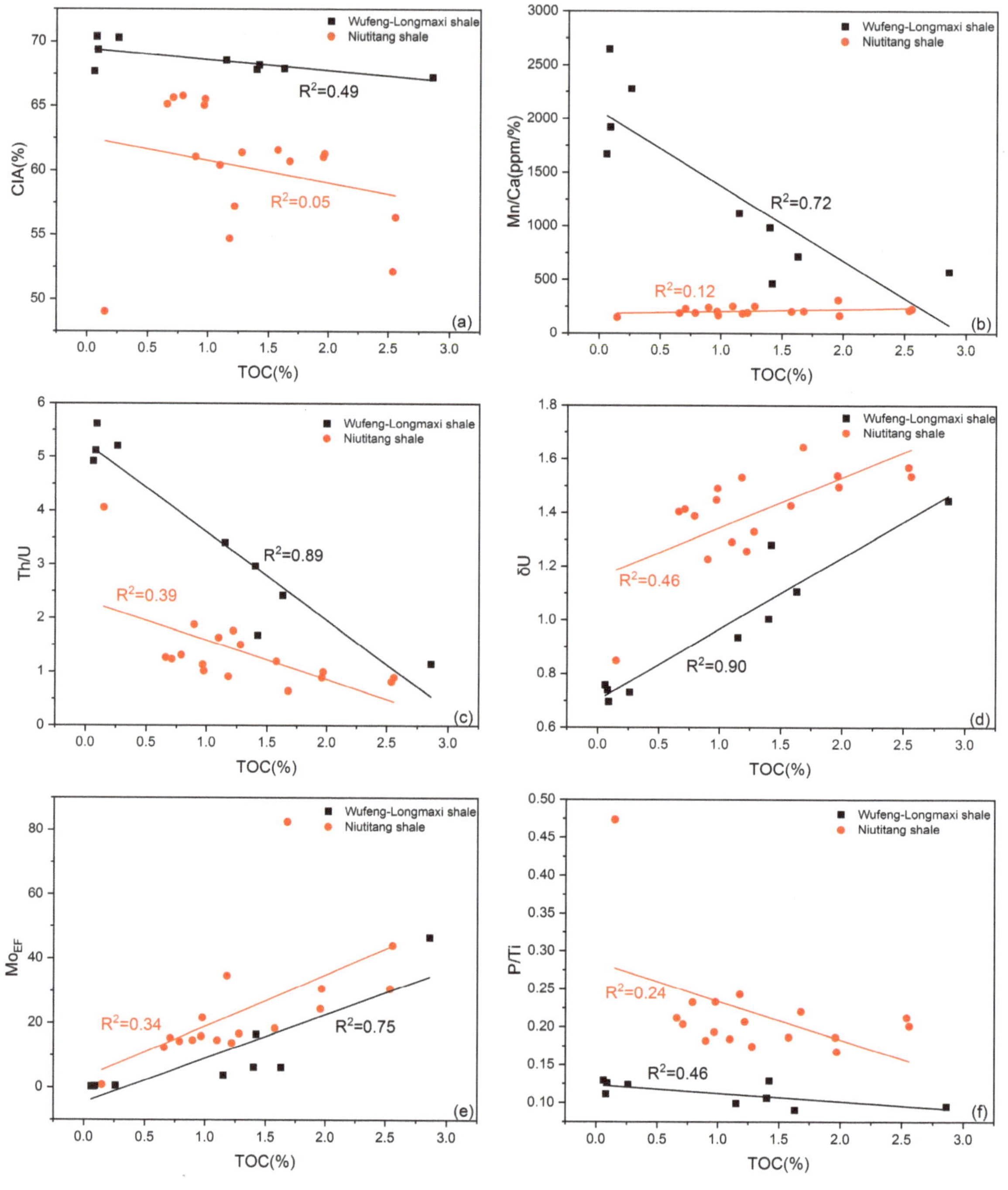

Figure 9. The relationship between different inorganic geochemical proxies and TOC contents.

5.6. Depositional and Organic Enrichment Models

Organic enrichment models mainly include preservation, production, and co-action models [66]. Cd/Mo and Co (ppm)× Mn (%) proxies have been proposed and applied to determine the dominant factor that impacts organic matter enrichment [66]. The results presented in Figure 10 indicate that the samples in the study well were mainly deposited in a restricted condition, based on Mo/TOC ratios. The Cd/Mn ratio reveals that organic matter enrichment of the Niutitang samples is mainly driven by preservation rather than productivity, while the dominating driving factor is the opposite for the Wufeng-Longmaxi samples.

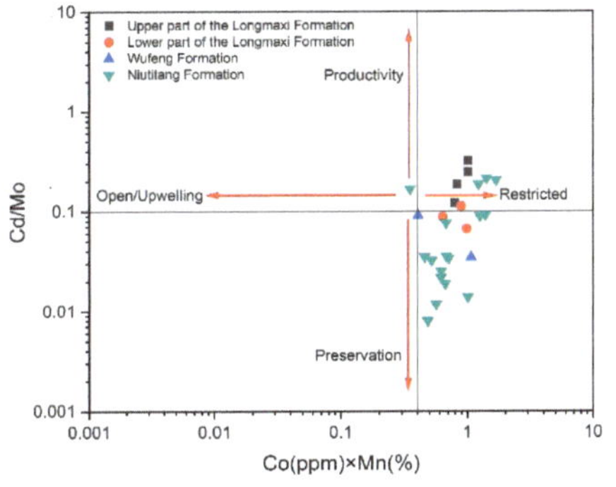

Figure 10. Cd/Mo versus Co (ppm) × Mn (%) for samples. The interpretation polygons are from Sweere et al. (2016) [66].

The Niutitang Formation was deposited in a moderately restricted basin with a euxinic reducing environment. This environment is considered to be the most suitable for preservation of organic matter (Figure 11a). The Wufeng Formation had similar reducing conditions and water mass restrictions to the Niutitang Formation (Figure 11b), but with increasing terrestrial influx intensity. As the water level decreased in the Longmaxi Formation, the terrestrial Influx steadily increased and diluted the richness of organic carbon. The hydrographic condition gradually became strongly restricted and the water environment shifted to dysoxic-oxic conditions due to a decrease in sea level, indicating that the paleo-productivity may have had a more significant impact on the organic matter enrichment compared to the paleo-redox condition (Figure 11c).

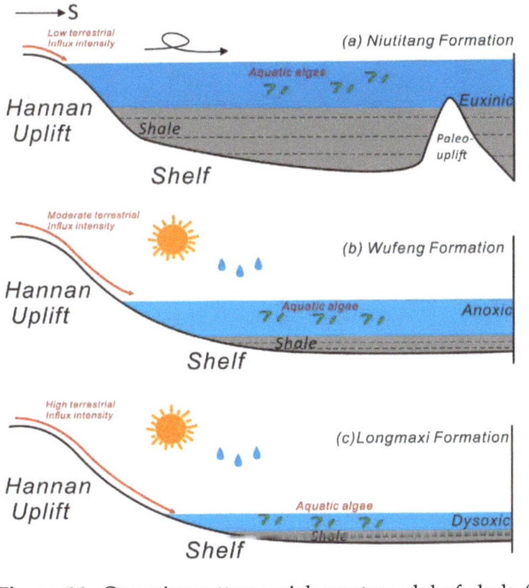

Figure 11. Organic matter enrichment model of shale from the Niutitang Formation (**a**), Wufeng Formation (**b**) and Longmaxi Formation (**c**).

6. Conclusions

The main conclusions from this study of the depositional environment and organic matter enrichment in Lower Paleozoic shale samples from the northeastern margin of the Yangtze Platform are as follows:

(1) Warm and humid conditions existed for deposition of the Wufeng-Longmaxi samples. In contrast, palaeoclimatic condition indicators suggest cold andarid conditions in the Niutitang samples.
(2) The terrestrial influx intensity in the Niutitang samples was relatively low. However, the terrestrial influx rate increased in the Wufeng and Longmaxi samples.
(3) Parameters that measure palaeoredox conditions indicate an anoxic environment in the Niutitang, Wufeng and Lower Longmaxi samples. In contrast, the paleo-redox conditions were more oxic during the sedimentation of the Upper Longmaxi samples.
(4) The P/Ti ratio reveals a low paleo-productivity for the Wufeng-Longmaxi and the Niutitang samples. Additionally, Wufeng-Longmaxi samples possess a lower paleo-productivity compared to the Niutitang samples in the well.
(5) An obvious difference exists in the organic matter enrichment of the shale samples from different formations. The organic matter enrichment for samples in this study can be jointly influenced by paleo-redox conditions and paleo-productivity. Organic matter enrichment of the Niutitang shale is mainly driven by preservation rather than productivity. The dominant driving factor is the opposite for the Wufeng-Longmaxi shale.

Author Contributions: Methodology, P.L.; Investigation, P.L.; Writing—original draft, P.L.; Writing—review & editing, C.L. and R.G.; Funding acquisition, C.L. and R.G. All authors have read and agreed to the published version of the manuscript.

Funding: This study was financially supported by the "100-Talent" program from Chinese Academy of Sciences and the Postdoctoral Science Foundation of China (No. 2022MD713802).

Institutional Review Board Statement: Not applicable.

Informed Consent Statement: Not applicable.

Data Availability Statement: The data presented in this study are available on request from the corresponding author.

Acknowledgments: We would like to thank the reviewers for their valuable comments to improve the quality of the paper.

Conflicts of Interest: The authors declare no conflict of interest.

References

1. Dong, D.; Shi, Z.; Guan, Q.; Jiang, S.; Zhang, M.; Zhang, C.; Wang, S.; Sun, S.; Yu, R.; Liu, D.; et al. Progress, challenges and prospects of shale gas exploration in the Wufeng–Longmaxi reservoirs in the Sichuan Basin. *Nat. Gas Ind. B* **2018**, *5*, 415–424. [CrossRef]
2. Zou, C.; Dong, D.; Wang, S.; Li, J.; Li, X.; Wang, Y.; Li, D.; Cheng, K. Geological characteristics and resource potential of shale gas in China. *Pet. Explor. Dev.* **2010**, *37*, 641–653. [CrossRef]
3. Guo, T. The Fuling Shale Gas Field—A highly productive Silurian gas shale with high thermal maturity and complex evolution history, southeastern Sichuan Basin, China. *Interpretation* **2015**, *3*, SJ25–SJ34. [CrossRef]
4. Chen, X.; Bao, S.; Zhai, G.; Zhou, Z.; Tong, C.; Wang, C. The discovery of shale gas within Lower Cambrian marine facies at Shan Nandi-1 well on the margin of Hannan palaeouplift. *China Geol.* **2018**, *45*, 412–413.
5. Jarvie, D.M.; Hill, R.J.; Ruble, T.E.; Pollastro, R.M. Unconventional shale-gas systems: The Mississippian Barnett Shale of north-central Texas as one model for thermogenic shale-gas assessment. *AAPG Bull.* **2007**, *91*, 475–499. [CrossRef]
6. Tissot, B.; Durand, B.; Espitalie, J.; Combaz, A. Influence of nature and diagenesis of organic matter in formation of petroleum. *AAPG Bull.* **1974**, *58*, 499–506.
7. Wang, Y.; Zhu, Y.; Liu, S.; Zhang, R. Methane adsorption measurements and modeling for organic-rich marine shale samples. *Fuel* **2016**, *172*, 301–309. [CrossRef]
8. Wu, Y.; Fan, T.; Jiang, S.; Yang, X.; Ding, H.; Meng, M.; Wei, D. Methane Adsorption Capacities of the Lower Paleozoic Marine Shales in the Yangtze Platform, South China. *Energy Fuels* **2015**, *29*, 4160–4167. [CrossRef]

9. Zou, J.; Rezaee, R.; Xie, Q.; You, L.; Liu, K.; Saeedi, A. Investigation of moisture effect on methane adsorption capacity of shale samples. *Fuel* **2018**, *232*, 323–332. [CrossRef]
10. Cox, G.M.; Jarrett, A.; Edwards, D.; Crockford, P.W.; Halverson, G.P.; Collins, A.S.; Poirier, A.; Li, Z.-X. Basin redox and primary productivity within the Mesoproterozoic Roper Seaway. *Chem. Geol.* **2016**, *440*, 101–114. [CrossRef]
11. Li, Y.; Liu, W.; Liu, P.; Luo, H.; Wang, X.; Zhang, D.; Yuan, Y. Paleoenvironment and Organic Matter Enrichment of the Middle Ordovician Marine Carbonates in the Ordos Basin of China: Evidence from Element Geochemistry. *ACS Earth Space Chem.* **2022**, *6*, 44–55. [CrossRef]
12. Mohialdeen, I.M.J.; Hakimi, M.H. Geochemical characterisation of Tithonian–Berriasian Chia Gara organic-rich rocks in northern Iraq with an emphasis on organic matter enrichment and the relationship to the bioproductivity and anoxia conditions. *J. Asian Earth Sci.* **2016**, *116*, 181–197. [CrossRef]
13. Schmid, S.; Kunzmann, M.; Pagès, A. Inorganic geochemical evaluation of hydrocarbon source rock potential of Neoproterozoic strata in the Amadeus Basin, Australia. *Mar. Pet. Geol.* **2017**, *86*, 1092–1105. [CrossRef]
14. Tribovillard, N.; Algeo, T.J.; Lyons, T.; Riboulleau, A. Trace metals as paleoredox and paleoproductivity proxies: An update. *Chem. Geol.* **2006**, *232*, 12–32. [CrossRef]
15. Choi, J.H.; Hariya, Y. Geochemistry and depositional environment of Mn oxide deposits in the Tokoro Belt, northeastern Hokkaido, Japan. *Econ. Geol.* **1992**, *87*, 1265–1274. [CrossRef]
16. Li, Y.; Zhang, T.; Ellis, G.S.; Shao, D. Depositional environment and organic matter accumulation of Upper Ordovician–Lower Silurian marine shale in the Upper Yangtze Platform, South China. *Palaeogeogr. Palaeoclimatol. Palaeoecol.* **2017**, *466*, 252–264. [CrossRef]
17. Rimmer, S.M. Geochemical paleoredox indicators in Devonian–Mississippian black shales, Central Appalachian Basin (USA). *Chem. Geol.* **2004**, *206*, 373–391. [CrossRef]
18. Pan, S.; Liao, Y.; Jiang, B.; Wan, Z.; Wang, F. Impact of natural weathering on source rocks: Organic and inorganic geochemical evidence from the Triassic Chang 7 outcrop profile in Tongchuan of the Southern Ordos Basin (China). *Int. J. Coal Geol.* **2022**, *263*, 104119. [CrossRef]
19. Liu, Z.; Yan, D.; Du, X.; Li, S. Organic matter accumulation of the early Cambrian black shales on the flank of Micangshan-Hannan Uplift, northern upper Yangtze Block, South China. *J. Pet. Sci. Eng.* **2021**, *200*, 108378. [CrossRef]
20. Chang, Y.; Xu, C.H.; Reiners, P.W. The exhumation evolution of the Micang Shan-Hannan uplift since Cretaceous: Evidence from apatite (U-Th)/He dating. *Acta Geophys. Sin.* **2010**, *53*, 912–919.
21. Chen, X.; Guo, T.; Shi, D.; Hou, Q.; Wang, C. Pore structure characteristics and adsorption capacity of Niutitang Formation shale in southern Shaanxi. *Lithol. Reserv* **2019**, *31*, 52–60.
22. Tian, T.; Yang, P.; Ren, Z.; Fu, D.; Zhou, S.; Yang, F.; Li, J. Hydrocarbon migration and accumulation in the Lower Cambrian to Neoproterozoic reservoirs in the Micangshan tectonic zone, China: New evidence of fluid inclusions. *Energy Rep.* **2020**, *6*, 721–733. [CrossRef]
23. Zhang, Y.; Jia, X.; Wang, Z. Palaeogeography and provenance analysis of Early Cambrian Xiannüdong Formation in the Micangshan area. *Acta Geol. Sin.* **2019**, *93*, 2904–2920.
24. Liu, W.; Yang, X.; Shu, S.; Liu, L.; Yuan, S. Precambrian Basement and Late Paleoproterozoic to Mesoproterozoic Tectonic Evolution of the SW Yangtze Block, South China: Constraints from Zircon U–Pb Dating and Hf Isotopes. *Minerals* **2018**, *8*, 333. [CrossRef]
25. Ma, R.; Xiao, Y.; Wei, X.; He, Z.; Li, Y. The magmatic activity and tectonic evolution in the micangshan area, china. *J. Mineral. Petrol.* **1997**, *17*, 76–82.
26. Li, Y.; Xiong, X.; Feng, Y. Mesozoic uplift of the Dabashan and Micangshan-Hannan Dome in the South Qinling orogenic belt. *Sci. China Earth Sci.* **2022**, *65*, 426–436. [CrossRef]
27. Tian, T.; Zhou, S.; Fu, D.; Yang, F.; Li, J. Calculation of the original abundance of organic matter at high-over maturity: A case study of the Lower Cambrian Niutitang shale in the Micangshan-Hannan Uplift, SW China. *J. Pet. Sci. Eng.* **2019**, *179*, 645–654. [CrossRef]
28. Tuo, J.; Wu, C.; Zhang, M. Organic matter properties and shale gas potential of Paleozoic shales in Sichuan Basin, China. *J. Nat. Gas Sci. Eng.* **2016**, *28*, 434–446. [CrossRef]
29. Wu, J.; Liang, C.; Jiang, Z.; Zhang, C. Shale reservoir characterization and control factors on gas accumulation of the Lower Cambrian Niutitang shale, Sichuan Basin, South China. *Geol. J.* **2019**, *54*, 1604–1616. [CrossRef]
30. Fan, J.; Melchin, M.J.; Chen, X.; Wang, Y.; Zhang, Y.; Chen, Q.; Chi, Z.; Chen, F. Biostratigraphy and geography of the Ordovician-Silurian Lungmachi black shales in South China. *Sci. China Earth Sci.* **2011**, *54*, 1854. [CrossRef]
31. Xiao, B.; Xiong, L.; Zhao, Z.; Fu, X. Sedimentary tectonic pattern of Wufeng and Longmaxi Formations in the northern margin of Sichuan Basin, South China. *Int. Geol. Rev.* **2022**, *64*, 2166–2185. [CrossRef]
32. McLennan, S.M. Rare earth elements in sedimentary rocks; influence of provenance and sedimentary processes. *Rev. Mineral. Geochem.* **1989**, *21*, 169–200.
33. Taylor, S.R.; McLennan, S.M. *The Continental Crust: Its Composition and Evolution*; Blackwell Scientific Pub.: Palo Alto, CA, USA, 1985.
34. McLennan, S.M.; Hemming, S.R.; Taylor, S.R.; Eriksson, K.A. Early Proterozoic crustal evolution: Geochemical and NdPb isotopic evidence from metasedimentary rocks, southwestern North America. *Geochim. Cosmochim. Acta* **1995**, *59*, 1153–1177. [CrossRef]
35. Nesbitt, H.W.; Young, G.M. Early Proterozoic climates and plate motions inferred from major element chemistry of lutites. *Nature* **1982**, *299*, 715–717. [CrossRef]

36. McLennan, S.M. Weathering and Global Denudation. *J. Geol.* **1993**, *101*, 295–303. [CrossRef]
37. Fedo, C.M.; Wayne Nesbitt, H.; Young, G.M. Unraveling the effects of potassium metasomatism in sedimentary rocks and paleosols, with implications for paleoweathering conditions and provenance. *Geology* **1995**, *23*, 921–924. [CrossRef]
38. Tao, S.; Xu, Y.; Tang, D.; Xu, H.; Li, S.; Chen, S.; Liu, W.; Cui, Y.; Gou, M. Geochemistry of the Shitoumei oil shale in the Santanghu Basin, Northwest China: Implications for paleoclimate conditions, weathering, provenance and tectonic setting. *Int. J. Coal Geol.* **2017**, *184*, 42–56. [CrossRef]
39. Goodarzi, F. Comparison of the geochemistry of lacustrine oil shales of Mississippian age from Nova Scotia and New Brunswick, Canada. *Int. J. Coal Geol.* **2020**, *220*, 103398. [CrossRef]
40. Goodarzi, F.; Goodarzi, N.N.; Malachowska, A. Elemental composition, environment of deposition of the lower Carboniferous Emma Fiord formation oil shale in Arctic Canada. *Int. J. Coal Geol.* **2021**, *244*, 103715. [CrossRef]
41. Hayashi, K.-I.; Fujisawa, H.; Holland, H.D.; Ohmoto, H. Geochemistry of ~1.9 Ga sedimentary rocks from northeastern Labrador, Canada. *Geochim. Cosmochim. Acta* **1997**, *61*, 4115–4137. [CrossRef]
42. Li, Y.; Sun, P.; Liu, Z.; Yao, S.; Xu, Y.; Liu, R.; Li, Y. Geochemistry of the Permian Oil Shale in the Northern Bogda Mountain, Junggar Basin, Northwest China: Implications for Weathering, Provenance, and Tectonic Setting. *ACS Earth Space Chem.* **2020**, *4*, 1332–1348. [CrossRef]
43. Yang, B.; Hu, B.; Bao, Z.; Zhang, Z. REE geochemical characteristics and depositional environment of the black shale-hosted Baiguoyuan Ag-V deposit in Xingshan, Hubei Province, China. *J. Rare Earths* **2011**, *29*, 499–506. [CrossRef]
44. Zhao, Z.H. *Trace Elements Geochemistry*, 2nd ed.; Science Press: Beijing, China, 2016.
45. Roser, B.P.; Korsch, R.J. Determination of Tectonic Setting of Sandstone-Mudstone Suites Using SiO_2 Content and K_2O/Na_2O Ratio. *J. Geol.* **1986**, *94*, 635–650. [CrossRef]
46. Hatch, J.R.; Leventhal, J.S. Relationship between inferred redox potential of the depositional environment and geochemistry of the Upper Pennsylvanian (Missourian) Stark Shale Member of the Dennis Limestone, Wabaunsee County, Kansas, U.S.A. *Chem. Geol.* **1992**, *99*, 65–82. [CrossRef]
47. Jones, B.; Manning, D.A.C. Comparison of geochemical indices used for the interpretation of palaeoredox conditions in ancient mudstones. *Chem. Geol.* **1994**, *111*, 111–129. [CrossRef]
48. Wei, Y.; Li, X.; Zhang, R.; Li, X.; Lu, S.; Qiu, Y.; Jiang, T.; Gao, Y.; Zhao, T.; Song, Z.; et al. Influence of a Paleosedimentary Environment on Shale Oil Enrichment: A Case Study on the Shahejie Formation of Raoyang Sag, Bohai Bay Basin, China. *Front. Earth Sci.* **2021**, *9*, 736054. [CrossRef]
49. Zhao, B.; Li, R.; Qin, X.; Wang, N.; Zhou, W.; Khaled, A.; Zhao, D.; Zhang, Y.; Wu, X.; Liu, Q. Geochemical characteristics and mechanism of organic matter accumulation of marine-continental transitional shale of the lower permian Shanxi Formation, southeastern Ordos Basin, north China. *J. Pet. Sci. Eng.* **2021**, *205*, 108815. [CrossRef]
50. Kimura, H.; Watanabe, Y. Oceanic anoxia at the Precambrian-Cambrian boundary. *Geology* **2001**, *29*, 995–998. [CrossRef]
51. Wignall, P.B.; Twitchett, R.J. Oceanic Anoxia and the End Permian Mass Extinction. *Science* **1996**, *272*, 1155–1158. [CrossRef]
52. Algeo, T.J.; Tribovillard, N. Environmental analysis of paleoceanographic systems based on molybdenum–uranium covariation. *Chem. Geol.* **2009**, *268*, 211–225. [CrossRef]
53. Zheng, Y.; Anderson, R.F.; van Geen, A.; Kuwabara, J. Authigenic molybdenum formation in marine sediments: A link to pore water sulfide in the Santa Barbara Basin. *Geochim. Cosmochim. Acta* **2000**, *64*, 4165–4178. [CrossRef]
54. Wang, C.; Wang, Q.; Chen, G.; He, L.; Xu, Y.; Chen, L.; Chen, D. Petrographic and geochemical characteristics of the lacustrine black shales from the Upper Triassic Yanchang Formation of the Ordos Basin, China: Implications for the organic matter accumulation. *Mar. Pet. Geol.* **2017**, *86*, 52–65. [CrossRef]
55. Tribovillard, N.; Algeo, T.J.; Baudin, F.; Riboulleau, A. Analysis of marine environmental conditions based onmolybdenum–uranium covariation—Applications to Mesozoic paleoceanography. *Chem. Geol.* **2012**, *324–325*, 46–58. [CrossRef]
56. Wu, Y.; Liu, C.; Liu, Y.; Gong, H.; Awan, R.S.; Li, G.; Zang, Q. Geochemical characteristics and the organic matter enrichment of the Upper Ordovician Tanjianshan Group, Qaidam Basin, China. *J. Pet. Sci. Eng.* **2022**, *208*, 109383. [CrossRef]
57. Algeo, T.J.; Lyons, T.W. Mo–total organic carbon covariation in modern anoxic marine environments: Implications for analysis of paleoredox and paleohydrographic conditions. *Paleoceanography* **2006**, *21*, PA1016. [CrossRef]
58. Gupta, L.P.; Kawahata, H. Downcore diagenetic changes in organic matter and implications for paleoproductivity estimates. *Glob. Planet. Chang.* **2006**, *53*, 122–136. [CrossRef]
59. Chen, G.; Gang, W.; Chang, X.; Wang, N.; Zhang, P.; Cao, Q.; Xu, J. Paleoproductivity of the Chang 7 unit in the Ordos Basin (North China) and its controlling factors. *Palaeogeogr. Palaeoclimatol. Palaeoecol.* **2020**, *551*, 109741. [CrossRef]
60. Xu, L.; Cheng, Y.; Zhang, J.; Liu, Y.; Yang, Y. Controls on the Organic Matter Accumulation of the Marine–Continental Transitional Shanxi Formation Shale in the Southeastern Ordos Basin. *ACS Omega* **2022**, *7*, 4317–4332. [CrossRef]
61. Redfield, A.C. The Biological Control of Chemical Factors in the Environment. *Am. Sci.* **1958**, *46*, 230A-221.
62. Paytan, A.; McLaughlin, K. The Oceanic Phosphorus Cycle. *Chem. Rev.* **2007**, *107*, 563–576. [CrossRef]
63. Schoepfer, S.D.; Shen, J.; Wei, H.; Tyson, R.V.; Ingall, E.; Algeo, T.J. Total organic carbon, organic phosphorus, and biogenic barium fluxes as proxies for paleomarine productivity. *Earth-Sci. Rev.* **2015**, *149*, 23–52. [CrossRef]
64. Yang, M.; Zuo, Y.; Fu, X.; Qiu, L.; Li, W.; Zhang, J.; Zheng, Z.; Zhang, J. Paleoenvironment of the Lower Ordovician Meitan Formation in the Sichuan Basin and Adjacent Areas, China. *Minerals* **2022**, *12*, 75. [CrossRef]

65. Dembicki, J.H. Chapter 3-Source Rock Evaluation. In *Practical Petroleum Geochemistry for Exploration and Production*; Dembicki, J.H., Ed.; Elsevier: Amsterdam, The Netherlands, 2017; pp. 61–133.
66. Sweere, T.; van den Boorn, S.; Dickson, A.J.; Reichart, G.-J. Definition of new trace-metal proxies for the controls on organic matter enrichment in marine sediments based on Mn, Co, Mo and Cd concentrations. *Chem. Geol.* **2016**, *441*, 235–245. [CrossRef]

Journal of
Marine Science and Engineering

MDPI

Review

Plastics and Five Heavy Metals from Sea Beaches: A Geographical Synthesis of the Literary Information

Anna V. Mikhailenko [1] and Dmitry A. Ruban [2,*]

1 Department of Physical Geography, Ecology, and Nature Protection, Institute of Earth Sciences, Southern Federal University, Zorge Street 40, Rostov-on-Don 344090, Russia
2 Department of Organization and Technologies of Service Activities, Institute of Tourism, Service and Creative Industries, Southern Federal University, 23-ja Linija Street 43, Rostov-on-Don 344019, Russia
* Correspondence: ruban-d@mail.ru

Abstract: Pollution of the oceans and seas, as well as their coastal zones, with plastics has become serious challenge, which is also related to the Anthropocene marine geology and geochemistry. Notably, plastics can bear heavy metals. The related knowledge is scattered through scientific publications, and, thus, it needs generalization. The present study synthesizes the published information about the geographical distribution of cadmium, chromium, mercury, nickel, and zinc associated with plastics on sea beaches. A bibliographical survey is undertaken, and the collected literary information is organized so as to document the principal localities of beached plastics bearing these metals. About twenty localities are established in many parts of the world, and the majority of them correspond to the coasts of the Atlantic and Indian oceans and their seas. Significant attention has been paid by the previous researchers to Northwest Europe and South and East Asia. The available information is enough to postulate the global extent of the problem of heavy metals association with plastics on beaches. Real or potential risks to the environment were reported in many cases. This review also shows that the industry of tourism, hospitality, and recreation contributes to the accumulation of such plastics. The analyzed literary information demonstrates several biases: for instance, Hg-bearing plastics on beaches were reported from the smaller number of localities relatively to the other metals, and the beaches and coastal recreational areas of many regions are yet to be studied in regard to heavy metals associated with plastics.

Keywords: Anthropocene; coasts; marine geochemistry; pollution; tourism

Citation: Mikhailenko, A.V.; Ruban, D.A. Plastics and Five Heavy Metals from Sea Beaches: A Geographical Synthesis of the Literary Information. *J. Mar. Sci. Eng.* **2023**, *11*, 626. https://doi.org/10.3390/jmse11030626

Academic Editor: Michele Arienzo

Received: 23 February 2023
Revised: 10 March 2023
Accepted: 15 March 2023
Published: 16 March 2023

1. Introduction

Plastics, which include macro-, meso-, and microplastics, have become a serious environmental challenge for contemporary society. Having various forms, compositions, and origin, plastics pollute landscapes on the planetary scale. Particularly, oceans and seas, as well as their coastal zones, have already become too affected by plastic pollution [1–8]. This challenge should not be simplified because plastics are not only "just litter", unacceptable aesthetically, and marking anthropogenic modifications of the natural environment [9–11], but are also a trigger of biological and chemical (often biochemical) processes with negative or unclear consequences [12–18], as well as a contributor to carbon cycling and climate changes [19–23]. The dynamics of plastics in marine environments conjugate with biological activities [24]. In other words, marine plastic pollution is a highly-complex problem with different aspects, each of which should be addressed with significant attention.

Heavy metals associate closely with plastics in marine and coastal environments. The former can be linked to additives [25–27] and adsorption [28–32]. Indeed, bioavailability also matters [33–35]. Such an association increases (if not multiplies) the danger from marine plastic pollution because elevated concentrations of such heavy metals as cadmium or mercury may have significant negative influences on the living nature [36–40]. Special attention should be paid to beaches. On the one hand, their recreational use often leads to a voluminous supply and accumulation of plastics [41–47], a part of which can be further transported to open marine environments. On the other hand, beaches can serve as barriers, where plastics (both touristic and non-touristic) accumulate, and, thus, contribute to the exposure of dangerous substances (also heavy metals) to visitors and biota. If so, one can expect the presence of not only plastics-related heavy metals on beaches, but also the related risks to their visitors and entire beach-based and marine ecosystems. Principally, beaches are within the global plastisphere (this has become a common term [48–50]), and their studies in regard to plastic-associated heavy metals are related to the focus of the Anthropocene geology and geochemistry [51–55]. These studies form a promising direction of geological oceanography.

Although research into beached plastics-related heavy metals started about fifteen years ago and intensified very recently, the related publications are already rather numerous, and, thus, the lines of evidence are scattered through the significant amount of sources (see review below). Taking into account the urgency of this information for environmental management, as well as for conceptual developments in the Anthropocene marine geology and geochemistry, synthesizing these lines of evidence appears to be very important. This corresponds to the current interest in environmental geochemistry (also marine and coastal) linked to tourism, hospitality, and recreation [56–58]. Indeed, such syntheses can address different aspects of the problem. The objective of the present contribution is to provide a geographical synthesis of the literary evidence of heavy metals associated with plastics on sea beaches. The need for such a geographical emphasis is dictated by two reasons. First, tracing the spatial distribution of the considered phenomenon reveals its scale. Second, the same analysis indicate the geographical biases in the knowledge of this phenomenon, which is important for the coordination of further investigations. This work does not pretend to be a comprehensive review of the particular environmental problem taken entirely, but it aims at documenting its geographical patterns in the light of the published knowledge.

2. Materials and Methods

From the various heavy metals which may be associated with plastics on sea beaches, attention is paid in this work to cadmium (Cd), chromium (Cr), mercury (Hg), nickel (Ni), and zinc (Zn). Initially, the literature was checked, and these metals were selected because they are among the best studied and widely documented. Moreover, negative environmental consequences and the related risks from pollution by cadmium [59], chromium [60], mercury [61], nickel [62], and zinc [63] in oceans and seas, as well as on beaches, are well-known.

The literary evidence, i.e., the information already published in international scientific journals, constitutes the material of the present study. It was collected by the "standard" procedures employed for the purposes of bibliographical surveys and reviews, which have become widespread in earth and environmental research [57,64–67]. The bibliographical database "Scopus", which boasts extensive coverage of the scientific literature [68,69], was used to collect the sources, and their subsequent check allowed information suitable for the present study to be established (Figure 1). It should be stressed that the majority of the leading environmental journals (if not all of them) are included in "Scopus", and this database offers quick indexation of their articles, which is important for such a quickly-grown research field as marine plastics. Moreover, "Scopus" offers a unique search engine for the quick and comprehensive collection of the literature.

A total of 38 articles devoted to plastics-associated cadmium, chromium, mercury, nickel, and zinc from beaches were found (Table 1). The majority of them were published in the major international journals ("Marine Pollution Bulletin" was most commonly selected by the authors). Notably, about two thirds of them appeared during the past five years (especially since the beginning of the 2020s). Among them, there are articles that considered several metals and a single metal. Although almost all the sources express environmental concerns, not all articles state beach pollution by heavy metals (some works simply note the presence of plastics-related heavy metals on beaches), and it is not easy to understand which particular metal is thought to be real or potential danger in some cases (several works consider heavy metals and specify the related risks in too general words). This seems to be a normal situation for the emerging research direction, and this is why this analysis focuses on the general geographical patterns.

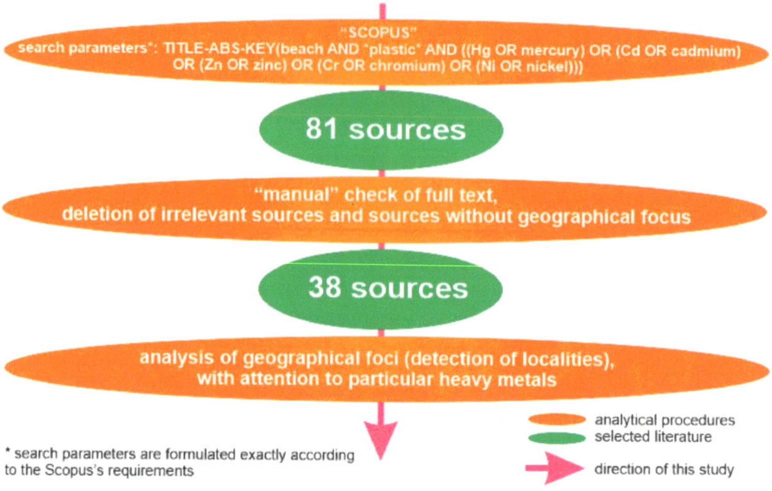

Figure 1. Procedures of collecting the literary information for this study.

Methodologically, the present analysis is framed by its objective, namely providing the geographical synthesis of the available knowledge. First of all, the geographical focus of the particular articles is established. Most commonly, this is a particular locality comprising one or several closely located beaches; the affinity to countries can be easily found. After the initial analysis of the literature, it is possible to attribute some very close, even overlapping, localities to a single locality. This is also possible when one work deals with several localities. These slight modifications are necessary to map the information on a global scale. Then, the localities are specified for each of the selected heavy metals. This permits the mapping of the global distribution of cadmium, chromium, mercury, nickel, and zinc in association with plastics on beaches. Indeed, this distribution reflects the only present state of the scientific knowledge. Taking into account the necessity of special, state-of-the-art techniques for investigation of this association, it is reasonable to believe that almost all (if not all) related information can be found in the only articles published in top international journals. It's possible incompleteness does not matter because it is among the aims of this study to judge about the degree of this incompleteness. In other words, this study focuses on mapping what is known from the literature and not what can take place in fact. The simple maps can be interpreted qualitatively to judge about the distribution of particular heavy metals associated with beached plastics. Attention is paid to two additional issues. First, the notions of environmental danger of plastics-associated heavy metals on beaches in the analyzed sources are considered. It should be stressed that the interpretations of this danger differ between the works: it is argued via comparison to some reference levels or via experiments or it is only hypothesized on the basis of too general assumptions (the latter occurs too often). This situation is expected because many works focus on the general issue of beach pollution by plastics, and heavy metals are considered "marginally". Moreover, it appears that the comprehensive studies of heavy metal cycling in different environments polluted by plastics are yet to be done, and, thus, the only preliminary judgments of the noted danger are possible. Second, the notions of whether the industry of tourism, hospitality, and recreation contributes to the accumulation of plastics and associated heavy metals are taken into account.

Table 1. The bibliographical information compiled for the purposes of this study.

Source	Heavy Metal(s) *	Locality	Dangerous Levels **	Tourism Impact ***
[70]	Cr, Hg, Ni	Cartagena (Colombia)	Yes	
[71]	Cd, Cr, Zn	Southwest England (UK)	Yes	
[72]	Hg	Persian Gulf and Kuwait Bay (Kuwait)	Yes	Yes
[73]	Cd, Cr, Zn	Guadeloupe (France)	Yes	
[74]	Cd, Cr, Zn	Guadeloupe (France)	Yes	
[75]	Cd, Cr, Ni	Persian Gulf (Iran)	Yes	
[76]	Cd, Zn	Guadeloupe (France)		
[77]	Cd, Hg, Zn	Vancouver (Canada)	Yes	Yes
[78]	Cd, Cr, Ni, Zn	Gulf of Guinea (Nigeria)		
[79]	Cd, Cr, Ni, Zn	Gulf of Guinea (Nigeria)	Yes	
[80]	Hg	Gdansk and Gdynia (Poland)	Yes	
[81]	Cd, Cr, Ni, Zn	Southwest England (UK)	Yes	
[82]	Hg	South Texas (USA)	Yes	
[83]	Cd, Cr, Ni, Zn	St Mary's Island (India)	Yes	Yes
[84]	Cd, Cr, Ni, Zn	Texel (the Netherlands)		
[85]	Cd, Ni, Zn	Hong Kong (China)	Yes	
[86]	Cd, Cr, Ni, Zn	Vis Island (Croatia)	Yes	
[87]	Cd	Southwest England (UK)	Yes	
[88]	Cd, Cr	Goto Islands (Japan)	Yes	
[89]	Cd, Cr, Ni, Zn	West Coast (Sri Lanka)	Yes	
[90]	Cr, Ni, Zn	Chennai (India)	Yes	Yes
[91]	Cr, Ni, Zn	Veracruz (Mexico)		Yes
[92]	Hg	Murcia (Spain)	Yes	Yes
[93]	Cr	West Coast (Sri Lanka)	Yes	Yes
[94]	Cd, Ni, Zn	Bahia (Brazil)		Yes
[95]	Cr	Chennai (India)	Yes	Yes
[96]	Cr	Claromecó (Argentina)		Yes
[97]	Cd, Cr, Hg, Ni, Zn	Southwest England (UK)	Yes	
[98]	Cd, Cr	Southwest England (UK)	Yes	
[99]	Zn	Southwest England (UK)		
[100]	Cd	Southwest England (UK)		
[101]	Cd, Cr, Hg, Zn	Northwest Europe (France, the Netherlands, UK)	Yes	
[102]	Cd, Cr, Ni, Zn	Southwest England (UK)	Yes	
[103]	Cr, Zn	São Paulo (Brazil)		
[104]	Zn	KwaZulu-Natal (South Africa)		Yes
[105]	Cd, Cr, Ni, Zn	Shenzhen (China)		Yes
[106]	Cd, Cr, Ni	Pearl River Estuary (China)	Yes	Yes
[107]	Cr, Ni, Zn	Kota Kinabalu (Malaysia)		

Notes: * from those heavy metals considered in this study (Cd, Cr, Hg, Ni, Zn); ** direct statement of real and hypothesized (potential) dangers from heavy metal(s) associated with plastics; *** direct statement of contribution of tourism, hospitality, and recreation to plastic supply/accumulation (in some papers, the origin of (micro)plastics is not discussed).

3. Results

3.1. Geographical Patterns

The association of the selected heavy metals with plastics on beaches was reported for many parts of the world. Our analysis of the published data permits the registration of thirteen localities belonging to twelve countries for cadmium (Table 2), seventeen localities belonging to fifteen countries for chromium (Table 3), nine localities in nine countries for mercury (Table 4), thirteen localities from twelve countries for nickel (Table 5), and sixteen localities belonging to fourteen countries for zinc (Table 6).

Table 2. Localities where cadmium associated with plastics have been studied.

ID	Locality	Sources
BA	Bahia (Brazil)	[94]
GO	Goto Islands (Japan)	[98]
GU	Guadeloupe (France)	[73,74,76]
GG	Gulf of Guinea (Nigeria)	[78,79]
HK	Hong Kong–Pearl River Estuary (China)	[85,105,106]
NF	Northern Coast (France)	[101]
PI	Persian Gulf (Iran)	[75]
SE	Southwest England (UK)	[71,81,87,97,98,100–102]
SM	St Mary's Island (India)	[83]
TE	Texel (the Netherlands)	[84,101]
VA	Vancouver (Canada)	[77]
VI	Vis Island (Croatia)	[86]
WS	West Coast (Sri Lanka)	[89]

Table 3. Localities where chromium associated with plastics have been studied.

ID	Locality	Sources
CA	Cartagena (Colombia)	[70]
CE	Chennai (India)	[90,95]
CL	Claromecó (Argentina)	[96]
GO	Goto Islands (Japan)	[88]
GU	Guadeloupe (France)	[73,74]
GG	Gulf of Guinea (Nigeria)	[78,79]
HK	Hong Kong–Pearl River Estuary (China)	[105,106]
KK	Kota Kinabalu (Malaysia)	[107]
NF	Northern Coast (France)	[101]
PI	Persian Gulf (Iran)	[75]
SP	São Paulo (Brazil)	[103]
SE	Southwest England (UK)	[71,81,97,98,101,102]
SM	St Mary's Island (India)	[83]
TE	Texel (the Netherlands)	[84,101]
VE	Veracruz (Mexico)	[91]
VI	Vis Island (Croatia)	[86]
WS	West Coast (Sri Lanka)	[89,93]

Table 4. Localities where mercury associated with plastics have been studied.

ID	Locality	Sources
CA	Cartagena (Colombia)	[70]
GD	Gdansk and Gdynia (Poland)	[80]
MU	Murcia (Spain)	[92]
NF	Northern Coast (France)	[101]
PK	Persian Gulf and Kuwait Bay (Kuwait)	[72]
SE	Southwest England (UK)	[97,101]
ST	South Texas (USA)	[82]
TE	Texel (the Netherlands)	[101]
VA	Vancouver (Canada)	[77]

Table 5. Localities where nickel associated with plastics have been studied.

ID	Locality	Sources
BA	Bahia (Brazil)	[94]
CA	Cartagena (Colombia)	[70]
CE	Chennai (India)	[90]
GG	Gulf of Guinea (Nigeria)	[78,79]
HK	Hong Kong–Pearl River Estuary (China)	[85,105,106]
KK	Kota Kinabalu (Malaysia)	[107]
PI	Persian Gulf (Iran)	[75]
SE	Southwest England (UK)	[81,97,102]
SM	St Mary's Island (India)	[83]
TE	Texel (the Netherlands)	[84]
VE	Veracruz (Mexico)	[91]
VI	Vis Island (Croatia)	[86]
WS	West Coast (Sri Lanka)	[89]

Mapping the collected literary information implies the wide geographical distribution of the beaches with plastics bearing the selected heavy metals (Figures 2–6). Although the most intensively studied locality is Southwest England in the United Kingdom, one should note the presence of such beaches in Africa, the Americas, Asia, and Europe. The previous studies were concentrated in West Europe and South and East Asia. Significant attention has been paid to South India and Sri Lanka. The majority of the localities correspond to the coasts of the Atlantic and Indian oceans. This is true for all five heavy metals (Figures 2, 3, 5 and 6), except for mercury, for which the localities tend to concentrate along the Atlantic coasts only (Figure 4). Zinc is the only considered heavy metal, for which an association with plastics is reported from more than one locality in Africa (Figure 6). Nonetheless, the observed association of cadmium, chromium, mercury, nickel, and zinc with beached plastics across the globe seems to be a global phenomenon because it is found in different geographical domains, and this is true even for mercury with the least number of localities (Figure 4).

In some cases, the concentrations of the selected heavy metals are found to be elevated enough to speculate about real or potential dangers to the local environment (Table 1). However, it is explained above that the bases of such judgments differ and are not always clear in some cases. It should also be noted that tourist activities may be responsible for the accumulation of plastics bearing heavy metals (Table 1).

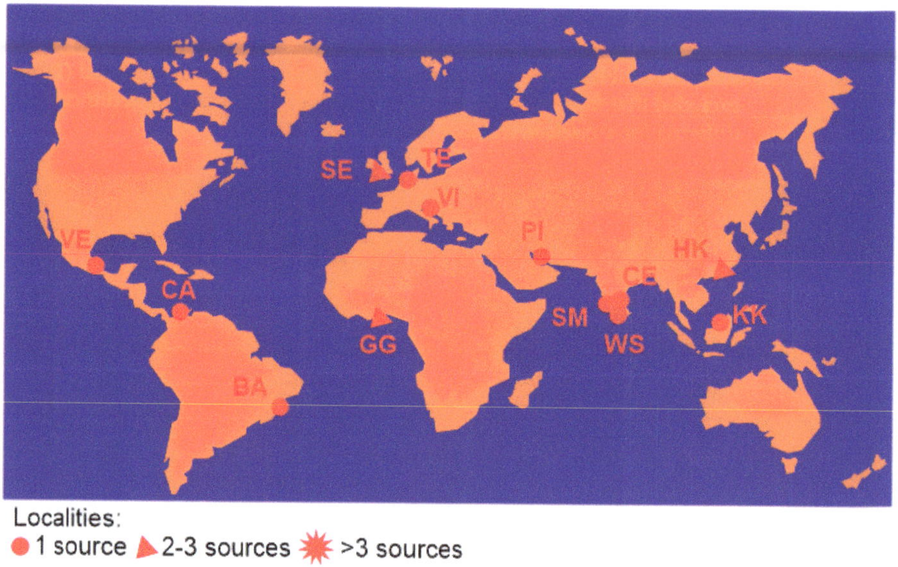

Figure 5. Geographical distribution of the beaches with Ni-bearing plastics in the light of the literary information (see Table 5 for IDs of localities).

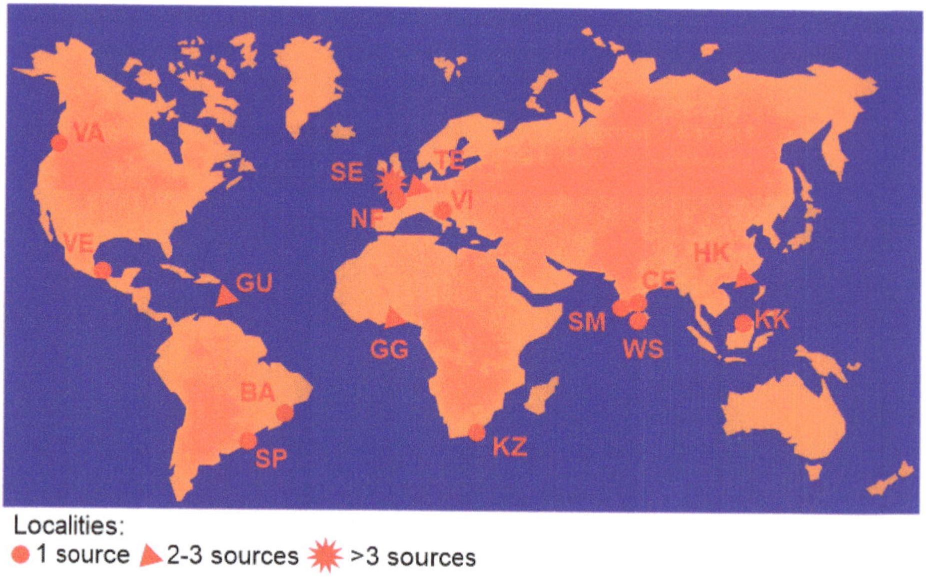

Figure 6. Geographical distribution of the beaches with Zn-bearing plastics in the light of the literary information (see Table 6 for IDs of localities).

3.2. Selected Examples

For cadmium, a representative example includes the beaches of Vancouver in Canada (Figure 2). There, Fernandes et al. [77] examined litter from the beach surface and, particularly, reported significant concentrations of cadmium (up to 32.3 $\mu g g^{-1}$) in candy wrappers from recreational areas. Even higher concentrations of this metal were found there in an unidentified fragment, presumably of industrial origin. Taking into account the significance of recreational activities on some studied beaches, where more than half of the plastic litter consists of food packaging [77], the related source of cadmium raises environmental concerns. A comparison with the "natural" levels of cadmium in beach sediments and nearby seawater is required to judge the pollution and true environmental risks.

For chromium, a notable example is the beaches of Chennai in India (Figure 3). Initially, Suman et al. [95] reported the absorption of trace metals, including chromium, on the surface of petroleum-based plastic debris collected from the beach surface and the upper layer of sediments. Its concentrations reach 1.11 μgg^{-1}, and they are relatively higher for polyamide polymers. Although the contribution of tourism and hotels may matter, the authors of this study indicated on the different origins of plastics in their case. A bit later, Ranjani et al. [90] undertook an extensive examination of microplastics on the southeast coast of India (Chennai coast) where the upper layer of sediments was sampled. Their study proved the adsorption of chromium on plastics and multiple sources of the latter. These specialists also stressed the major role of the industry of tourism, hospitality, and recreation in the supply of plastics and its seasonal variations on several beaches. Importantly, they take into account the concentrations of heavy metals in coastal environments to be compared from those in plastics.

As an example for mercury, the Mediterranean beaches of Murcia in Spain can be considered. According to Santos-Echeandía et al. [92] who examined different parts of these beaches, the concentration of mercury exceeds 8200 nmol kg^{-1}. This metal tends to associate with polyurethane and polyvinyl chloride and with black plastics. In regard to what is reported in this study, the association can be explained by both additives and absorption. Although different sources of plastics are proposed, beach recreation is among them. Importantly, it appears that plastics are able to increase concentrations of mercury higher than those in the aquatic environment, with the related effects on exposure of organisms to Hg-pollution.

A typical example for nickel can be found in Kota Kinabalu in Malaysia, where beached plastics were studied by Zahari et al. [107] who sampled the upper layer of beach sediments. Their investigation established the concentrations of nickel up to 20 mg kg^{-1}. Interestingly, this metal associates with different kinds of plastics on different beaches: polypropylene in one case and polystyrene in the other case. The interpretations of the sources of plastics presented in the work by Zahari et al. [107] leave the impression that these sources are diverse, being linked to local household activities and delivery by currents and waves; nonetheless, the contribution of beach recreation cannot be excluded. Indeed, a comparison to nickel concentrations in beach sediments and seawater will provide new insights.

For zinc, a representative example is the beaches of Shenzhen, which is attributed to the Hong Kong–Pearl River Estuary in South China. Xiang et al. [105] reported the relatively high concentrations of zinc (>500 mg kg^{-1}) from plastics collected on these beaches. The experiments showed that heavy metals (including zinc) from plastics do not pose negative health effects. However, the same example implies that recreational activities contribute substantially to accumulation of Zn-rich plastics.

4. Discussion

4.1. Biases and Related Interpretations

The registered geographical patterns (Figures 2–6) do not reflect true differences in the distribution of plastics-associated heavy metals, but they reflect differences in the scope of the international research and the analytical opportunities. There are evident geographical biases. Of the biggest concern is the absence of information (at least, it was not found with this bibliographical survey) from Australia, with its long shoreline and numerous beaches. The localities tend to concentrate along the coasts of the Atlantic and Indian oceans, whereas those of the other oceans are too rare or absent. The scarcity of information from the Mediterranean appears to be intriguing.

Speaking about the noted biases highlights another, research-related issue. One would hypothesize that investigations of the association of heavy metals with plastics requires advanced analytics and significant funding, and, thus, they should concentrate in the developed countries. However, the outcomes of the present analysis of the literary information (Tables 2–6) totally disprove such a hypothesis. Many localities are found in the developing countries, and they were investigated by the research teams from these countries to mark their big success in high-class research. In contrast, the records from some developed countries are scarce or absent, which questions the research agenda there. Most probably, the geographical biases should be explained by the very sporadic interest of the entire international research community to the urgent problem, which coexists on the agenda together with many other problems. Anyway, the intense studies of this problem in Southwest England [71,81,87,97–102], Guadeloupe [73,74,76], and South China [85,105,106] demonstrate how this research theme can rise in relation to the particular areas.

The present synthesis implies that plastics bearing heavy metals are often supplied by the industry of tourism, hospitality, and recreation (Table 7). The analysis of the content of the collected literature (Table 1) also suggests that this industry is significant, but usually only one of many (local and remote) sources of plastics, and it dominates rather rarely (see also the examples given above). Moreover, the literature often suggests that the proper management of beaches, including regular cleaning and restrictions for visitors, minimizes the role of this source to a significant degree (for instance, see [107] about the efficacy of such measures in Kota Kinabalu in Malaysia). Nonetheless, the other research has shown that the industry of tourism, hospitality, and recreation can produce voluminous amounts of plastics on beaches, which also actively "catch" plastics delivered by currents and waves. The related lines of evidence were obtained from Bali in Indonesia [108], California in the USA [109], Northern Sardinia in Italy [110], Phuket in Thailand [111], and South Australia [112]. If so, one can assume that the knowledge of heavy metals associated with plastics on beaches due to touristic activities (Table 7) remains very incomplete, and the problem may be significantly stronger than one can imagine. A specific issue deserving attention is the creation of artificial beaches and the modification of natural beaches for recreational purposes. Apparently, they can become new barriers for the accumulation of plastics bearing heavy metals, and, thus, the industry of tourism, hospitality, and recreation contributes to the problem indirectly, i.e., even without the supply of plastics.

Table 7. Localities where plastics bearing heavy metals accumulate due to tourist activities.

Locality	Heavy Metals *	Sources
Bahia (Brazil)	Cd, Ni, Zn	[94]
Chennai (India)	Cr, Ni, Zn	[90,95]
Claromecó (Argentina)	Cr	[96]
Hong Kong–Pearl River Estuary (China)	Cd, Cr, Ni, Zn	[105,106]
KwaZulu-Natal (South Africa)	Zn	[104]
Murcia (Spain)	Hg	[92]
Persian Gulf and Kuwait Bay (Kuwait)	Hg	[72]
St Mary's Island (India)	Cd, Cr, Ni, Zn	[83]
Vancouver (Canada)	Cd, Hg, Zn	[77]
Veracruz (Mexico)	Cr, Ni, Zn	[91]
West Coast (Sri Lanka)	Cr	[93]

Note: * all considered heavy metals reported in association with plastics from a given locality are indicated.

The present geographical synthesis proves the global scale of the problem with the presence of heavy metals associated with beached plastics. Although its studies have remained scarce, and the number of the related localities remains rather limited (Tables 2–6), cadmium (Figure 2), chromium (Figure 3), mercury (Figure 4), nickel (Figure 5), and zinc (Figure 6) are associated with plastics on the beaches in many parts of the world. Irrespectively of whether heavy metals occur as additives or are adsorbed from outside (also from seawater), they often tend to concentrate on beaches thanks to the accumulation of various plastics. If so, the geochemical peculiarities of coastal environments experience changes. Taking into account the noted global scale of these processes, it is possible to state their contribution to the Anthropocene modification of the planetary environment. This geological interval is marked by geochemical perturbations, including those linked to heavy metals [113,114] and plastics [115]. It should be stressed that beaches are actively visited by definition, as well as requiring cleanness and preservation of their natural state. Therefore, the noted geochemical changes mark the situation when the specific Anthropocene dynamics affect the specific anthropogenic environment.

4.2. Implications

Although the present work focuses on the geographical synthesis of the literary information, its outcomes also permit outlining some practical implications. The association of heavy metals (especially with dangerous concentrations) with plastics on sea beaches is a serious challenge to the industry of tourism, hospitality, and recreation, for which beaches constitute a crucial resource. The interests of beach visitors should be reflected by policy-makers and the media. These stakeholders can pose tasks (these can be simplified as the cleanness and environmental safety of beaches) and provide funding to researchers, who can only realize the scale of the problem on the global, national, and local scales. The outcomes of the related research projects should be communicated to the noted stakeholders to prescribe regulations, rules, and recommendations to organizations of tourism, hospitality, and recreation, as well as to the other stakeholders. The considered environmental challenge seems to be too complex and still too poorly understood (for instance, see the geographical biases identified above) that it cannot be addressed without "deep" involvement of the international research community (Figure 7). As demonstrated by the present study, research teams from many countries are able to provide the related service to the contemporary society.

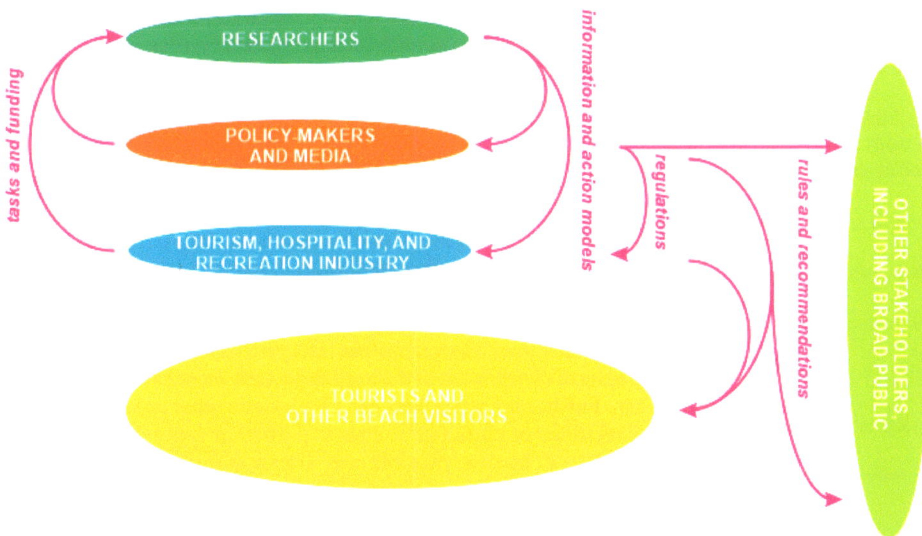

Figure 7. Proposed action framework for dealing with the studied problem.

The undertaken analysis of the available literature has faced certain difficulties. The main difficulty is the significant heterogeneity of the published scientific information. The problem is not that the different articles consider different heavy metals and different localities. More important is that the outcomes of the particular studies are reported differently. For instance, the concentrations of heavy metals are stated either numerically in the text or only graphically on the figures, the links to different plastic types are either

22. Kvale, K. Implications of plastic pollution on global marine carbon cycling and climate. *Emerg. Top. Life Sci.* **2022**, *6*, 359–369. [PubMed]
23. Smeaton, C. Augmentation of global marine sedimentary carbon storage in the age of plastic. *Limnol. Oceanogr. Lett.* **2021**, *6*, 113–118. [CrossRef]
24. Li, J.; Shan, E.; Zhao, J.; Teng, J.; Wang, Q. The factors influencing the vertical transport of microplastics in marine environment: A review. *Sci. Total Environ.* **2023**, *870*, 161893. [CrossRef] [PubMed]
25. Turner, A. Black plastics: Linear and circular economies, hazardous additives and marine pollution. *Environ. Int.* **2018**, *117*, 308–318. [CrossRef]
26. Turner, A.; Filella, M. Polyvinyl chloride in consumer and environmental plastics, with a particular focus on metal-based additives. *Environ. Sci. Process. Impacts* **2021**, *23*, 1376. [CrossRef]
27. Zha, F.; Shang, M.; Ouyang, Z.; Guo, X. The aging behaviors and release of microplastics: A review. *Gondwana Res.* **2022**, *108*, 60–71. [CrossRef]
28. Cássio, F.; Batista, D.; Pradhan, A. Plastic Interactions with Pollutants and Consequences to Aquatic Ecosystems: What We Know and What We Do Not Know. *Biomolecules* **2022**, *12*, 798. [CrossRef]
29. Goh, P.B.; Pradit, S.; Towatana, P.; Khokkiatiwong, S.; Chuan, O.M. Laboratory Experiment on Copper and Lead Adsorption Ability of Microplastics. *Sains Malays.* **2022**, *51*, 993–1004. [CrossRef]
30. Liu, Y.; Zhang, K.; Xu, S.; Yan, M.; Tao, D.; Chen, L.; Wei, Y.; We, C.; Liu, G.; Lam, P.K.S. Heavy metals in the "plastisphere" of marine microplastics: Adsorption mechanisms and composite risk. *Gondwana Res.* **2022**, *108*, 171–180. [CrossRef]
31. Khalid, N.; Aqeel, M.; Noman, A.; Khan, S.M.; Akhter, N. Interactions and effects of microplastics with heavy metals in aquatic and terrestrial environments. *Environ. Pollut.* **2021**, *290*, 118104. [CrossRef] [PubMed]
32. Torres, F.G.; Dioses-Salinas, D.C.; Pizarro-Ortega, C.I.; De-la-Torre, G.E. Sorption of chemical contaminants on degradable and non-degradable microplastics: Recent progress and research trends. *Sci. Total Environ.* **2021**, *757*, 143875. [CrossRef] [PubMed]
33. Chen, G.; Fu, Q.; Tan, X.; Yang, H.; Luo, Y.; Shen, M.; Gu, Y. Speciation and release risk of heavy metals bonded on simulated naturally-aged microplastics prepared from artificially broken macroplastics. *Environ. Pollut.* **2022**, *295*, 118695. [CrossRef] [PubMed]
34. Li, X.; Chen, Y.; Zhang, S.; Dong, Y.; Pang, Q.; Lynch, I.; Xie, C.; Guo, Z.; Zhang, P. From marine to freshwater environment: A review of the ecotoxicological effects of microplastics. *Ecotoxicol. Environ. Saf.* **2023**, *251*, 114564. [CrossRef] [PubMed]
35. Xie, Q.; Li, H.; Li, Z.; Zhang, H.; Yuan, M.; Wu, M.; Li, H.; Xu, X. Accumulation, chemical speciation and ecological risks of heavy metals on expanded polystyrene microplastics in seawater. *Gondwana Res.* **2022**, *108*, 181–192. [CrossRef]
36. Bernhoft, R.A. Mercury toxicity and treatment: A review of the literature. *J. Environ. Public Health* **2012**, *2012*, 460508. [CrossRef]
37. Das, P.; Samantaray, S.; Rout, G.R. Studies on cadmium toxicity in plants: A review. *Environ. Pollut.* **1997**, *98*, 29–36. [CrossRef]
38. Satarug, S.; Baker, J.R.; Urbenjapol, S.; Haswell-Elkins, M.; Reilly, P.E.B.; Williams, D.J.; Moore, M.R. A global perspective on cadmium pollution and toxicity in non-occupationally exposed population. *Toxicol. Lett.* **2003**, *137*, 65–83. [CrossRef]
39. Scheuhammer, A.M. The chronic toxicity of aluminium, cadmium, mercury, and lead in birds: A review. *Environ. Pollut.* **1987**, *46*, 263–295. [CrossRef]
40. Wren, C.D. A review of metal accumulation and toxicity in wild mammals. I. Mercury. *Environ. Res.* **1986**, *40*, 210–244. [CrossRef]
41. Bozzeda, F.; Zangaro, F.; Colangelo, M.A.; Pinna, M. Relationships between size and abundance in beach plastics: A power-law approach. *Mar. Pollut. Bull.* **2021**, *173*, 113005. [CrossRef] [PubMed]
42. Corcoran, P.L.; Biesinger, M.C.; Grifi, M. Plastics and beaches: A degrading relationship. *Mar. Pollut. Bull.* **2009**, *58*, 80–84. [CrossRef] [PubMed]
43. Feld, L.; da Silva, V.H.; Strand, J. Characterization of foamed plastic litter on Danish reference beaches—Pollution assessment and multivariate exploratory analysis. *Mar. Pollut. Bull.* **2022**, *180*, 113774. [CrossRef] [PubMed]
44. Fruergaard, M.; Laursen, S.N.; Larsen, M.N.; Niebe, K.B.; Bentzon-Tarp, A.; Svenningsen, S.K.; Acevedo, N.L.I.; Trinh, B.-S.; Tran-Thi, P.T.; Doan-Nhu, H.; et al. Abundance and sources of plastic debris on beaches in a plastic hotspot, Nha Trang, Viet Nam. *Mar. Pollut. Bull.* **2023**, *186*, 114394. [CrossRef] [PubMed]
45. Kutralam-Muniasamy, G.; Pérez-Guevara, F.; Shruti, V.C. (Micro)plastics: A possible criterion for beach certification with a focus on the Blue Flag Award. *Sci. Total Environ.* **2022**, *803*, 150051. [CrossRef]
46. Moreira, F.T.; Balthazar-Silva, D.; Barbosa, L.; Turra, A. Revealing accumulation zones of plastic pellets in sandy beaches. *Environ. Pollut.* **2016**, *218*, 313–321. [CrossRef]
47. Turra, A.; Manzano, A.B.; Dias, R.J.S.; Mahiques, M.M.; Barbosa, L.; Balthazar-Silva, D.; Moreira, F.T. Three-dimensional distribution of plastic pellets in sandy beaches: Shifting paradigms. *Sci. Rep.* **2014**, *4*, 4435. [CrossRef]
48. Amaral-Zettler, L.A.; Zettler, E.R.; Mincer, T.J. Ecology of the plastisphere. *Nat. Rev. Microbiol.* **2020**, *18*, 139–151. [CrossRef]
49. Dąbrowska, A. A roadmap for a Plastisphere. *Mar. Pollut. Bull.* **2021**, *167*, 112322. [CrossRef]
50. Yu, Y.; Miao, L.; Adyel, T.M.; Waldschlager, K.; Wu, J.; Hou, J. Aquatic plastisphere: Interactions between plastics and biofilms. *Environ. Pollut.* **2023**, *322*, 121196. [CrossRef]
51. Bank, M.S.; Hansson, S.V. The Plastic Cycle: A Novel and Holistic Paradigm for the Anthropocene. *Environ. Sci. Technol.* **2019**, *53*, 7177–7179. [CrossRef] [PubMed]
52. De-la-Torre, G.E.; Dioses-Salinas, D.C.; Pizarro-Ortega, C.I.; Santillán, L. New plastic formations in the Anthropocene. *Sci. Total Environ.* **2021**, *754*, 142216. [CrossRef] [PubMed]

53. Porta, R. Anthropocene, the plastic age and future perspectives. *FEBS Open Bio.* **2021**, *11*, 948–953. [CrossRef]
54. Taffel, S. Technofossils of the Anthropocene: Media, geology, and plastics. *Cult. Politics* **2016**, *12*, 355–375. [CrossRef]
55. Zalasiewicz, J.; Waters, C.N.; Ivar do Sul, J.A.; Corcoran, P.L.; Barnosky, A.D.; Cearreta, A.; Edgeworth, M.; Galuszka, A.; Jeandel, C.; Leinfelder, R.; et al. The geological cycle of plastics and their use as a stratigraphic indicator of the Anthropocene. *Anthropocene* **2016**, *13*, 4–17. [CrossRef]
56. de Souza, A.M.; Rocha, D.S.; Guerra, J.V.; Cunha, B.A.; Martins, M.V.A.; Geraldes, M.C. Metal concentrations in marine sediments of the Rio de Janeiro Coast (Brazil): A proposal to establish new acceptable levels of contamination. *Mar. Pollut. Bull.* **2021**, *165*, 112113. [CrossRef] [PubMed]
57. Mikhailenko, A.V.; Ruban, D.A.; Ermolaev, V.A.; van Loon, A.J. Cadmium pollution in the tourism environment: A literature review. *Geoscience* **2020**, *10*, 242. [CrossRef]
58. Nour, H.E.; Helal, S.A.; Wahab, M.A. Contamination and health risk assessment of heavy metals in beach sediments of Red Sea and Gulf of Aqaba, Egypt. *Mar. Pollut. Bull.* **2022**, *177*, 113517. [CrossRef]
59. Gu, X.; Xu, L.; Wang, Z.; Ming, X.; Dang, P.; Ouyang, W.; Lin, C.; Liu, X.; He, M.; Wang, B. Assessment of cadmium pollution and subsequent ecological and health risks in Jiaozhou Bay of the Yellow Sea. *Sci. Total Environ.* **2021**, *774*, 145016. [CrossRef]
60. Tumolo, M.; Ancona, V.; De Paola, D.; Losacco, D.; Campanale, C.; Massarelli, C.; Uricchio, V.F. Chromium pollution in European water, sources, health risk, and remediation strategies: An overview. *Int. J. Environ. Res. Public Health* **2020**, *17*, 5438. [CrossRef]
61. Bonsignore, M.; Manta, D.S.; Barsanti, M.; Conte, F.; Delbono, I.; Horvat, M.; Quinci, E.M.; Schirone, A.; Shlyapnikov, Y.; Sprovieri, M. Mercury isotope signatures in sediments and marine organisms as tracers of historical industrial pollution. *Chemosphere* **2020**, *258*, 127435. [CrossRef]
62. Gauthier, P.T.; Blewett, T.A.; Garman, E.R.; Schlekat, C.; Middleton, E.T.; Suominen, E.; Crémazy, A. Environmental risk of nickel in aquatic Arctic ecosystems. *Sci. Total Environ.* **2021**, *797*, 148921. [CrossRef]
63. Kim, S.-M.; Choi, Y. Assessment of Lead (Pb) and Zinc (Zn) contamination in beach sands by hot spot analysis. *J. Coast. Res.* **2019**, *91*, 321–325. [CrossRef]
64. de Almeida Ribeiro Carvalho, M.; Botero, W.G.; de Oliveira, L.C. Natural and anthropogenic sources of potentially toxic elements to aquatic environment: A systematic literature review. *Environ. Sci. Pollut. Res.* **2022**, *29*, 51318–51338. [CrossRef]
65. Gopal, M.; Lemu, H.G.; Gutema, E.M. Sustainable Additive Manufacturing and Environmental Implications: Literature Review. *Sustainability* **2023**, *15*, 504. [CrossRef]
66. Rajeswari, S.; Saravanan, P.; Linkesver, M.; Rajeshkannan, R.; Rajasimman, M. Identifying global status and research hotspots of heavy metal remediation: A phase upgrade study. *J. Environ. Manag.* **2022**, *324*, 116265. [CrossRef] [PubMed]
67. Xu, Z.; Shao, T.; Dong, Z.; Li, S. Research progress of heavy metals in desert—Visual analysis based on CiteSpace. *Environ. Sci. Pollut. Res.* **2022**, *29*, 43648–43661. [CrossRef] [PubMed]
68. Pranckutė, R. Web of science (WoS) and Scopus: The titans of bibliographic information in today's academic world. *Publications* **2021**, *9*, 12. [CrossRef]
69. Singh, V.K.; Singh, P.; Karmakar, M.; Leta, J.; Mayr, P. The journal coverage of Web of Science, Scopus and Dimensions: A comparative analysis. *Scientometrics* **2021**, *126*, 5113–5142. [CrossRef]
70. Acosta-Coley, I.; Mendez-Cuadro, D.; Rodriguez-Cavallo, E.; de la Rosa, J.; Olivero-Verbel, J. Trace elements in microplastics in Cartagena: A hotspot for plastic pollution at the Caribbean. *Mar. Pollut. Bull.* **2019**, *139*, 402–411. [CrossRef]
71. Ashton, K.; Holmes, L.; Turner, A. Association of metals with plastic production pellets in the marine environment. *Mar. Pollut. Bull.* **2010**, *60*, 2050–2055. [CrossRef]
72. Bu-Olayan, A.H.; Thomas, B.V. Combined toxicity of mercury and plastic wastes to crustacean and gastropod inhabiting the waters in Kuwait. *J. Environ. Biol.* **2015**, *36*, 1291–1296. [PubMed]
73. Catrouillet, C.; Davranche, M.; Khatib, I.; Fauny, C.; Wahl, A.; Gigault, J. Metals in microplastics: Determining which are additive, adsorbed, and bioavailable. *Environ. Sci. Process. Impacts* **2021**, *23*, 553–558. [CrossRef]
74. Cormier, B.; Gambardella, C.; Tato, T.; Perdriat, Q.; Costa, E.; Veclin, C.; Le Bihanic, F.; Grassl, B.; Dubocq, F.; Karrman, A.; et al. Chemicals sorbed to environmental microplastics are toxic to early life stages of aquatic organisms. *Ecotoxicol. Environ. Saf.* **2021**, *208*, 111665. [CrossRef]
75. Dobaradaran, S.; Schmidt, T.C.; Nabipour, I.; Khajeahmadi, N.; Tajbakhsh, S.; Saeedi, S.; Mohammadi, M.J.; Keshtkar, M.; Khorsand, M.; Faraji Ghasemi, F. Characterization of plastic debris and association of metals with microplastics in coastline sediment along the Persian Gulf. *Waste Manag.* **2018**, *78*, 649–658. [CrossRef] [PubMed]
76. El Hadri, H.; Gigault, J.; Mounicou, S.; Grassl, B.; Reynaud, S. Trace element distribution in marine microplastics using laser ablation-ICP-MS. *Mar. Pollut. Bull.* **2020**, *160*, 111716. [CrossRef]
77. Fernandes, S.; Farzaneh, S.; Bendell, L.I. Abundance and distribution of beach litter with acutely toxic metal concentrations. *Mar. Pollut. Bull.* **2020**, *159*, 111479. [CrossRef] [PubMed]
78. Fred-Ahmadu, O.H.; Ayejuyo, O.O.; Benson, N.U. Dataset on microplastics and associated trace metals and phthalate esters in sandy beaches of tropical Atlantic ecosystems, Nigeria. *Data Brief* **2020**, *31*, 105755. [CrossRef]
79. Fred-Ahmadu, O.H.; Ayejuyo, O.O.; Tenebe, I.T.; Benson, N.U. Occurrence and distribution of micro(meso)plastic-sorbed heavy metals and metalloids in sediments, Gulf of Guinea coast (SE Atlantic). *Sci. Total Environ.* **2022**, *813*, 152650. [CrossRef]
80. Graca, B.; Bełdowska, M.; Wrzesień, P.; Zgrundo, A. Styrofoam debris as a potential carrier of mercury within ecosystems. *Environ. Sci. Pollut. Res.* **2014**, *21*, 2263–2271. [CrossRef]

81. Holmes, L.A.; Turner, A.; Thompson, R.C. Adsorption of trace metals to plastic resin pellets in the marine environment. *Environ. Pollut.* **2012**, *160*, 42–48. [CrossRef] [PubMed]
82. Jiang, X.; Conner, N.; Lu, K.; Tunnell, J.W.; Liu, Z. Occurrence, distribution, and associated pollutants of plastic pellets (nurdles) in coastal areas of South Texas. *Sci. Total Environ.* **2022**, *842*, 156826. [CrossRef]
83. Khaleel, R.; Valsan, G.; Rangel-Buitrago, N.; Warrier, A.K. Hidden problems in geological heritage sites: The microplastic issue on Saint Mary's Island, India, Southeast Arabian Sea. *Mar. Pollut. Bull.* **2022**, *182*, 114043. [CrossRef] [PubMed]
84. Kühn, S.; van Oyen, A.; Booth, A.M.; Meijboom, A.; van Franeker, J.A. Marine microplastic: Preparation of relevant test materials for laboratory assessment of ecosystem impacts. *Chemosphere* **2018**, *213*, 103–113. [CrossRef] [PubMed]
85. Li, W.; Lo, H.-S.; Wong, H.-M.; Zhou, M.; Wong, C.-Y.; Tam, N.F.-Y.; Cheung, S.-G. Heavy metals contamination of sedimentary microplastics in Hong Kong. *Mar. Pollut. Bull.* **2020**, *153*, 110977. [CrossRef] [PubMed]
86. Maršić-Lučić, J.; Lušić, J.; Tutman, P.; Varezic Dubravka, D.; Šiljić, J.; Pribudić, J. Levels of trace metals on microplastic particles in beach sediments of the island of Vis, Adriatic Sea, Croatia. *Mar. Pollut. Bull.* **2018**, *137*, 231–236. [CrossRef]
87. Massos, A.; Turner, A. Cadmium, lead and bromine in beached microplastics. *Environ. Pollut.* **2017**, *227*, 139–145. [CrossRef]
88. Nakashima, E.; Isobe, A.; Kako, S.; Itai, T.; Takahashi, S. Quantification of toxic metals derived from macroplastic litter on Ookushi Beach, Japan. *Environ. Sci. Technol.* **2012**, *46*, 10099–10105. [CrossRef]
89. Perera, U.L.H.P.; Subasinghe, H.C.S.; Ratnayake, A.S.; Weerasingha, W.A.D.B.; Wijewardhana, T.D.U. Maritime pollution in the Indian Ocean after the MV X-Press Pearl accident. *Mar. Pollut. Bull.* **2022**, *185*, 114301. [CrossRef]
90. Ranjani, M.; Veerasingam, S.; Venkatachalapathy, R.; Jinoj, T.P.S.; Guganathan, L.; Mugilarasan, M.; Vethamony, P. Seasonal variation, polymer hazard risk and controlling factors of microplastics in beach sediments along the southeast coast of India. *Environ. Pollut.* **2022**, *305*, 119315. [CrossRef]
91. Sánchez-Hernández, L.J.; Ramírez-Romero, P.; Rodríguez-González, F.; Ramos-Sanchez, V.H.; Maquez Montes, R.A.; Romero-Paredes Rubio, H.; Sujitha, S.B.; Jonathan, M.P. Seasonal evidences of microplastics in environmental matrices of a tourist dominated urban estuary in Gulf of Mexico, Mexico. *Chemosphere* **2021**, *277*, 130261. [CrossRef]
92. Santos-Echeandía, J.; Rivera-Hernández, J.R.; Rodrigues, J.P.; Moltó, V. Interaction of mercury with beached plastics with special attention to zonation, degradation status and polymer type. *Mar. Chem.* **2020**, *222*, 103788. [CrossRef]
93. Sewwandi, M.; Amarathunga, A.A.D.; Wijesekara, H.; Mahatantila, K.; Vithanage, M. Contamination and distribution of buried microplastics in Sarakkuwa beach ensuing the MV X-Press Pearl maritime disaster in Sri Lankan sea. *Mar. Pollut. Bull.* **2022**, *184*, 114074. [CrossRef] [PubMed]
94. Souza, L.A.; Santos, A.C.S.S.; Leão, J.M.; Schaeppi, O.C.; Hatje, V. Occurrence and contents of trace metals and rare earth elements on plastic pellets. *Mar. Pollut. Bull.* **2022**, *185*, 114261. [CrossRef] [PubMed]
95. Suman, T.Y.; Li, W.-G.; Alif, S.; Faris, V.R.P.; Amarnath, D.J.; Ma, J.-G.; Pei, D.-S. Characterization of petroleum-based plastics and their absorbed trace metals from the sediments of the Marina Beach in Chennai, India. *Environ. Sci. Eur.* **2020**, *32*, 110. [CrossRef]
96. Truchet, D.M.; López, A.D.F.; Ardusso, M.G.; Rimondino, G.N.; Buzzi, N.S.; Malanca, F.E.; Spetter, C.V.; Severini, M.D.F. Microplastics in bivalves, water and sediments from a touristic sandy beach of Argentina. *Mar. Pollut. Bull.* **2021**, *173*, 113023. [CrossRef]
97. Turner, A. Heavy metals, metalloids and other hazardous elements in marine plastic litter. *Mar. Pollut. Bull.* **2016**, *111*, 136–142. [CrossRef]
98. Turner, A. Trace elements in fragments of fishing net and other filamentous plastic litter from two beaches in SW England. *Environ. Pollut.* **2017**, *224*, 722–728. [CrossRef]
99. Turner, A. Polystyrene foam as a source and sink of chemicals in the marine environment: An XRF study. *Chemosphere* **2021**, *263*, 128087. [CrossRef]
100. Turner, A.; Arnold, R.; Williams, T. Weathering and persistence of plastic in the marine environment: Lessons from LEGO. *Environ. Pollut.* **2020**, *262*, 114299. [CrossRef]
101. Turner, A.; Wallerstein, C.; Arnold, R. Identification, origin and characteristics of bio-bead microplastics from beaches in western Europe. *Sci. Total Environ.* **2019**, *664*, 938–947. [CrossRef]
102. Twyford, S.I.; Turner, A. Association of metals with expanded polystyrene in the marine environment. *Sci. Total Environ.* **2023**, *871*, 161920. [CrossRef]
103. Vedolin, M.C.; Teophilo, C.Y.S.; Turra, A.; Figueira, R.C.L. Spatial variability in the concentrations of metals in beached microplastics. *Mar. Pollut. Bull.* **2018**, *129*, 487–493. [CrossRef] [PubMed]
104. Vetrimurugan, E.; Jonathan, M.P.; Sarkar, S.K.; Roy, P.D.; Velumani, S.; Sakthi, J.S. Occurrence, distribution and provenance of micro plastics: A large scale quantitative analysis of beach sediments from southeastern coast of South Africa. *Sci. Total Environ.* **2020**, *746*, 141103. [CrossRef] [PubMed]
105. Xiang, J.-Y.; Wang, Q.; Shao, M.-S.; Wang, Q.; Wu, H.-N.; Xu, Q.-Y. Assessment of beach plastic waste and its heavy metal pollution in Shenzhen. *China Environ. Sci.* **2020**, *40*, 3097–3105.
106. Xie, Q.; Li, H.-X.; Lin, L.; Li, Z.-L.; Huang, J.-S.; Xu, X.-R. Characteristics of expanded polystyrene microplastics on island beaches in the Pearl River Estuary: Abundance, size, surface texture and their metals-carrying capacity. *Ecotoxicology* **2021**, *30*, 1632–1643. [CrossRef] [PubMed]
107. Zahari, N.Z.; Tuah, P.M.; Junaidi, M.R.; Ali, S.A.M. Identification, Abundance, and Chemical Characterization of Macro-, Meso-, and Microplastics in the Intertidal Zone Sediments of Two Selected Beaches in Sabah, Malaysia. *Water* **2022**, *14*, 1600. [CrossRef]

108. Suteja, Y.; Atmadipoera, A.S.; Riani, E.; Nurjaya, N.I.; Nugroho, D.; Purwiyanto, A.I.S. Stranded marine debris on the touristic beaches in the south of Bali Island, Indonesia: The spatiotemporal abundance and characteristic. *Mar. Pollut. Bull.* **2021**, *173*, 113026. [CrossRef] [PubMed]
109. Horn, D.; Miller, M.; Anderson, S.; Steele, C. Microplastics are ubiquitous on California beaches and enter the coastal food web through consumption by Pacific mole crabs. *Mar. Pollut. Bull.* **2019**, *139*, 231–237. [CrossRef]
110. Corbau, C.; Lazarou, A.; Buoninsegni, J.; Oilivo, E.; Gazale, V.; Nardin, W.; Simeoni, U.; Carboni, D. Linking marine litter accumulation and beach user perceptions on pocket beaches of Northern Sardinia (Italy). *Ocean. Coast. Manag.* **2023**, *232*, 106442. [CrossRef]
111. Akkajit, P.; Tipmanee, D.; Cherdsukjai, P.; Suteerasak, T.; Thongnonghin, S. Occurrence and distribution of microplastics in beach sediments along Phuket coastline. *Mar. Pollut. Bull.* **2021**, *169*, 112496. [CrossRef] [PubMed]
112. Hayes, A.; Kirkbride, P.; Leterme, S.C. Variation in polymer types and abundance of microplastics from two rivers and beaches in Adelaide, South Australia. *Mar. Pollut. Bull.* **2021**, *172*, 112842. [CrossRef] [PubMed]
113. Aliff, M.N.; Reavie, E.D.; Post, S.P.; Zanko, L.M. Anthropocene geochemistry of metals in sediment cores from the Laurentian Great Lakes. *PeerJ* **2020**, *2020*, e9034. [CrossRef]
114. Marx, S.K.; McGowan, H.A.; Kamber, B.S.; Knight, J.M.; Denholm, J.; Zawadzki, A. Unprecedented wind erosion and perturbation of surface geochemistry marks the Anthropocene in Australia. *J. Geophys. Res. Earth Surf.* **2014**, *119*, 45–61. [CrossRef]
115. Waters, C.N.; Zalasiewicz, J.; Summerhayes, C.; Barnosky, A.D.; Poirier, C.; Galuszka, A.; Cearreta, A.; Edgeworth, M.; Ellis, E.C.; Ellis, N.; et al. The Anthropocene is functionally and stratigraphically distinct from the Holocene. *Science* **2016**, *351*, aad2622. [CrossRef] [PubMed]
116. Kleppin, L.; Schröder, W.; Pesch, R.; Schmidt, G. Development and testing of a metadata- and WebGIS-application for the exposure monitoring by mosses in Germany. A contribution to the LTER-network. *Umweltwiss. Und Schadst. -Forsch.* **2008**, *20*, 38–48.
117. Rocca, J.D.; Simonin, M.; Blaszczak, J.R.; Ernakovich, J.G.; Gibbons, S.M.; Midani, F.S.; Washburne, A.D. The Microbiome Stress Project: Toward a global meta-analysis of environmental stressors and their effects on microbial communities. *Front. Microbiol.* **2019**, *10*, 3272. [CrossRef]
118. Patil, S.; Bafana, A.; Naoghare, P.K.; Krishnamurthi, K.; Sivanesan, S. Environmental prevalence, fate, impacts, and mitigation of microplastics—A critical review on present understanding and future research scope. *Environ. Sci. Pollut. Res.* **2021**, *28*, 4951–4974. [CrossRef]
119. Puri, M.; Gandhi, K.; Kumar, M.S. Emerging environmental contaminants: A global perspective on policies and regulations. *J. Environ. Manag.* **2023**, *332*, 117334. [CrossRef]

Journal of
Marine Science and Engineering

MDPI

Article

Punctiform Breakup and Initial Oceanization in the Central Red Sea Rift

Ya-Di Sang [1], Bakhit M. T. Adam [1], Chun-Feng Li [1,2,3,*], Liang Huang [1], Yong-Lin Wen [1], Jia-Ling Zhang [1,4] and Yu-Tao Liu [1,5]

1 Ocean College, Zhejiang University, Zhoushan 316021, China
2 Hainan Institute, Zhejiang University, Sanya 572025, China
3 Laboratory for Marine Mineral Resources, Qingdao National Laboratory for Marine Science and Technology, Qingdao 266237, China
4 Key Laboratory of Submarine Geoscience, Second Institute of Oceanography, Ministry of Natural Resources, Hangzhou 310012, China
5 Science & Technology on Integrated Information System Laboratory, Institute of Software, Chinese Academy of Sciences, Beijing 100190, China
* Correspondence: cfli@zju.edu.cn

Abstract: The Central Red Sea Rift is a natural laboratory to study the transition from rifting to spreading. Based on new reflection seismic profiles and gravity modeling, we examined the crustal structure, tectonic evolution, breakup mechanism, and future evolution of the Central Red Sea Rift. Along this rift axis, the breakup of continental lithosphere is discontinuous and the oceanic crust is limited to the axial deeps. The punctiform breakup and formation of deeps is assisted by mantle upwelling and topographic uplift, but the nucleation is directly controlled by the normal-fault system. The discontinuities spaced between axial deeps within the relatively continuous central troughs are presently axial domes or highs and will evolve into new deeps with tectonic subsidence. Isolated deeps will grow and connect with each other to become a continuous central trough, before transitioning into a unified spreading center.

Keywords: Central Red Sea Rift; continental breakup; mantle upwelling; fault nucleation; rift to drift transition; initial seafloor spreading

Citation: Sang, Y.-D.; Adam, B.M.T.; Li, C.-F.; Huang, L.; Wen, Y.-L.; Zhang, J.-L.; Liu, Y.-T. Punctiform Breakup and Initial Oceanization in the Central Red Sea Rift. *J. Mar. Sci. Eng.* **2023**, *11*, 808. https://doi.org/10.3390/jmse11040808

Academic Editors: George Kontakiotis, Assimina Antonarakou and Dmitry A. Ruban

Received: 9 March 2023
Revised: 6 April 2023
Accepted: 7 April 2023
Published: 10 April 2023

1. Introduction

The mechanism forming the very first piece of oceanic crust during continental breakup has yet to be better understood [1–6]. The Central Red Sea Rift is transitional between the southern Red Sea Rift, which has developed typical seafloor spreading with continuous seafloor magnetic stripes, and the northern part, which lacks a rift valley and magnetic anomalies (Figure 1a) [1,7–9]. The Central Red Sea Rift provides a unique window to understanding the initial formation of oceans, as it is now undergoing the final breakup and initial oceanization [10–13].

The Red Sea Rift connects with the East Africa Rift System and the Gulf of Aden at the Afar Triple Junction (Figure 1a). It developed on the broad Precambrian Arabian–Nubian shield that was strongly influenced by the Neoproterozoic Pan-African collisions that formed NW-striking shear systems, suture zones, and lineaments [1,14]. The Arabian–Nubian shield, as a part of Gondwanaland, has drifted northwards since the early Paleozoic and collided with Eurasia by the end of the Eocene to Oligocene, forming the Bitlis–Zagros Thrust (Figure 1a) [15–17]. The Arabian block relatively moved counter-clockwise with respect to the Nubia block (Figure 1a), and the Cenozoic NW-striking intercontinental rift systems started to develop along Pan-African inherited weak zones, in coincidence with Afar volcanism (~30 Ma) [17–21].

The Central Red Sea Rift has developed a series of spaced deeps with hypothesized formation of oceanic crust along the rift axis. Magnetic anomalies and high seismic velocity imply massive basaltic intrusions in the axial trough [22,23], and fresh basaltic rocks were collected at Deep Sea Drilling Project (DSDP) Site 226 in the Atlantis II Deep (Figure 1b) [24–26]. The deeps become sparser, narrower, shorter, and shallower northwards (Figure 1b) [1,27–29]. Groups of relatively closely spaced deeps form central troughs, separated by inter-trough zones (ITZs) [22,27,30,31]. Within the troughs, deeps are further spaced by second-order discontinuities in axial domes or highs (Figure 1b). There exist marked differences in magnetic and gravity anomalies and structure and composition between the deeps and inter-trough zones [22,30,32–36]. Similar geomorphic and structural segmentation is also found in other propagation tips of spreading centers, such as the Ostler fault zone in New Zealand [37], the Cocos–Nazca spreading ridge [38], and the Woodlark spreading ridge [39], reflecting the mechanisms of initiation and propagation of mid-ocean ridges. The formation of the deeps and breakup mechanism of these initial spreading centers need to be clarified [37–39].

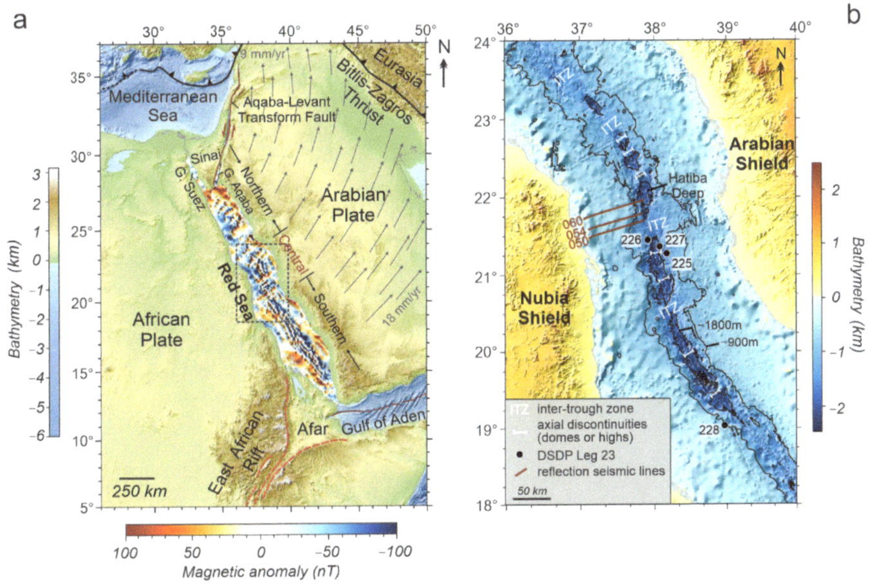

Figure 1. (**a**) Tectonic settings and the magnetic anomaly map of the Red Sea Rift. Bathymetry data are from the GEBCO Compilation Group 2021 (https://www.gebco.net, accessed on 3 September 2021); magnetic anomaly data are from EMAG2 (https://ngdc.noaa.gov/geomag/emag2.html, accessed on 2 March 2020). Grey arrows show the plate movement of the Arabian Plate relative to the Nubia Plate, and the direction and velocity are calculated from the model MORVEL [40]. Grey dashed box shows the area of Figure 1b. (**b**) Bathymetry of the Central Red Sea. Red lines show the locations of the reflection seismic profiles used in this study, the black dots show sites 225–228 on the DSDP Leg 23 [24,41–43], and the white lines represent the "discontinuities" between axial deeps within the trough.

In the Central Red Sea Rift, the nature of crust outside the deeps and the forces controlling continental lithosphere breakup are still controversial. It was proposed that the deeps are nuclei oceanic crusts formed by hotspot [25,27], or edge-driven mantle convection [10,18], surrounded by continental crust with basaltic intrusions [11,18,22,44–46]. The deeps may also be pull-apart basins under regional pure shear crustal stretching, in the early stage of punctiform fracturing [31,47,48]. In a two-stage seafloor spreading model (41–34 and 4–5 Ma, respectively), Girdler and Styles [19] suggested that the oceanic crust is not limited within the main trough, but even extend over the whole shelves of the Red Sea. Finally, the deeps may be just "windows" of the underlying well-developed spreading center, exposed by the dissolution of salt deposits [5,32,49–52].

Existing models fail to reach a consensus on the distribution of oceanic crust in the Red Sea Rift, and often neglect the interaction between mantle activities and tectonism. To study the rifting process and the formation of the axial deeps, we interpret three new reflection seismic profiles perpendicular to the rift axis. The eastern ends of the three profiles reach different axial zones of the Central Red Sea Rift (Figure 1b), providing a chance to reveal evolution from axial deeps to the inter-trough zones. We establish the initial oceanization model of the Central Red Sea Rift to explain how a continuous spreading center forms and how the rift to drift transition will be achieved.

2. Materials and Methods

2.1. Reflection Seismic Data Acquisition and Interpretation

The reflection seismic lines extend from the west shelf, crossing the main trough margin, to Red Sea Rift axis, for more than 80 km in distance (Figure 1b). They were acquired and processed during the CPOC08 survey in 2008 by the Oil Exploration and Production Authority, Sudan Ministry of Energy. The data acquisition parameters are listed in Table 1. The data processing steps include low-cut band pass filtering, auto despike/manual bad trace editing, F–K filtering, wave equation multiple reduction, geometrical spreading compensation and exponential gain, surface consistent amplitude correction, Tau-P deconvolution, surface related multiple elimination, high resolution demultiple filtering, prestack time migration (PSTM), poststack zero phase deconvolution, FX-DCON, AGC, time variant band pass, and equalization.

Table 1. Reflection seismic data acquisition parameters.

Parameters		Values	Units
Recording	Record length	8	s
	Sampling rate	2	ms
	Shot point interval	25	m
Streamer	Number	1	-
	Streamer length	8100	m
	Operating depth	7	m
	Channels	648	-
	Group length	12.5	m
Source	Airgun array volume	2960	cubic inch
	Source depth	7	m

Based on the extensive geological studies and DSDP drill results [17,24,41–43,45,46,48,53–59], three regional seismic unconformities in addition to the acoustic basement can be identified: the top and bottom of salt deposition, and the reflector S, which can be traced in the whole rift basin and is regarded as the transition from rifting to drifting at 5 Ma in the Southern Red Sea Rift [16,17,19,22,60] (Figure 2).

From the Oligocene to the Middle Miocene, the Red Sea Rift experienced constant extension and subsidence [15–17]. The seawater channels at the northern and southern ends of the Red Sea rift opened successively and the sedimentary facies evolved from terrestrial to hemipelagic-deep marine deposits (Figure 2) [61,62]. Due to the reorganization of the plate kinematics along the Aqaba–Levant transform boundary at ~14 Ma, the seawater exchange with the Mediterranean was restricted [12,16,62–65]. In addition, the climate became arid to semiarid [32]. From the middle of Middle Miocene to the end of Middle Miocene, evaporites were widely deposited across the whole Red Sea (Figure 2), with thickness varying from tens of meters to thousands of meters [32,59,66]. After the long period of tectonic quiescence and saline lake sediment (Figure 2) [25,60,67,68], during the late Miocene, the Red Sea Rift basin reconnected with the Indian Ocean [62,69], forming layered evaporite and shallow marine deposits (Figure 2). The top of this sequence is truncated by the reflector S. During the Plio–Pleistocene, the Red Sea Rift Basin maintained a shallow marine environment (Figure 2). The sedimentary unit deposited in a stable and relatively low-energy hydrodynamic environment.

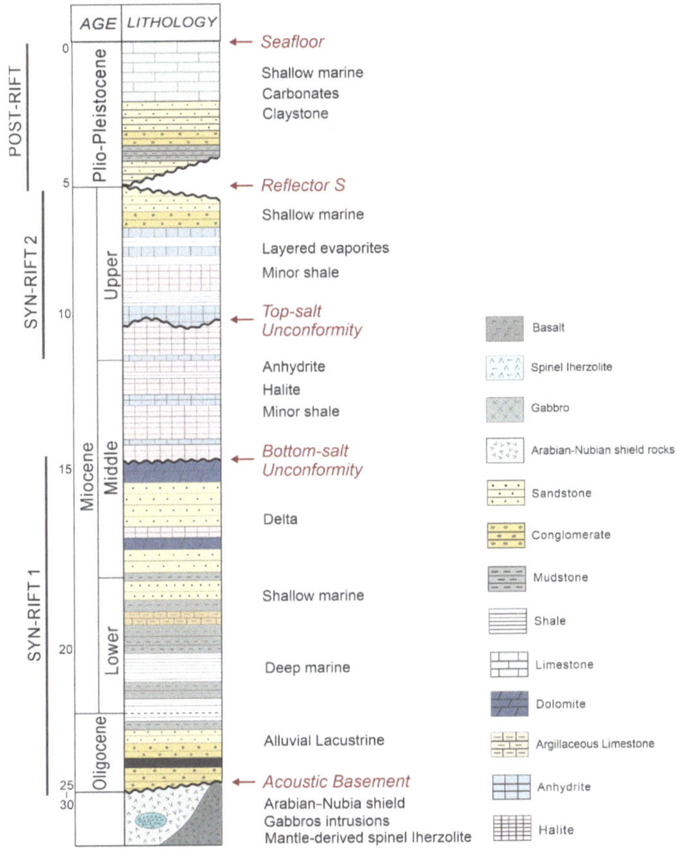

Figure 2. Stratigraphy of the Red Sea. Basement composition and stratigraphy refer to previous geological studies and DSDP drill results [17,45,46,48,54–59].

2.2. Free-Air Gravity Modeling

To study the nature of crust and deep structure of the Central Red Sea Rift, we apply gravity modeling to explore the density structure of the study area.

The Moho reflector is unrecognizable in our profiles, so we calculated the Moho depths through gravity inversion based on the Parker–Oldenburg method [70,71], following the procedures of Bai et al. [72]. We estimated the mantle residual anomaly from the Bouguer gravity anomaly model WGM2012 [73], after sediment thickness correction based on Crust 1.0 [74] and GlobSed [75], and lithospheric thermal correction based on the global age model of oceanic crust [76,77].

Figure 3a shows the calculated Moho depth in the Central Red Sea region. We compared our results with some published density and velocity structure profiles around the study area to test the reliability of our results (Figure 3b) [16,35,47,78,79]. Our calculated values are close to previously observed results especially near the rift axis (Figure 3b). The errors in the continental margins were caused by thick and variable sediments, and the gravity effects of these sediments may have been inadequately corrected. Moho depths vary from a maximum of >20 km at the continental margin to about 5 km at the southern rift axis (Figure 3a). Along the rift axis, Moho rises under the troughs and deepens under inter-trough zones. Moho also deepens northwards, consistent with the northward propagation of the Red Sea Rift (Figure 3a).

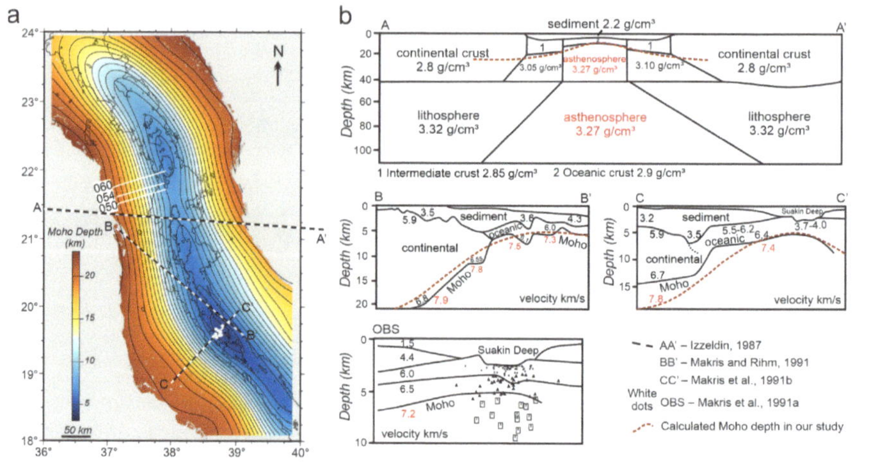

Figure 3. Calculated Moho depth and comparison with published data. (**a**) Moho depth from gravity inversion. Black lines show the isobaths of the calculated Moho depth. Grey lines along the rift axis represent the isobaths of −900 m and −1800 m bathymetries. White lines indicate reflection seismic profiles in this study. Black and white lines (AA′ [78], BB′ [47], CC′ [79]) represent locations of the published density and velocity structures. White dots on the line CC′ show the distribution of OBS stations within the Suakin Deep [9]. (**b**) Published density and velocity structures [47,78,79] and microearthquakes records [9]. Red dashed lines indicate calculated Moho depths along published profiles.

Initial density models were built based on reflection seismic data. In gravity modeling, we update the model until the error between calculated values and observed free-air gravity anomaly is acceptable. The structural layers and parameters we set are shown in Table 2, with references to previously published results [7,9,47,78–81].

Table 2. Structural layers and parameters in the gravity model.

Layers		Density (g/cm^3)	Velocity (km/s)
	Water	1.06	1.5
Sediment	S1 (postsalt sediments)	2.3	2.8
	S2 (salt)	2.2	4.0
	S3 (presalt sediments)	2.4	3.5
Crust	Continental	2.8	6.2
	Transitional	2.85	6.1
	Oceanic	2.9	6.0
Mantle	Mantle	3.3	7.5
	Upwelled mantle	3.1	7.6
	Asthenosphere	3.27	7.4

3. Tectonic Evolution and Density Structure of the Central Red Sea Rift

3.1. Tectonic Evolution of the Central Red Sea Rift

Referring to the sedimentary evolution of the Red Sea (Figure 2) [17,24,41–43,45,46,48,53–59], we can identify four major unconformities in the reflection seismic profiles. The top of the acoustic basement reflectors are indistinct, but can be identified as the bottom of the sediment cover. The top and bottom of evaporite deposition are mainly traced according to the different reflector characteristics between evaporite deposition and sub- and supraevaporite seismic units (Figures 4–6). Evaporite deposition shows chaotic and blank internal reflection and minor high amplitude reflectors from thin beds of anhydrite, shale, mudstone, and siltstone [56,57], and the sub- and supraevaporite strata show moderately continuous reflectors of intermediate to high amplitudes (Figures 4–6). The reflector S is a significant regional unconformity, showing continuous strong reflectors. Faults are marked by offsets in sequences, continuous reflectors, and major interpreted horizons (Figures 4–6).

3.1.1. Differential Sedimentary and Tectonic Evolution between the Axial Deep and ITZ

In the center of the Hatiba Deep, the acoustic basement is exposed to the seafloor with a strong reflector distinguishable from the reflectors in the marginal area (Figure 4). In profiles 054 and 050, the basement close to the southern inter-trough zone deepens and flattens again (Figures 5 and 6). Intense fault activities concentrated at the boundaries of axial deeps along the central trough, forming step-fault zones (Figure 4). On the contrary, in the inter-trough zones, basement faults occurred on the rift axis before the late Miocene (Figure 6), and the inter-trough zones experienced more uniform subsidence without active fault activities in the later stage (Figures 5 and 6). Salt tectonics developed in the Central Red Sea Rift are mainly salt domes, salt walls, diapirism, salt pillows, and salt anticlines, usually forming angular unconformities and halokinetic sequences with the suprasalt strata (Figures 4–6). Reflection seismic data reveals that evaporites were only exposed at the boundary fault escarpments of the axial deep, and they uniformly deposited at the ITZ, but were absent at the center of the deep (Figures 4–6).

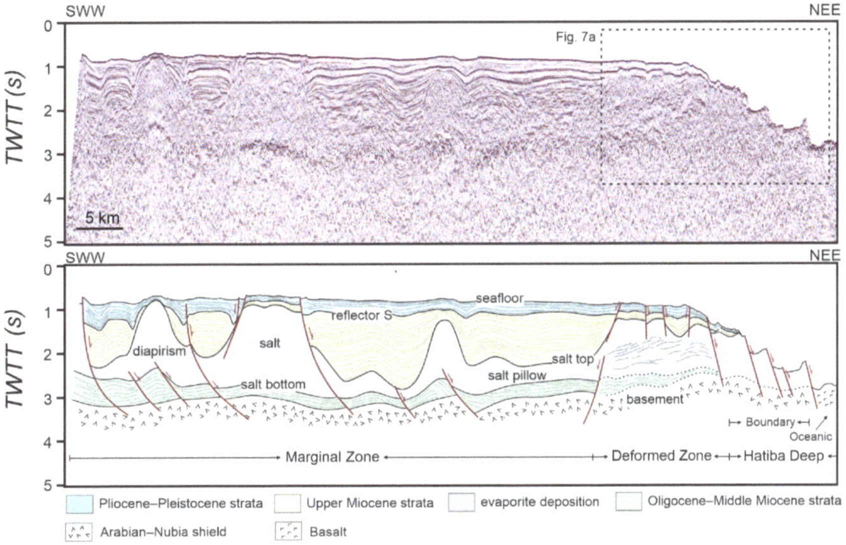

Figure 4. Reflection seismic profile 060 and interpretations. Dashed box shows the area of Figure 7a.

During the Plio–Pleistocene, the southern part of the Red Sea Rift started seafloor spreading and experienced the post-rift stage, while the central Red Sea Rift was still in the syn-rift stage. Intense fault activities during the third rifting stage only concentrated at the boundary of axial deep, shaping the step-fault zone and axial deep (Figure 4), leaving the southern inter-trough zone subsided uniformly (Figure 6). The development and structural style of salt tectonics correspond to the tectonic evolution of the whole rift basin. In the Central Red Sea Rift, salt flowage towards the axis driven by heterogeneous gravity load can strongly shape the geomorphology of the central trough [5,32,49–51]. The distribution of evaporites along the rift axis reflects potential differential salt movement towards the axial deep and ITZ influenced by the basement topography [5,32], or dissolution of the salt deposits in the axial deep associated with the hydrothermal circulation [32,52]. The differential tectonism and evaporite deposition between the axial deep and the ITZ shaped the rift axis with geomorphic segmentation.

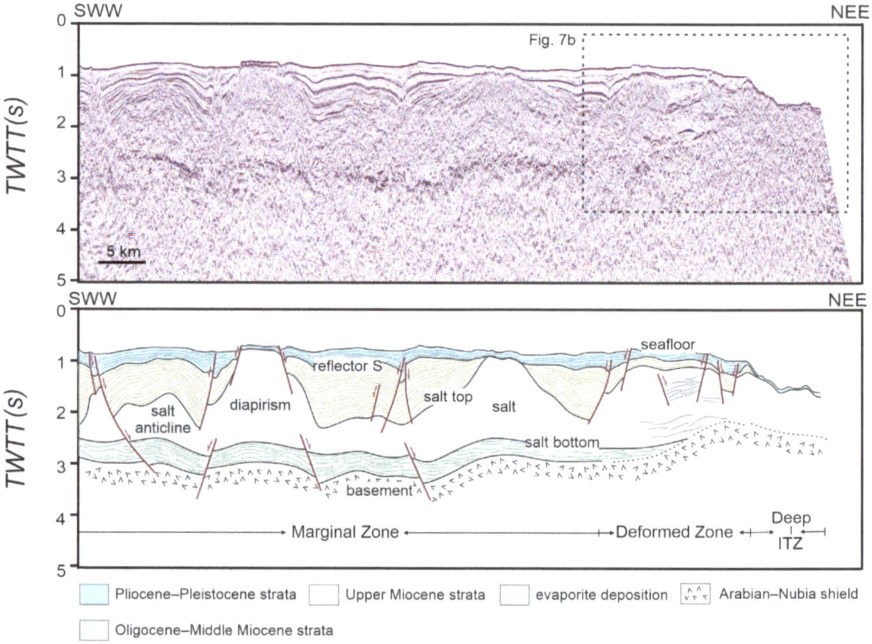

Figure 5. Reflection seismic profile 054 and interpretations. Dashed box shows the area of Figure 7b.

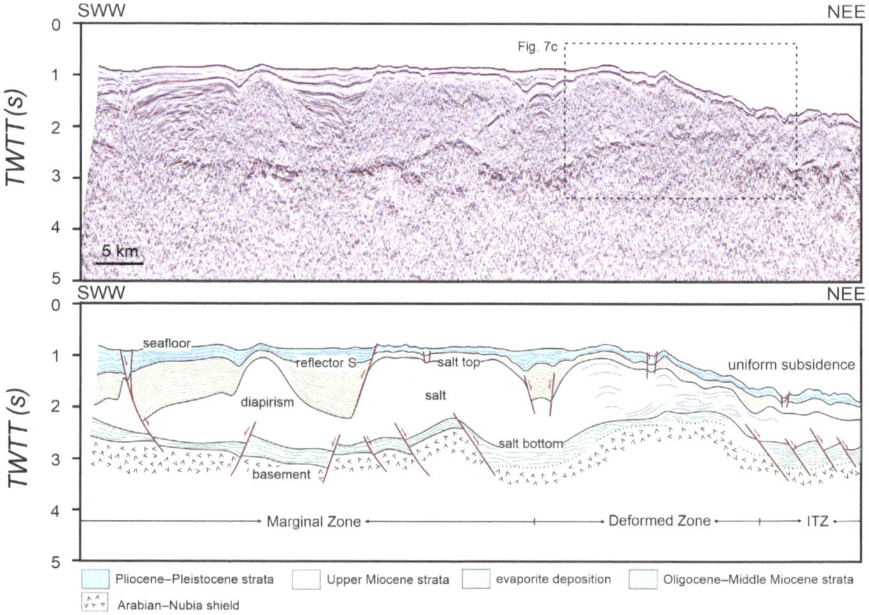

Figure 6. Reflection seismic profile 050 and interpretations. Dashed box shows the area of Figure 7c.

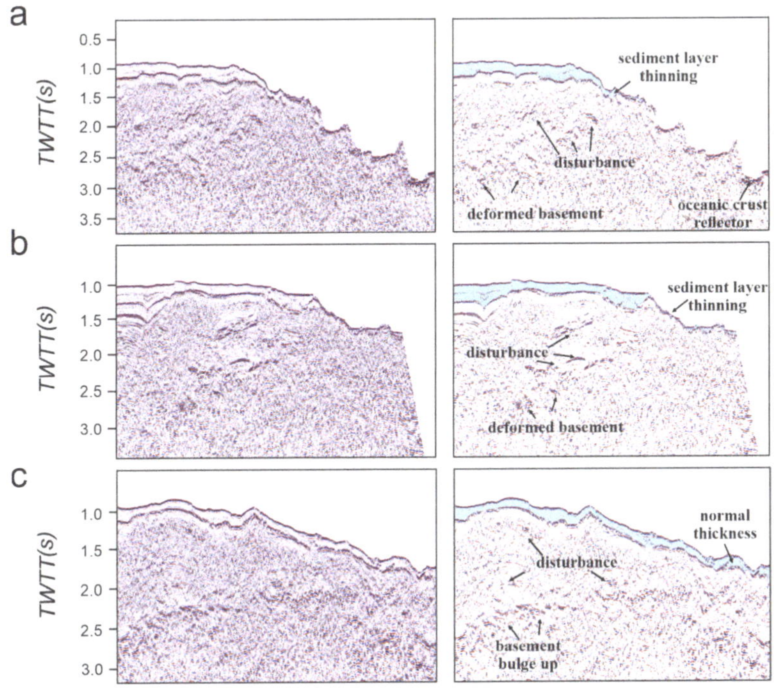

Figure 7. Deformed zones. Indicators of regional uplift: deformed basement, interrupted Pliocene–Pleistocene sediments, and disturbances within the evaporite deposition. Differential crustal vertical movement between the axial deep and the inter-trough zone. (**a**) Profile 060. (**b**) Profile 054. (**c**) Profile 050.

3.1.2. Deformed Zone and Postevaporite Uplift

The acoustic basement in the wide marginal area has reflectors with moderate amplitude and continuity, and has gentle topography and forms horsts and grabens during the development of the continental margin (Figures 4–6). Close to the rift axis, the basement becomes less continuous, intensively disturbed, and tends to bulge upward (Figures 4–6) compared with acoustic basement reflectors in the wide marginal area. Evaporite deposition in our reflection seismic profiles shows chaotic or blank internal reflection (Figures 4–6); while corresponding to the deformation of the basement, discontinuous strong reflectors were formed within the evaporites (Figure 7). The Pliocene–Pleistocene strata significantly thinned around the deep and are absent in the center of the Hatiba Deep (Figures 4 and 5), but thickened again in the inter-trough zone (Figure 6). Evaporites that flow into the rift axis uniformly deposited at the ITZ, but were absent at the center of the deep (Figures 4–6).

Based on the observations, we define the domain where the basement formed a dome-like structure and was significantly deformed as the deformed zone (Figures 4–6). The up-bulge basement, significant thinning of Pliocene–Pleistocene strata, and the disturbance in the evaporites (Figures 4–7) imply crustal vertical movements along the Central Red Sea Rift axis after the evaporite deposition. Similar basement deformation and regional sedimentary interruption were reported near the axial deeps, and are thought to be influenced by crustal vertical movements during lithospheric thinning [19,22,27,32,82].

3.1.3. Formation of the Axial Deep

Our reflection seismic data further revealed the differential sediment and tectonic evolution between the Hatiba Deep and the southern ITZ, suggesting differential vertical crustal movements between the axial deep and the inter-trough zone. Thinner Pliocene–Pleistocene strata around the deep than the inter-trough zone and the absence of evaporites in the center of the deep (Figure 7) indicate a high topography around the deep before subsidence. Evaporites uniformly deposited at the ITZ, but were absent at the center of the deep, implying the differential uplift that can influence the evaporite flowage into the rift axis. The high topography around the axial deep can obstruct salt flowage towards the deep [16,32], and the lack of the overlying Pliocene–Pleistocene strata can induce the dissolution of evaporites at the center of the deep [32]. Without the uplift and higher topography, the axial deep should be invaded by evaporites and covered by Pliocene–Pleistocene strata, like the inter-trough zone (Figure 6).

Therefore, we proposed that the axial deep was the center of the "Postevaporite Uplift" before collapsing to form deeps. After the evaporite was deposited, regional uplift occurred along the rift axis. The basement was deformed and bulged upward, overlying evaporite deposition was disturbed, and the Pliocene–Pleistocene sediments pinched out towards the rift axis (Figure 8a). During the third rifting stage, intense fault activities focused on the stress concentrations of early uplift, forming isolated axial deeps (Figures 4 and 8b). However, the subsidence outside of the centers of uplift was uniform, forming gentle inter-trough zones without the development of the normal-fault system (Figure 7c).

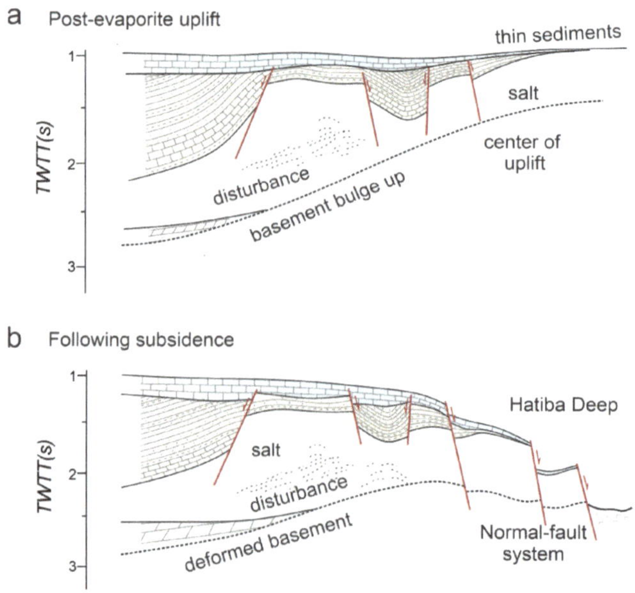

Figure 8. Sediment and tectonic evolution of the Hatiba Deep based on profile 060. (**a**) As the center of the Pliocene–Pleistocene uplift. (**b**) Intense tectonic subsidence controlled the formation of the deep. Lithologic patterns and unconformities refer to Figure 2.

3.2. Gravity Modeling Results

From the continental margin to the rift axis, the free-air gravity anomalies first reach a low (~−20 mGal) at the model distance of 20 km due to thick deposition of low-density salt, then increase to the maxima (~20 mGal) at the model distance of 60–80 km at the deformed zones, and finally decrease towards the rift axis (Figure 9). In the deformed zones, the free-air gravity anomaly highs in all three models are too large to be caused only by the bulge of the basement that we observed in the reflection seismic profiles (Figures 4–6), even taking the high-density oceanic crust into account. Denser materials are needed under the deformed zones.

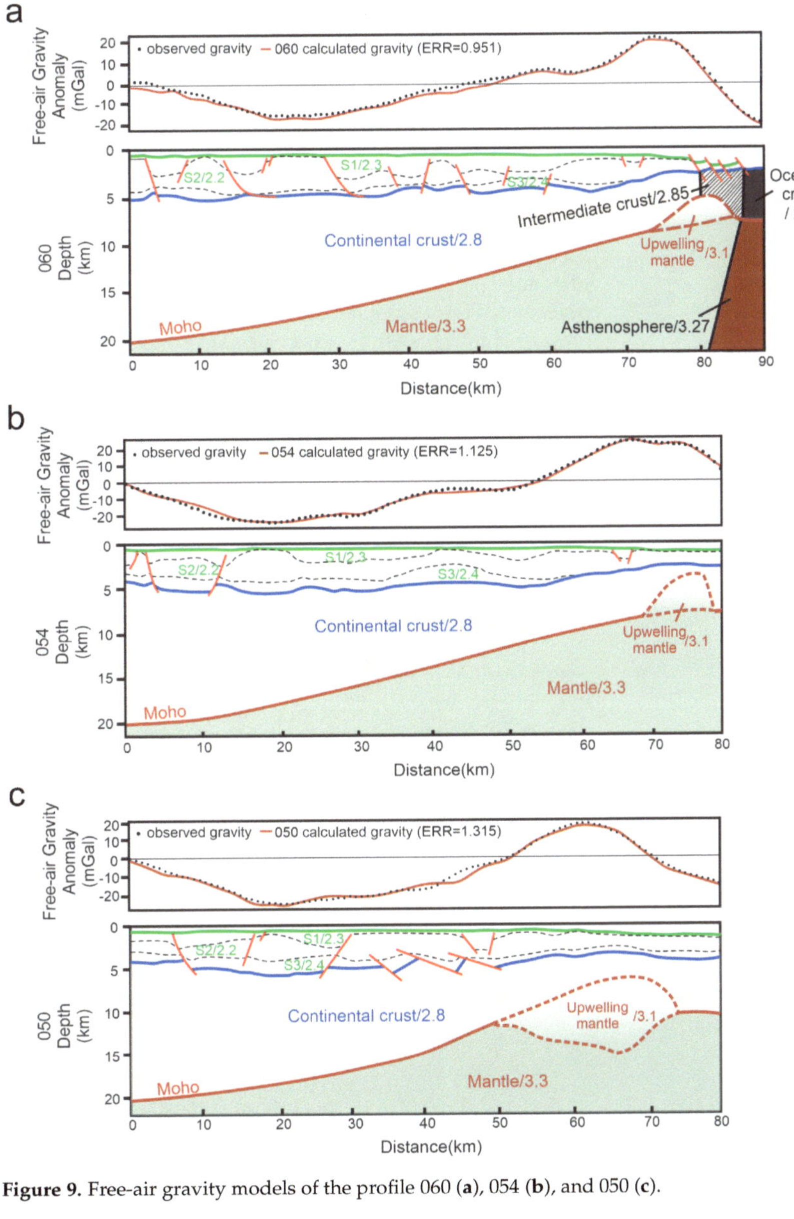

Figure 9. Free-air gravity models of the profile 060 (**a**), 054 (**b**), and 050 (**c**).

In the Central Red Sea Rift, high-density anomaly at the depths of 8–15 km can be caused by newborn oceanic crust under the whole main trough [22], prespreading igneous intrusions to the Precambrian Arabian–Nubian shield [11], mantle diapirs into the crust [27], or other processes related to the initial oceanization [83,84]. We have tested the newborn oceanic crust or igneous intrusion with densities ranging from 2.9 to 3.0 g/cm^3 [80], but the calculated gravity anomalies were not high enough to fit the free-air gravity anomalies at the deformed zones, especially in profile 050. Tramontini and Davis [22] also noticed the density of the high-velocity layer under the axial trough almost needs to exceed the maximum density of igneous rocks to fit the free-air gravity anomaly and suggested the contribution of the mantle. We interpret the dense materials as upwelled mantle rocks (Figure 9). High average heat-flow value in the Red Sea and the formation of hot hydrothermal brine indicate a relatively shallower heat source [85]. Initial oceanic basalts collected in the rift axis reveal a high temperature but low-pressure melting, implying ascending mantle during the initial burst of oceanic crust [10]. Blocks of mantle-derived peridotite support the involvement of mantle upwelling during the evolution of the Central Red Sea Rift [45,46,53]. Mantle upwelling is a possible driven force for the postevaporite uplift we interpreted (Figure 8).

We confirm that the oceanic crust is only limited to the center of the axial deep. In profile 060 (Figure 9a), the low relief at the eastern end would have given much lower free-air gravity values than observation if the underlying crust was not oceanic. We model the range of the intermediate crust and find it is limited to the boundary of the deep, where the normal fault system is active. On the contrary, in profiles 054 and 050 (Figure 9b,c), the nature of the crust at the eastern ends of these profiles can not be oceanic or intermediate with the low free-air gravity anomalies, and the crust even thickens again to about 8 km in profile 050 (Figure 9c). The extremely thinned crust above the upwelled mantle in profiles 054 and 050 cannot be intermediate or/and oceanic crust either, or the calculated gravity values would exceed the observed values (Figure 9b,c).

3.3. *Tectono-Geomorphic Segmentation of the Central Red Sea Rift*

The Red Sea is a narrow young ocean with a burst of oceanic crust younger than 3–5 Ma in its southern part. Since ~5 Ma, the third rifting stage has strongly shaped the Central Red Sea Rift valley (Figures 4–6). However, the Central Red Sea Rift does not have a typical spreading center yet, but formed segmentations of geomorphology, tectonism, and deep structure.

Geomorphic segmentation of the rift axis developed in two orders. The first-order segmentation consists of relatively continuous central troughs and inter-trough zones; within the troughs, there are still second-order discontinuities (axial domes or highs) between deeps (Figure 10). Salt movement and local dissolution influenced the geomorphology along the rift axis [5,32,49–51], but a simple sedimentary genesis cannot explain the regular segmentation.

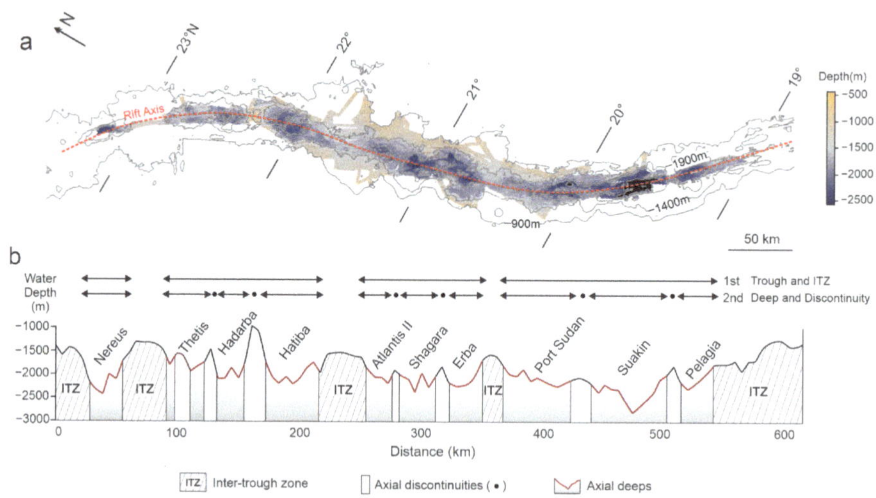

Figure 10. Segmentation of the Central Red Sea Rift axis. (**a**) Bathymetric map of the rift axis; data from Augustin et al. [34]. Grey lines are isobaths of −900 m, −1400 m, and −1900 m, respectively; data from the GEBCO Compilation Group 2021 (https://www.gebco.net, accessed on 3 September 2021). Red dash line represents the axis of the Central Red Sea Rift. (**b**) Water depth variation along the red dash line in (**a**). The red solid lines are indicators of the oceanic crust in axial deeps. Longer arrows represent the first-order segments, central troughs spaced by ITZs; shorter ones represent the second-order segments, axial deeps spaced by the discontinuities (represented by the black dots).

The seismic reflection data reveal the differences in sedimentary and tectonic evolution between the axial deep and the inter-trough zone. Close to the axial deep, the reflector S was interrupted, and Pliocene–Pleistocene strata were quite thinned or even absent. The border faults controlled the significant tectonic subsidence (Figure 4). While the inter-trough zone had a variable tectonic condition, subsidence was uniform and hardly influenced by the fault activities during the third rifting stage, and the basement was deeply buried by thick sediments (Figures 5 and 6). Correspondingly, differences in the nature of the crust between the axial deeps and inter-trough zones were further revealed by gravity models. The oceanic crust is only limited in the center of deeps bounded by the normal fault systems that developed during the third rifting stage (Figure 9), whereas in the inter-trough zone the continental crust thickens (Figure 9). Besides, the Moho tends to rise under the main troughs and deepens under the inter-trough zones (Figure 3). Bouguer gravity anomaly is higher in the axial deeps than the inter-trough zone and discontinuities [35], indicating the deep structure segmentation along the rift axis.

4. Punctiform Breakup and Initial Oceanization Mechanism

In contrast to previous views of a continuously developed spreading center or oceanic crust extending coast to coast [19,51], we argue that the Central Red Sea Rift is still at its initial oceanization. Differential evolution between the axial deeps and the intervals/segmentations support the punctiform breakup, which formed sparse oceanization windows interweaving with continental crust along the rift axis at first.

After the evaporite deposition in the middle Miocene, a regional uplift occurred, causing the basement bulge, sediment deformation, and anomalous thinning of the Pliocene–Pleistocene strata (Figures 4–6). Gravity modeling suggests that the driving force of the uplift is mantle upwelling (Figure 9). Huismans et al. [86] also proposed the occurrence of asthenosphere upwelling and surface doming during passive extension of the intraplate rift after the end of syn-rift through numerical models. However, mantle upwelling did not directly cause the breakup of the continental lithosphere, contradicting previously proposed mantle-dominate breakup models [11,18,27,87,88]. Mantle upwelling thinned the continental crust to <5 km in thickness, but did not change the nature of the crust (Figure 9).

Mantle upwelling and regional uplift also concentrate stress. By comparing the effects of the uplift between the deep and the southern inter-trough zone (Figure 7), we conclude that the centers of the uplift were located in the present axial deeps. Consequently, the fault activities during the third rifting stage concentrated at the deep boundary controlled significant tectonic subsidence, while leaving the inter-trough zones tectonically quiescent (Figures 4–6). It was the nucleation of the fault system that directly led to the punctiform breakup and the initial oceanization limited in the center of the deep (Figure 9).

Here we propose an initial oceanization model of the Central Red Sea Rift (Figure 11). Before 5 Ma, the Central Red Sea Rift already experienced two rifting phases, forming horst-graben systems in a wide zone and the main depression in the rift axis (Figures 4, 5, 6 and 11a). Along with the constant continental thinning and rifting under far-field stress, pre-existing weak zones developed during the early Arabian–Nubian shield evolution could reactivate and influence the upper mantle evolution. The inherited basement structure has an impact on the localization and orientation of rifting [89–91] and reactivates rheological heterogeneity [92]. The lithosphere mantle in pre-existing weak zones is less viscous and easier to trigger small-scale mantle upwelling and be eroded by convective asthenosphere [92,93]. The passive mantle upwelling can cause topographic uplift and thinning and weakening of the lithosphere [38,93]. In the Central Red Sea Rift, mantle upwelling did not induce massive volcanism and the direct breakup of the continental lithosphere. The rising mantle can weaken the overlying thin lithosphere due to a higher geothermal gradient, and translithospheric faults developed around the boundaries of the rift axis under the long-term extension may induce partial melting of the lithospheric mantle (Figure 11b) [11,94,95]. The underplating of high-density materials derived from melts occurred in the lower continental crust (Figure 11c). After the mantle upwelling, the centers of uplift were prone to concentrating extensional stress in the next stage. The onset of fault nucleation in the third rifting phase was limited in isolated deeps (Figure 11c). The well-developed normal-fault systems at the boundaries of the deeps controlled the extreme tectonic subsidence there (Figure 4), and the continental lithosphere finally broke up and seafloor spreading started (Figure 9a).

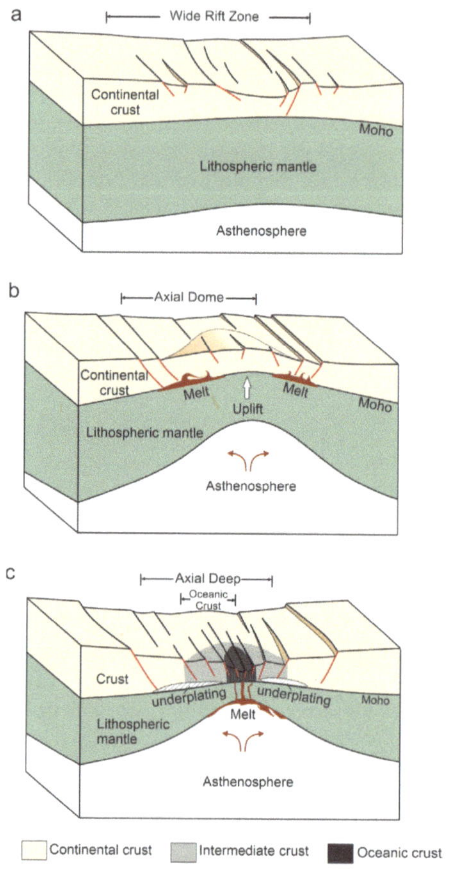

Figure 11. Conceptual model of initial oceanization. (**a**) Early rift initiation, after the first two rifting stages (30–35 Ma). (**b**) Mantle upwelling and topographic uplift concentrated in the weak zones. (**c**) Nucleated normal-fault systems controlled the final breakup and formation of the oceanic crust.

5. Future Evolution to a Continuous Spreading Center

Axial deeps first achieved the initial oceanization, and constituted the second-order segments of the Central Red Sea Rift with the axial discontinuities (Figure 10). Once the oceanic crust was generated sparsely along the rift axis, the instability of the lithosphere between the deeps and discontinuities could trigger small-scale mantle convection at the discontinuities. The discontinuities are at present axial domes or highs (Figure 10), with high free-air gravity anomalies (Figure 12a). Geochemistry also suggests relatively hot materials underlying the discontinuities [96,97]. These discontinuities are probably in the earlier stage of the continental breakup experiencing mantle upwelling and consequent topographic uplift (Figure 11b), and will evolve into new deeps after the stress concentration and fault nucleation (Figure 11c). Inside the relatively continuous central trough in the southern part of the Central Red Sea Rift, the free-air anomalies still show alternating highs and lows along the axial trough (Figure 12a), implying that the deeps and discontinuities subsided successively before forming troughs.

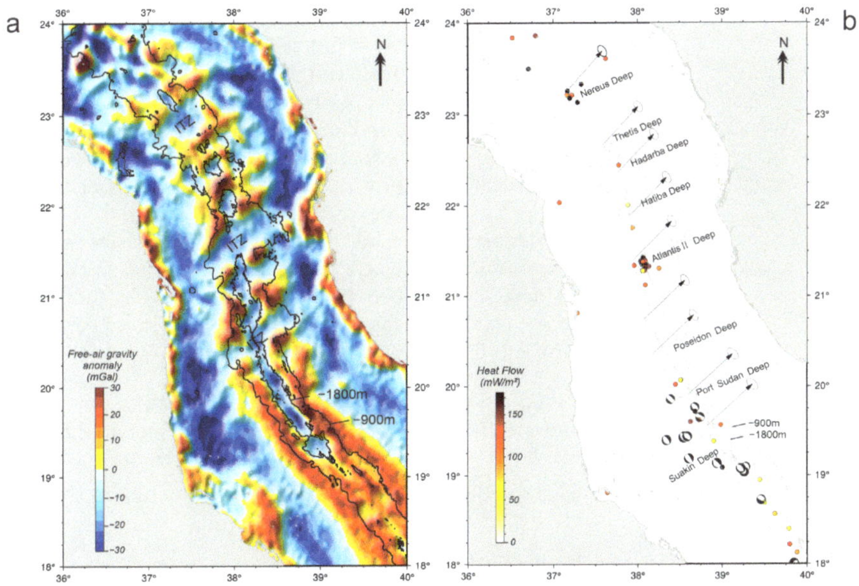

Figure 12. Geophysical and kinematic characteristics of the Central Red Sea Rift. (**a**) Free-air gravity anomalies of the Central Red Sea Rift [98]. Black lines along the rift axis indicate the isobaths of −900 m and −1800 m. (**b**) Plate kinematic, heat flow, and earthquakes in the Central Red Sea Rift. Black arrows represent relative motion angular velocities between the Nubia and Arabia, calculated according to the model MORVEL [40]. Grey lines indicate the isobaths of −900 m and −1800 m. Focal mechanism solutions of the earthquakes from 1976 to present are from the CMT catalog [99,100]. Dots are heat flow sites [101].

The growth of the normal-fault systems, from the nucleation to the growth and linkage, will also play an important role in future evolution. North of 21.5° N, where the deeps are more isolated (Figure 12b), the normal faults are spatially confined around the axial deep and did not propagate to the inter-trough zone (Figures 4–6), and the normal-fault systems are in the stage of initial nucleation. The most active and latest tectonism localizes in the propagation tips of isolated fault systems, giving extremely high heat flow in the Atlantis II Deep and the northmost deep in the Central Red Sea Rift (Figure 12b). With the continued growth of the normal fault systems, early formed depocenters will enlarge by interacting and linking with adjacent faults segments [102]. South of 20.5° N, the more mature and relatively continuous central trough has been bounded by mature and active normal fault systems according to earthquake focal mechanisms (Figure 12b). The normal fault system will propagate northwards under the developing far-field stress of divergence between the African and the Arabian Plates (Figure 12b).

The propagation mechanism of the Central Red Sea Rift can be applied to other propagation tips of spreading centers that form isolated axial deeps [37–39]. When the newly formed segments propagate in their preferred orientation controlled by the pre-existing basement structures, the rheological heterogeneity of the lithosphere will trigger mantle upwelling and activation of the inherited basement structure under far-field stress [38,39,89–94]. Isolated deeps form after the fault rupture and bounding fault systems gradually grow and line up [37]. The interaction and linkage between major fault segments may induce the offset between two adjacent major segments and the initiation of the transform faults [31].

6. Conclusions

The ongoing breakup of the continental lithosphere in the Central Red Sea Rift is discontinuous in time and space. At present, the punctiform breakup of the lithosphere and newborn oceanic crust are limited in the center of the axial deeps.

Two factors played important roles in the formation of the axial deeps, that is, mantle upwelling and normal-fault nucleation. Mantle upwelling in the Central Red Sea Rift was punctiform, triggered by rheological heterogeneity of lithosphere caused by pre-existing basement structure and newly formed weak zones. Mantle upwelling caused topographic uplift, high-density material underplating, and stress concentration, rather than direct continental breakup. It was the normal-fault system nucleated in the third rifting stage that shaped the deeps in the rift axis and controlled the final breakup.

Driven by the instability of the lithosphere after the discontinuous breakup, the discontinuities between the deeps are experiencing mantle upwelling and uplift now, and are destined to evolve into the future deeps by fault nucleation. The originally isolated normal-fault systems will gradually grow and interact and link with each other. Finally, the Central Red Sea Rift will evolve into a continuous newborn spreading center.

Author Contributions: Conceptualization, Y.-D.S. and C.-F.L.; methodology, Y.-L.W. and J.-L.Z.; software, L.H.; data curation, B.M.T.A.; writing—original draft preparation, Y.-D.S.; supervision, C.-F.L.; visualization, Y.-T.L. All authors have read and agreed to the published version of the manuscript.

Funding: This research was funded by National Natural Science Foundation of China, grant number 41776057 and 42176055.

Institutional Review Board Statement: Not applicable.

Informed Consent Statement: Not applicable.

Data Availability Statement: The bathymetry data used in this study are available at the GEBCO Compilation Group (2021) GEBCO 2021 Grid (https://www.gebco.net/data_and_products/gridded_bathymetry_data/, accessed on 3 September 2021). The magnetic anomaly data are available at the EMAG2 (https://geomag.colorado.edu/emag2-earth-magnetic-anomaly-grid-2-arc-minute-resolution.html, accessed on 2 March 2020). The earthquake events from 1976 to present used in this study are available in the CMT catalog (https://www.globalcmt.org/CMTfiles.html, accessed on 9 January 2022). The Model MORVEL (http://www.geology.wisc.edu/~chuck/MORVEL/motionframe_mrvl.html, accessed on 6 January 2022) was used for calculating relative motion angular velocities between the Nubia and Arabia. Sandwell et al. (2014) for the free-air gravity anomaly data (https://doi.org/10.1126/science.1258213, accessed on 13 April 2020), Bonvalot et al. (2012) for the WGM2012 global Bouguer gravity model (https://ccgm.org/en/catalogue, accessed on 4 May 2020), Augustin et al. (2016) for the high-resolution bathymetry data (https://doi.org/10.1016/j.geomorph.2016.08.028, accessed on 9 March 2022), and Pollack et al. (1993) for the heat flow data (https://doi.org/10.1029/93RG01249, accessed on 18 December 2020).

Acknowledgments: The authors thank the Sudan Ministry of Energy for providing the reflection seismic data used in this study available for the academic research purpose. The authors thank editors and reviewers for their comments and constructive suggestions.

Conflicts of Interest: The authors declare no conflict of interest.

References

1. Guennoc, P.; Pautot, G.; Coutelle, A. Surficial structures of the northern Red Sea axial valley from 23° N to 28°N: Time and space evolution of neo-oceanic structures. *Tectonophysics* **1988**, *153*, 1–23. [CrossRef]
2. Merle, O. A simple continental rift classification. *Tectonophysics* **2011**, *513*, 88–95. [CrossRef]
3. Keir, D.; Bastow, I.D.; Pagli, C.; Chambers, E.L. The development of extension and magmatism in the Red Sea rift of Afar. *Tectonophysics* **2013**, *607*, 98–114. [CrossRef]
4. Buck, W. The Dynamics of Continental Breakup and Extension. In *Treatise on Geophysics*, 2nd ed.; Elsevier: Amsterdam, The Netherlands, 2015; Volume 6, pp. 325–379. [CrossRef]
5. Augustin, N.; Devey, C.W.; van der Zwan, F.M.; Feldens, P.; Tominaga, M.; Bantan, R.A.; Kwasnitschka, T. The rifting to spreading transition in the Red Sea. *Earth Planet. Sci. Lett.* **2014**, *395*, 217–230. [CrossRef]

6. Gillard, M.; Autin, J.; Manatschal, G.; Sauter, D.; Munschy, M.; Schaming, M. Tectonomagmatic evolution of the final stages of rifting along the deep conjugate Australian-Antarctic magma-poor rifted margins: Constraints from seismic observations. *Tectonics* **2015**, *34*, 753–783. [CrossRef]
7. Cochran, J.R.; Martinez, F. Evidence from the northern Red Sea on the transition from continental to oceanic rifting. *Tectonophysics* **1988**, *153*, 25–53. [CrossRef]
8. Lowell, G.J.G.J.D. Sea-Floor Spreading and Structural Evolution of Southern Red Sea. *AAPG Bull.* **1972**, *56*, 247–259. [CrossRef]
9. Makris, J.; Franke, M.; Grützner, J. Microearthquake activity in the Suakin Deep, Red Sea and tectonic implications. *Tectonophysics* **1991**, *198*, 411–420. [CrossRef]
10. Ligi, M.; Bonatti, E.; Rasul, N.M.A. Seafloor Spreading Initiation: Geophysical and Geochemical Constraints from the Thetis and Nereus Deeps, Central Red Sea. In *The Red Sea. Springer Earth System Sciences*; Rasul, N., Stewart, I., Eds.; Springer: Berlin/Heidelberg, Germany, 2015; pp. 79–98. [CrossRef]
11. Ligi, M.; Bonatti, E.; Bosworth, W.; Cai, Y.; Cipriani, A.; Palmiotto, C.; Ronca, S.; Seyler, M. Birth of an ocean in the Red Sea: Oceanic-type basaltic melt intrusions precede continental rupture. *Gondwana Res.* **2018**, *54*, 150–160. [CrossRef]
12. Bosworth, W.; Stockli, D.F.; Helgeson, D.E. Integrated outcrop, 3D seismic, and geochronologic interpretation of Red Sea dike-related deformation in the Western Desert, Egypt—The role of the 23Ma Cairo "mini-plume". *J. Afr. Earth Sci.* **2015**, *109*, 107–119. [CrossRef]
13. Mohriak, W.U.; Leroy, S. Architecture of rifted continental margins and break-up evolution: Insights from the South Atlantic, North Atlantic and Red Sea–Gulf of Aden conjugate margins. *Geol. Soc. Lond. Spéc. Publ.* **2012**, *369*, 497–535. [CrossRef]
14. Girdler, R.W. The relationship of the red sea to the East African rift system. *Q. J. Geol. Soc.* **1958**, *114*, 79–105. [CrossRef]
15. Kusky, T.M.; Abdelsalam, M.; Tucker, R.D.; Stern, R.J. Evolution of the East African and related orogens, and the assembly of Gondwana. *Precambrian Res.* **2003**, *123*, 81–85. [CrossRef]
16. Almalki, K.A.; Betts, P.G.; Ailleres, L. The Red Sea—50 years of geological and geophysical research. *Earth-Sci. Rev.* **2015**, *147*, 109–140. [CrossRef]
17. Bosworth, W.; Huchon, P.; McClay, K. The Red Sea and Gulf of Aden Basins. *J. Afr. Earth Sci.* **2005**, *43*, 334–378. [CrossRef]
18. Ligi, M.; Bonatti, E.; Tontini, F.C.; Cipriani, A.; Cocchi, L.; Schettino, A.; Bortoluzzi, G.; Ferrante, V.; Khalil, S.; Mitchell, N.; et al. Initial burst of oceanic crust accretion in the Red Sea due to edge-driven mantle convection. *Geology* **2011**, *39*, 1019–1022. [CrossRef]
19. Girdler, R.W.; Styles, P. Two Stage Red Sea Floor Spreading. *Nature* **1974**, *247*, 7–11. [CrossRef]
20. Ghebreab, W. Tectonics of the Red Sea region reassessed. *Earth-Sci. Rev.* **1998**, *45*, 1–44. [CrossRef]
21. Mohr, P. Structure of Yemeni Miocene dike swarms and emplacement of coeval granite plutons. *Tectonophysics* **1991**, *198*, 203–221. [CrossRef]
22. Tramontini, C.; Davies, D. A Seismic Refraction Survey in The Red Sea. *Geophys. J. Int.* **1969**, *17*, 225–241. [CrossRef]
23. Drake, C.L.; Girdler, R.W. A Geophysical Study of the Red Sea. *Geophys. J. Int.* **1964**, *8*, 473–495. [CrossRef]
24. Whitmarsh, R.B.; Ross, D.A.; Ali, S.; Boudreaux, J.E.; Coleman, R.; Fleisher, R.L.; Girdler, R.W.; Manheim, F.T.; Matter, A.; Nigrini, C.; et al. 16. SITE 226. Deep Sea Drill. *Proj. Initial Rep.* **1974**, *23*, 595–600.
25. Girdle, R.R.W. 28. miocene evaporites in red sea cores, their relevance to the problem of the width and age of oceanic crust beneath the red sea. *Deep Sea Drill. Proj. Initial Rep.* **1974**, *23*, 913–921.
26. le Quentrec, M.; Sichler, B. 3-D inversion of deep tow magnetic data on the Atlantis II Deep (Red Sea): Hydrothermal and geodynamic interpretation. *Tectonophysics* **1991**, *198*, 421–439. [CrossRef]
27. Bonatti, E. Punctiform initiation of seafloor spreading in the Red Sea during transition from a continental to an oceanic rift. *Nature* **1985**, *316*, 33–37. [CrossRef]
28. Ehrhardt, A.; Hübscher, C.; Gajewski, D. Conrad Deep, Northern Red Sea: Development of an early stage ocean deep within the axial depression. *Tectonophysics* **2005**, *411*, 19–40. [CrossRef]
29. El Khrepy, S.; Koulakov, I.; Al-Arifi, N. Crustal and uppermost mantle structure beneath the continental rifting area of the Gulf of Suez from earthquake tomography. *Tectonophysics* **2016**, *668-669*, 92–104. [CrossRef]
30. Searle, R.C.; Ross, D.A. A Geophysical Study of the Red Sea Axial Trough between 20*5 and 22 N. *Geophys. J. Int.* **1975**, *43*, 555–572. [CrossRef]
31. Izzeldin, A.Y. Transverse structures in the central part of the Red Sea and implications on early stages of oceanic accretion. *Geophys. J. Int.* **1989**, *96*, 117–129. [CrossRef]
32. Mitchell, N.C.; Ligi, M.; Ferrante, V.; Bonatti, E.; Rutter, E. Submarine salt flows in the central Red Sea. *GSA Bull.* **2009**, *122*, 701–713. [CrossRef]
33. Mitchell, N.C.; Park, Y. Nature of crust in the central Red Sea. *Tectonophysics* **2014**, *628*, 123–139. [CrossRef]
34. Augustin, N.; van der Zwan, F.M.; Devey, C.W.; Ligi, M.; Kwasnitschka, T.; Feldens, P.; Bantan, R.A.; Basaham, A.S. Geomorphology of the central Red Sea Rift: Determining spreading processes. *Geomorphology* **2016**, *274*, 162–179. [CrossRef]
35. Makris, J.; Henke, C.; Egloff, F.; Akamaluk, T. The gravity field of the Red Sea and East Africa. *Tectonophysics* **1991**, *198*, 369–381. [CrossRef]
36. Shi, W.; Mitchell, N.C.; Kalnins, L.M.; Izzeldin, A. Oceanic-like axial crustal high in the central Red Sea. *Tectonophysics* **2018**, *747-748*, 327–342. [CrossRef]

37. Davis, K.; Burbank, D.W.; Fisher, D.; Wallace, S.; Nobes, D. Thrust-fault growth and segment linkage in the active Ostler fault zone, New Zealand. *J. Struct. Geol.* **2005**, *27*, 1528–1546. [CrossRef]
38. Smith, D.K.; Schouten, H.; Parnell-Turner, R.; Klein, E.M.; Cann, J.; Dunham, C.; Alodia, G.; Blasco, I.; Wernette, B.; Zawadzki, D.; et al. The Evolution of Seafloor Spreading Behind the Tip of the Westward Propagating Cocos-Nazca Spreading Center. *Geochem. Geophys. Geosyst.* **2020**, *21*. [CrossRef]
39. Little, T.A.; Baldwin, S.L.; Fitzgerald, P.G.; Monteleone, B. Continental rifting and metamorphic core complex formation ahead of the Woodlark spreading ridge, D'Entrecasteaux Islands, Papua New Guinea. *Tectonics* **2007**, *26*. [CrossRef]
40. DeMets, C.; Gordon, R.G.; Argus, D.F. Geologically current plate motions. *Geophys. J. Int.* **2010**, *181*, 1–80. [CrossRef]
41. Whitmarsh, R.B.; Ross, D.A.; Ali, S.; Boudreaux, J.E.; Coleman, R.; Fleisher, R.L.; Girdler, R.W.; Manheim, F.T.; Matter, A.; Nigrini, C.; et al. 15. SITE 225. Deep Sea Drill. *Proj. Initial Rep.* **1974**, *23*, 539–594.
42. Whitmarsh, R.B.; Ross, D.A.; Ali, S.; Boudreaux, J.E.; Coleman, R.; Fleisher, R.L.; Girdler, R.W.; Manheim, F.T.; Matter, A.; Nigrini, C.; et al. 17. SITE 227. Deep Sea Drill. *Proj. Initial Rep.* **1974**, *23*, 601–676.
43. Whitmarsh, R.B.; Ross, D.A.; Ali, S.; Boudreaux, J.E.; Coleman, R.; Fleisher, R.L.; Girdler, R.W.; Manheim, F.T.; Matter, A.; Nigrini, C.; et al. 18. SITE 228. Deep Sea Drill. *Proj. Initial Rep.* **1974**, *23*, 677–751.
44. Bonatti, E.; Simmons, E.C.; Breger, D.; Hamlyn, P.; Lawrence, J. Ultramafic rock/seawater interaction in the oceanic crust: Mg-silicate (sepiolite) deposit from the Indian Ocean floor. *Earth Planet. Sci. Lett.* **1983**, *62*, 229–238. [CrossRef]
45. Bonatti, E.; Clocchiatti, R.; Colantoni, P.; Gelmini, R.; Marinelli, G.; Ottonello, G.; Santacroce, R.; Taviani, M.; Abdel-Meguid, A.A.; Assaf, H.S.; et al. Zabargad (St. John's) Island: An uplifted fragment of sub-Red Sea lithosphere. *J. Geol. Soc.* **1983**, *140*, 677–690. [CrossRef]
46. Bonatti, E.; Ottonello, G.; Hamlyn, P.R. Peridotites from the Island of Zabargad (St. John), Red Sea: Petrology and geochemistry. *J. Geophys. Res. Atmos.* **1986**, *91*, 599–631. [CrossRef]
47. Makris, J.; Rihm, R. Shear-controlled evolution of the Red Sea: Pull apart model. *Tectonophysics* **1991**, *198*, 441–466. [CrossRef]
48. Voggenreiter, W.; Hötzl, H.; Mechie, J. Low-angle detachment origin for the Red Sea Rift System? *Tectonophysics* **1988**, *150*, 51–75. [CrossRef]
49. Coleman, A.J.; Jackson, C.A.-L.; Duffy, O.B. Balancing sub- and supra-salt strain in salt-influenced rifts: Implications for extension estimates. *J. Struct. Geol.* **2017**, *102*, 208–225. [CrossRef]
50. Pilcher, R.S.; Blumstein, R.D. Brine volume and salt dissolution rates in Orca Basin, northeast Gulf of Mexico. *AAPG Bull.* **2007**, *91*, 823–833. [CrossRef]
51. Augustin, N.; van der Zwan, F.M.; Devey, C.W.; Brandsdóttir, B. 13 million years of seafloor spreading throughout the Red Sea Basin. *Nat. Commun.* **2021**, *12*, 1–10. [CrossRef] [PubMed]
52. Levi, S. and Riddihough, R. Why are marine magnetic anomalies suppressed over sedimented spreading centres? *Geology* **1986**, *14*, 651–654. [CrossRef]
53. Bonatti, E.; Hamlyn, P.; Ottonello, G. Upper mantle beneath a young oceanic rift: Peridotites from the island of Zabargad (Red Sea). *Geology* **1981**, *9*, 474–479. [CrossRef]
54. Coleman, R. *Geologic Background of the Red Sea*; U.S. Geological Survey: Asheville, NC, USA, 1974; Volume 23. [CrossRef]
55. Supko, P.; Stoffers, P.; Coplen, T.; Whitmarsh, R.; Weser, O.; Ross, D. Petrography and Geochemistry of Red Sea Dolomite. Deep Sea Drill. *Proj. Initial Rep.* **1974**, *23*. [CrossRef]
56. Stoffers, P.; Kuhn, R.; Whitmarsh, R.; Weser, O.; Ross, D. Red Sea Evaporites: A Petrographic and Geochemical Study. Deep Sea Drill. *Proj. Initial Rep.* **1974**, *23*. [CrossRef]
57. Stoffers, P.; Ross, D.; Whitmarsh, R.; Weser, O. Sedimentary History of the Red Sea. Deep Sea Drill. *Proj. Initial Rep.* **1974**, *23*. [CrossRef]
58. Bunter, M.A.G.; Magid, A.E.M.A. The sudanese red sea: 1. new developments in stratigraphy and petroleum-geological evolution. *J. Pet. Geol.* **1989**, *12*, 145–166. [CrossRef]
59. Segev, A.; Avni, Y.; Shahar, J.; Wald, R. Late Oligocene and Miocene different seaways to the Red Sea–Gulf of Suez rift and the Gulf of Aqaba–Dead Sea basins. *Earth-Sci. Rev.* **2017**, *171*, 196–219. [CrossRef]
60. Almalki, K.A.; Betts, P.G.; Ailleres, L. Episodic sea-floor spreading in the Southern Red Sea. *Tectonophysics* **2014**, *617*, 140–149. [CrossRef]
61. Hughes, G.W.A.G.; Johnson, R.S. Lithostratigraphy of the Red Sea Region. *Geoarabia* **2005**, *10*, 49–126. [CrossRef]
62. Hughes, G.W.; Beydoun, Z.R. The red sea—Gulf of aden: Biostratigraphy, lithostratigraphy and palaeoenvironments. *J. Pet. Geol.* **1992**, *15*, 135–156. [CrossRef]
63. Goldberg, M.; Beyth, M. Tiran island: An internal block at the junction of the Red Sea rift and Dead Sea transform. *Tectonophysics* **1991**, *198*, 261–273. [CrossRef]
64. Amer, R.; Sultan, M.; Ripperdan, R.; Encarnación, J. Structural Architecture for Development of Marginal Extensional Sub-Basins in the Red Sea Active Rift Zone. *Int. J. Geosci.* **2012**, *3*, 133–152. [CrossRef]
65. Beydoun, Z.R. Evolution and development of the Levant (Dead Sea Rift) Transform System: A historical-chronological review of a structural controversy. *Geol. Soc. Lond. Spéc. Publ.* **1999**, *164*, 239–255. [CrossRef]
66. Adam, B.M.T.; Chun-Feng, L.; Wadi, D. Recognition of geological structures in the Red Sea Basin based on potential field data and 2D seismic profiles: Insights from Block 13, Sudan. *Arab. J. Geosci.* **2022**, *15*, 1–21. [CrossRef]

67. Lazar, M.; Ben-Avraham, Z.; Garfunkel, Z. The Red Sea—New insights from recent geophysical studies and the connection to the Dead Sea fault. *J. Afr. Earth Sci.* **2012**, *68*, 96–110. [CrossRef]
68. Blanchette, A.R.; Klemperer, S.L.; Mooney, W.D.; Zahran, H.M. Two-stage Red Sea rifting inferred from mantle earthquakes in Neoproterozoic lithosphere. *Earth Planet. Sci. Lett.* **2018**, *497*, 92–101. [CrossRef]
69. Hutchinson, R.W.; Engels, G.G. Tectonic Evolution in the Southern Red Sea and Its Possible Significance to Older Rifted Continental Margins. *GSA Bull.* **1972**, *83*, 2989–3002. [CrossRef]
70. Parker, R.L. The Rapid Calculation of Potential Anomalies. *Geophys. J. Int.* **1973**, *31*, 447–455. [CrossRef]
71. Oldenburg, D.W. The inversion and interpretation of gravity anomalies. *Geophysics* **1974**, *39*, 526–536. [CrossRef]
72. Bai, Y.; Williams, S.; Müller, D.; Liu, Z.; Hosseinpour, M. Mapping crustal thickness using marine gravity data: Methods and uncertainties. *Geophysics* **2014**, *79*, G27–G36. [CrossRef]
73. Bonvalot, S.; Balmino, G.; Briais, A.; Kuhn, M.; Peyrefitte, A.; Vales, N.; Biancale, R.; Gabalda, G.; Moreaux, G.; Reinquin, F.; et al. *World Gravity Map. Commission for the Geological Map of the World*; BGI-CGMW-CNES-IRD: Paris, France, 2012. Available online: https://ccgm.org/en/catalogue (accessed on 4 May 2020).
74. Laske, G.; Masters, G.; Ma, Z.; Pasyanos, M. Update on CRUST1.0—A 1-degree Global Model of Earth's Crust. In Proceedings of the EGU General Assembly 2013, Vienna, Austria, 7–12 April 2013; *Abstracts*, 15, Abstract EGU2013–2658.
75. Straume, E.O.; Gaina, C.; Medvedev, S.; Hochmuth, K.; Gohl, K.; Whittaker, J.M.; Fattah, R.A.; Doornenbal, J.C.; Hopper, J.R. GlobSed: Updated Total Sediment Thickness in the World's Oceans. *Geochem. Geophys. Geosyst.* **2019**, *20*, 1756–1772. [CrossRef]
76. Seton, M.; Müller, R.D.; Zahirovic, S.; Williams, S.; Wright, N.M.; Cannon, J.; Whittaker, J.M.; Matthews, K.J.; McGirr, R. A Global Data Set of Present-Day Oceanic Crustal Age and Seafloor Spreading Parameters. *Geochem. Geophys. Geosyst.* **2020**, *21*. [CrossRef]
77. Müller, R.D.; Zahirovic, S.; Williams, S.E.; Cannon, J.; Seton, M.; Bower, D.J.; Tetley, M.G.; Heine, C.; Le Breton, E.; Liu, S.; et al. A Global Plate Model Including Lithospheric Deformation Along Major Rifts and Orogens Since the Triassic. *Tectonics* **2019**, *38*, 1884–1907. [CrossRef]
78. Izzeldin, A. Seismic, gravity and magnetic surveys in the central part of the Red Sea: Their interpretation and implications for the structure and evolution of the Red Sea. *Tectonophysics* **1987**, *143*, 269–306. [CrossRef]
79. Makris, J.; Tsironidis, J.; Richter, H. Heatflow density distribution in the Red Sea. *Tectonophysics* **1991**, *198*, 383–393. [CrossRef]
80. Press, F. Section 9: Seismic Velocities. In *Handbook of Physical Constants*; Geological Society of America: Boulder, CO, USA, 1966; Volume 97, pp. 195–218. [CrossRef]
81. Cochran, J.R. Effects of finite rifting times on the development of sedimentary basins. *Earth Planet. Sci. Lett.* **1983**, *66*, 289–302. [CrossRef]
82. Coutelle, A.; Pautot, G.; Guennoc, P. The structural setting of the Red Sea axial valley and deeps: Implications for crustal thinning processes. *Tectonophysics* **1991**, *198*, 395–409. [CrossRef]
83. Nicolas, A.; Violette, J. Mantle flow at oceanic spreading centers: Models derived from ophiolites. *Tectonophysics* **1982**, *81*, 319–339. [CrossRef]
84. Lizarralde, D.; Axen, G.J.; Brown, H.E.; Fletcher, J.M.; González-Fernández, A.; Harding, A.J.; Holbrook, W.S.; Kent, G.M.; Paramo, P.; Sutherland, F.; et al. Variation in styles of rifting in the Gulf of California. *Nature* **2007**, *448*, 466–469. [CrossRef] [PubMed]
85. Langseth, M.G.; Taylor, P.T. Recent heat flow measurements in the Indian Ocean. *J. Geophys. Res. Atmos.* **1967**, *72*, 6249–6260. [CrossRef]
86. Huismans, R.S.; Podladchikov, Y.; Cloetingh, S. Transition from passive to active rifting: Relative importance of asthenospheric doming and passive extension of the lithosphere. *J. Geophys. Res. Solid Earth* **2001**, *106*, 11271–11291. [CrossRef]
87. Flament, N.; Gurnis, M.; Müller, R.D. A review of observations and models of dynamic topography. *Lithosphere* **2013**, *5*, 189–210. [CrossRef]
88. Ziegler, P.A.; Cloetingh, S. Dynamic processes controlling evolution of rifted basins. *Earth-Sci. Rev.* **2004**, *64*, 1–50. [CrossRef]
89. Phillips, T.B.; Jackson, C.A.-L.; Bell, R.E.; Duffy, O.B.; Fossen, H. Reactivation of intrabasement structures during rifting: A case study from offshore southern Norway. *J. Struct. Geol.* **2016**, *91*, 54–73. [CrossRef]
90. Peng, X.; Li, C.-F.; Shen, C.; Liu, Y.; Shi, H. Intra-basement structures and their implications for rifting of the northeastern South China Sea margin. *J. Asian Earth Sci.* **2022**, *225*, 105073. [CrossRef]
91. Salomon, E.; Koehn, D.; Passchier, C. Brittle reactivation of ductile shear zones in NW Namibia in relation to South Atlantic rifting. *Tectonics* **2015**, *34*, 70–85. [CrossRef]
92. Petersen, K.D.; Schiffer, C. Wilson cycle passive margins: Control of orogenic inheritance on continental breakup. *Gondwana Res.* **2016**, *39*, 131–144. [CrossRef]
93. Sacek, V. Post-rift influence of small-scale convection on the landscape evolution at divergent continental margins. *Earth Planet. Sci. Lett.* **2017**, *459*, 48–57. [CrossRef]
94. Chu, G.; Chen, H.; Falloon, T.; Han, J.; Zhang, S.; Cheng, J.; Zhang, X. Early Cretaceous mantle upwelling and melting of juvenile lower crust in the Middle-Lower Yangtze River Metallogenic Belt: Example from Tongshankou Cu-(Mo W) ore deposit. *Gondwana Res.* **2020**, *83*, 183–200. [CrossRef]
95. Chenin, P.; Jammes, S.; Lavier, L.L.; Manatschal, G.; Picazo, S.; Müntener, O.; Karner, G.D.; Figueredo, P.H.; Johnson, C. Impact of Mafic Underplating and Mantle Depletion on Subsequent Rifting: A Numerical Modeling Study. *Tectonics* **2019**, *38*, 2185–2207. [CrossRef]
96. Vita-Finzi, C.; Spiro, B. Isotopic indicators of deformation in the Red Sea. *J. Struct. Geol.* **2006**, *28*, 1114–1122. [CrossRef]

97. Haase, K.; Mühe, R.; Stoffers, P. Magmatism during extension of the lithosphere: Geochemical constraints from lavas of the Shaban Deep, northern Red Sea. *Chem. Geol.* **2000**, *166*, 225–239. [CrossRef]
98. Sandwell, D.T.; Müller, R.D.; Smith, W.H.F.; Garcia, E.; Francis, R. New global marine gravity model from CryoSat-2 and Jason-1 reveals buried tectonic structure. *Science* **2014**, *346*, 65–67. [CrossRef] [PubMed]
99. Dziewonski, A.M.; Chou, T.-A.; Woodhouse, J.H. Determination of earthquake source parameters from waveform data for studies of global and regional seismicity. *J. Geophys. Res. Atmos.* **1981**, *86*, 2825–2852. [CrossRef]
100. Ekström, G.; Nettles, M.; Dziewoński, A. The global CMT project 2004–2010: Centroid-moment tensors for 13,017 earthquakes. *Phys. Earth Planet. Inter.* **2012**, *200–201*, 1–9. [CrossRef]
101. Pollack, H.N.; Hurter, S.J.; Johnson, J.R. Heat flow from the Earth's interior: Analysis of the global data set. *Rev. Geophys.* **1993**, *31*, 267–280. [CrossRef]
102. Soliva, R.; Benedicto, A. A linkage criterion for segmented normal faults. *J. Struct. Geol.* **2004**, *26*, 2251–2267. [CrossRef]

Journal of
Marine Science and Engineering

MDPI

Article

New Insights into the Seamount Structure of the Northern Part of the Ninetyeast Ridge (Indian Ocean) through the Integrated Analysis of Geophysical Data

Vsevolod Yutsis [1,*], Oleg Levchenko [2], Alexander Ivanenko [2], Ilya Veklich [2], Nataliya Turko [3] and Yulia Marinova [2]

1 División de Geociencias Aplicadas, Instituto Potosino de Investigación Científica y Tecnológica (IPICYT), San Luis Potosí 78216, SLP, Mexico
2 Shirshov Institute of Oceanology, Russian Academy of Sciences, Moscow 119991, Russia
3 Geological Institute, Russian Academy of Sciences, Moscow 119017, Russia
* Correspondence: vsevolod.yutsis@ipicyt.edu.mx; Tel.: +52-444-8342000

Abstract: The linear Ninetyeast Ridge (NER) is the longest oceanic intraplate volcanic edifice and main feature in the Eastern Indian Ocean. Many seamounts are located on the ridge, whose origin and age remain unclear due to the lack of samples of the bedrock of which they are composed. Carbonate sedimentary caps on these seamounts prevent their direct geological sampling by dredging, therefore indirect geophysical methods are an alternative. Such integrated geophysical studies (the main methods are multibeam bathymetry and magnetic surveys) were carried out in cruise #42 of the *R/V Akademik Boris Petrov* in 2017 on a large seamount at the base of the NER's western slope near 0.5° S. The collected data also includes seismic reflection data that reveal morphology, fault tectonics, depth structure, and an assumed origin of this volcanic feature. The Ninetyeast Ridge was formed by the Kerguelen plume magmatism at 50° S in the giant N-S fault. The seamount studied in cruise #42 of the *R/V Akademik Boris Petrov* was formed mainly to the north as a result of two-stage magmatism in a transverse strike-slip fault. The first stage (47 Ma) formed the main western part of the seamount at 20° S. The second stage (23 Ma) formed its eastern part at 8° S. The time intervals between the formation of the main massif of the Ninetyeast Ridge and the stages of subsequent magmatism that formed the western and eastern parts of the seamount are approximately 31 and 55 Ma, respectively.

Keywords: seamount; Ninetyeast Ridge; bathymetry; fault; high-resolution seismic data; reflector; magnetic field; anomaly

Citation: Yutsis, V.; Levchenko, O.; Ivanenko, A.; Veklich, I.; Turko, N.; Marinova, Y. New Insights into the Seamount Structure of the Northern Part of the Ninetyeast Ridge (Indian Ocean) through the Integrated Analysis of Geophysical Data. *J. Mar. Sci. Eng.* **2023**, *11*, 924. https://doi.org/10.3390/jmse11050924

Academic Editors: George Kontakiotis, Assimina Antonarakou and Dmitry A. Ruban

Received: 8 April 2023
Revised: 24 April 2023
Accepted: 24 April 2023
Published: 26 April 2023

1. Introduction

Bathymetric surveys in the world's oceans revealed a large number of seamounts, the vast majority of which were formed at the boundaries of the lithospheric plate, where more than 95% of volcanic activity is concentrated. However, some seamounts were formed away from these boundaries. The explanation of intraplate volcanism required mechanisms and structures other than those that define the tectonics of the lithospheric plates (spreading, subduction, and transform faults). It was suggested that intraplate seamounts are the surface trace of the "hot" mantle plume and can therefore be considered as windows into the Earth's interior, helping to explain various deep processes in the lithosphere [1]. The most important information on the composition and age of rocks composing oceanic structures and, globally, on the chemical and physical properties of the oceanic lithosphere is obtained by analyzing samples raised by dredging and drilling. However, for intraplate seamounts in the world's oceans, such data are very limited; therefore, various geophysical methods are widely used to study them. In the Indian Ocean, there are many seamounts of different ages, the formation of which cannot be explained by a single mechanism [2–7]. The central structure of the Eastern Indian Ocean is the Ninetyeast Ridge (NER), formed according to the "hot spot trace" hypothesis as a result of the Indian Plate drift over the

Kerguelen mantle plume [8–13]. This is evidenced by the results of seven DSDP and ODP deep-water drilling sites due to a regular increase in the age of the basalts along the ridge from south to north from approximately 40 to 80 million years [14,15]. In the existing evolutionary models regarding the nature of the NER, there is generally some disagreement, with the exception of its ancient northern segment, the formation of which is unanimously attributed to intraplate volcanism in the Indian Plate away from the spreading ridge [8,9,13,16,17].

On the main massive body of the ridge, many seamounts were observed along its entire extent, including to the north of the equator [18]. Multibeam echosounder surveys revealed their detailed morphology in different segments of the NER [19–21]. Kopf et al. (2001) proposed that two seamounts in the southern segment of the NER near 17° S are presumably 6 Ma younger than the adjacent parts of the ridge [19]. These conclusions were obtained by correlating seismic profiles through these seamounts with the nearest Site 757 ODP. Attempts to dredge some seamounts in the northern segment on the NER' top in cruise KNOX06RR of the R/V *Roger Revelle* in 2007 were unsuccessful due to sedimentary caps [20,21]. Only from one volcano in the lower part of the western slope of the NER at 10° S in the #55 cruise of the R/V *Vityaz* in 1976 were basalts ~8 million years old dredged [22]. Thus, a young, late Miocene volcanism potentially existed 160 miles north of DSDP Site 214, where the absolute age of the basalts was determined to be between 52.9 and 59.0 Ma, late Paleocene–early Eocene [14,15], which, according to the above authors, cannot be explained within the Kerguelen hotspot model of NER formation or generally from the plume theory.

The sparse grid of vessel magnetic survey profiles determines the weak study of the anomalous magnetic field of the Ninetyeast Ridge, whose structure remains debatable to date. In contrast to the regular series of linear magnetic anomalies in the basins adjacent to the Ninetyeast Ridge [8,9], such single anomalies are rarely identified on the NER rather conventionally due to the limited amount and poor quality of the available magnetic data.

Ideas about the structure and nature of the NER are largely based on the results obtained in regular cruises of the Shirshov Institute of Oceanology (IO RAS) in the 70–80 years of the last century [13,23,24]. In winter 2017, after a long break, cruise #42 of the R/V *Akademik Boris Petrov* resumed geological and geophysical work in the Indian Ocean [25]. One of the tasks of the cruise was to investigate seamounts on the Ninetyeast Ridge. One of them was investigated at the polygon in the area of 0.5° N and 88.8° E (Figure 1). This seamount was selected from the ship's route on the General Bathymetric Chart of the Oceans (GEBCO) Gridded Bathymetry Data [26]. The polygon is located in the northern segment of the NER, which morphologically is a series of echeloned massifs, presumably tectonic blocks, into which it is broken by SW-NW strike-slip faults (Figure 1a) [21–23]. The research included a bathymetric survey using a multibeam echosounder, high-resolution seismic profiling with a parametric profiler, and a hydromagnetic survey. The survey on the polygon was carried out at a speed of 9 knots simultaneously by these three devices on seven profiles with a total length of about 120 km (Figure 1b).

The purpose of this article is to summarize the results from the first detailed geophysical survey of the seamount on the NER to assume the time and origin of this seamount. This allows for the generation of new ideas about the morphology of the seamount based on the new detailed bathymetric survey with a multibeam, as well as new ideas about the deep structure and nature of the mountain based on a detailed magnetic survey performed for the first time. Together, these new results allow us to make assumptions about the tectonic evolution of this NER segment. The new high-resolution seismic profiling data evidence the recent tectonic processes in this area. They confirm the high tectonic activity, which is manifested in the high modern seismicity [27–31].

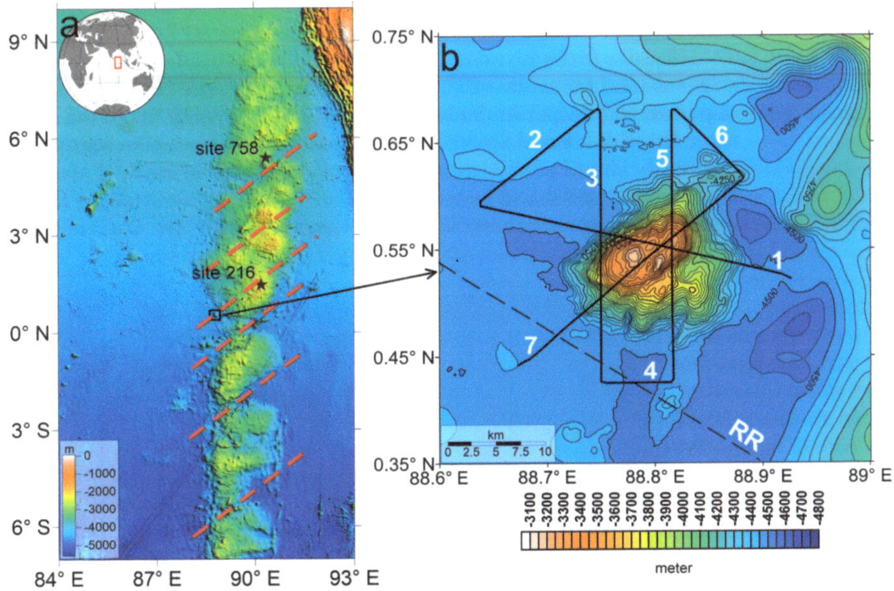

Figure 1. Scheme of the study area: (**a**) echeloned morphology of the northern segment of the Ninetyeast Ridge (position in the inset), GEBCO map base (2019), red dotted line—assumed tectonic faults, and star—deep ocean drilling sites DSDP and ODP; (**b**) bathymetric map with transects of geophysical survey of cruise #42 of R/V *Akademik Boris Petrov* (solid line). Dotted line—profile of the cruise RR1510 R/V *Roger Revelle* (2015).

2. Methods and Results

2.1. Bathymetry

2.1.1. Methods

A bathymetric survey of the seamount with continuous area coverage was performed using the ATLAS HYDROSWEEP DS-2 deep-water multibeam echosounder (ATLAS HYDROGRAPHIC, Bremen, Germany), operating at a frequency of 15.5 kHz. A depth measurement with an accuracy of 0.5% of its value was carried out by 240 beams in wide sector (up to 120°), with the acquisition bandwidth reaching 370% of the depth.

The Hydromap software version 4.2 system data acquisition, processing, and visualization software allowed the survey process to be controlled. During post-processing with the MB-System package [32], the raw multibeam echosounder data were processed (filtered, smoothed, and corrected) and exported from the internal format into ASCII format to calculate a digital elevation model (DEM).

Subsequently, the data for the DEM calculation were supplemented from the GEODAS database [33] with data from the expedition RR1510 of the R/V *Roger Revelle* (2015) that passed through this area [27]. The DEM was used for the construction of the bathymetric map, as well as the calculation of morphometric characteristics, allocation of geomorphological units, etc., in the SAGA GIS program [34]. For the analysis of the relief given below, the data from the seismic profiling by the ATLAS PARASOUND DS2 parametric profiler were also used.

2.1.2. Results

The seamount is located in the lower part of the western slope of the NER at a depth of >4500 m. Its relative height is ~1400 m and the minimum mark is 3093 m (Figures 1 and 2). The base of the seamount on the isobath of 4500 m has an almost rectangular shape of approximately 17 × 15 km. Its longer sides have a SW-NE strike (azimuth ~53°), and

have similar depositional environments; therefore, we assumed that the sedimentology and physical properties did not vary significantly.

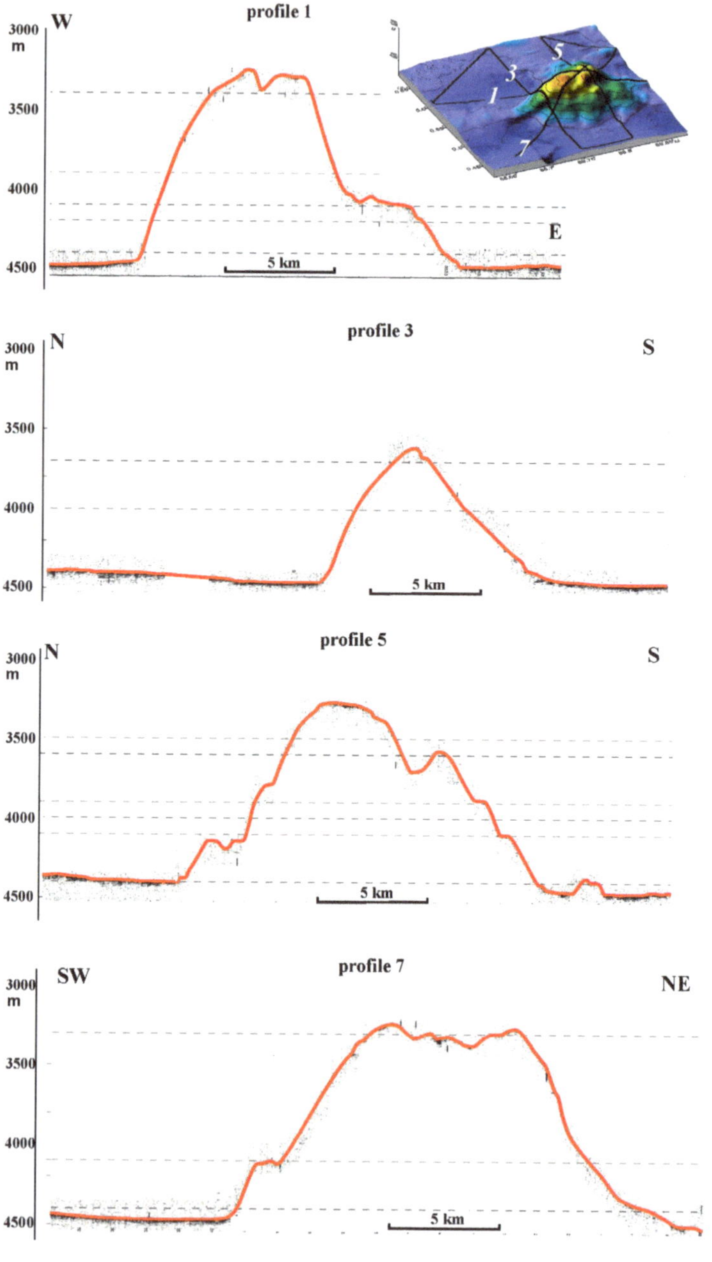

Figure 3. Seismic profiles of the ATLAS PARASOUND DS2 parametric sub-bottom profiler on a very compressed scale through the seamount. The seafloor is highlighted in red. The location of the profiles is shown in the inset above and in Figure 1b. More detailed portions of some seismic profiles are shown in Figure 5a (Profile 5) and Figure 6b (Profile 3).

2.2.2. Results

The visible thickness of the sedimentary cover according to seismic data is not constant. It varies from a few meters to 80–100 m in the northwestern part of the study area. The four seismic profiles crossing the seamount show no reflection boundaries in the sedimentary cover on the seamount (Figure 3). The sediment structure with internal reflectors below the bottom surface is really not visible due to the very compressed and very small scales of the PARASOUND records. Furthermore, volcanic rocks composing the seamount are exposed directly in the bottom surface, and acoustical seismic signals do not penetrate normally below the seafloor here. Unconsolidated loose sediments with a thickness of several meters are found here only in small depressions on the seamount. This pattern is typical for seismic reflection profiles through seamounts. Therefore, the records of the sub-bottom profiler PARASOUND with a high frequency of 4 kHz look similar to a recording of a single-beam echosounder.

The visible thickness of the sedimentary cover areas off the seamount slopes reaches 80–100 m (Figures 4–6). An analysis of the wave pattern of the seismic records in the northwestern part of the study area made it possible to identify three seismo-stratigraphic units, whose boundaries correspond to horizons named A, B, and C (Figure 4). Prominent seismic reflectors/horizons were traced visually based on their characteristic reflection pattern, amplitude, and continuity.

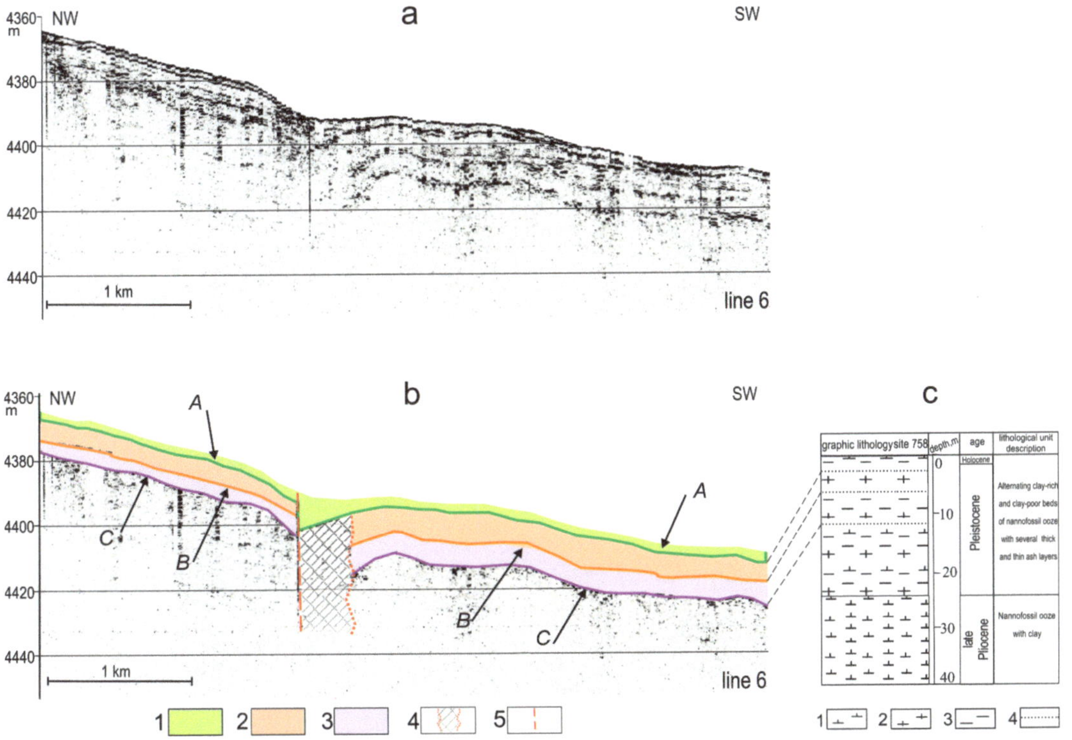

Figure 4. Seismic line 6 north to the seamount: (**a**) original uninterpreted and (**b**) interpreted. Reflectors A, B, and C are shown by green, orange, and violet lines. 1—seismo-stratigraphic unit 1; 2—seismo-stratigraphic unit 2; 3—seismo-stratigraphic unit 3; 4—disturbed zone with a chaotic acoustic structure and no of correlation of reflectors; and 5—fault. It was not possible to identify the seismo-stratigraphic unit below reflector C. The position of line 6 is shown in Figure 1a. (**c**)—lithostratigraphic summary of Site 758 ODP. 1—clayey nannofossil ooze; 2—nannofossil ooze; 3—clay; and 4—volcanic ash.

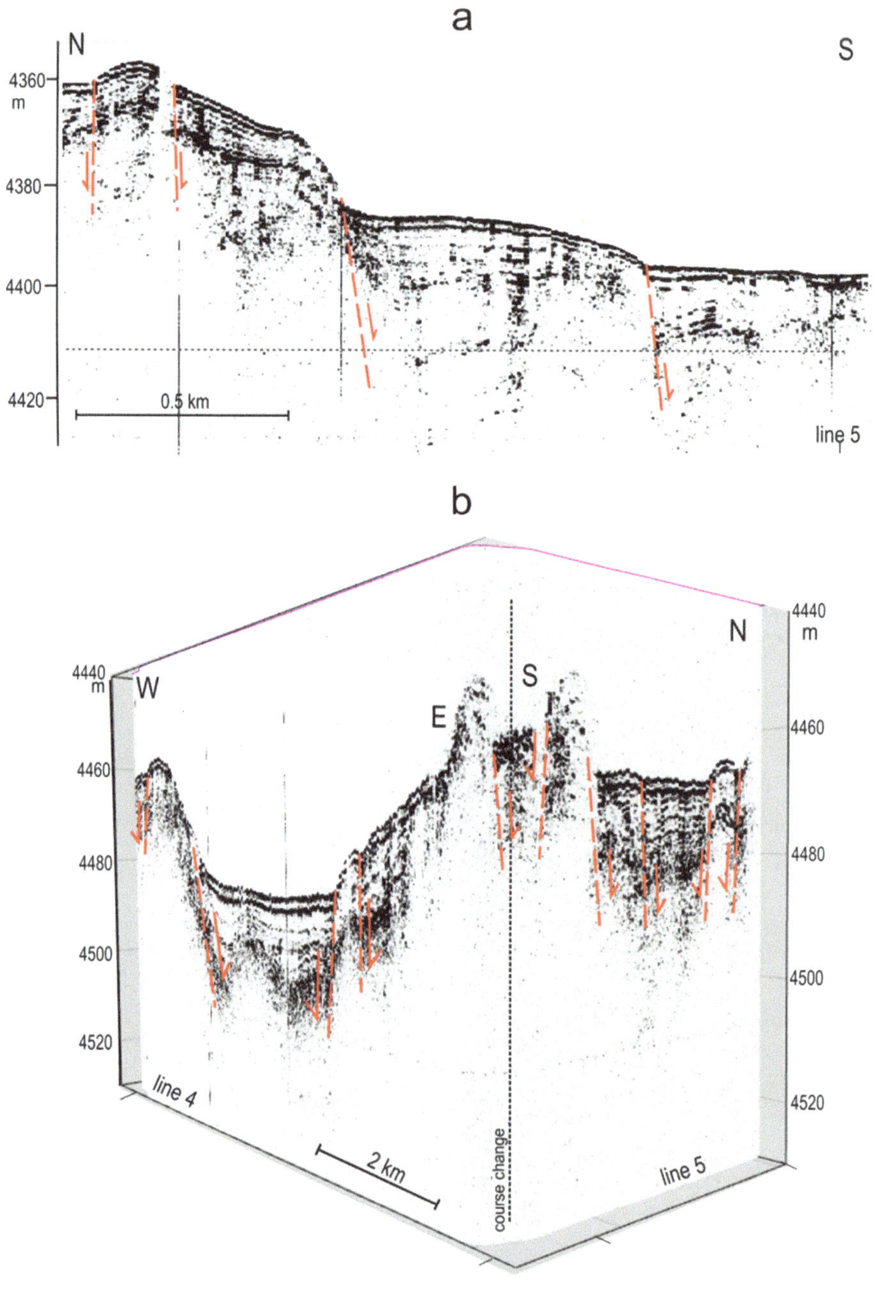

Figure 5. The PARASOUND profiles near the seamount slopes showing fault possible activity. (**a**) Profile 5 north to the seamount. (**b**) Profiles 4 and 5 south to the seamount. Possible faults (red dotted line) in the upper part of the sedimentary section, coming to the bottom surface. The red arrows show the direction of displacement along the fault. The position of the profiles is shown in Figures 1b and 3.

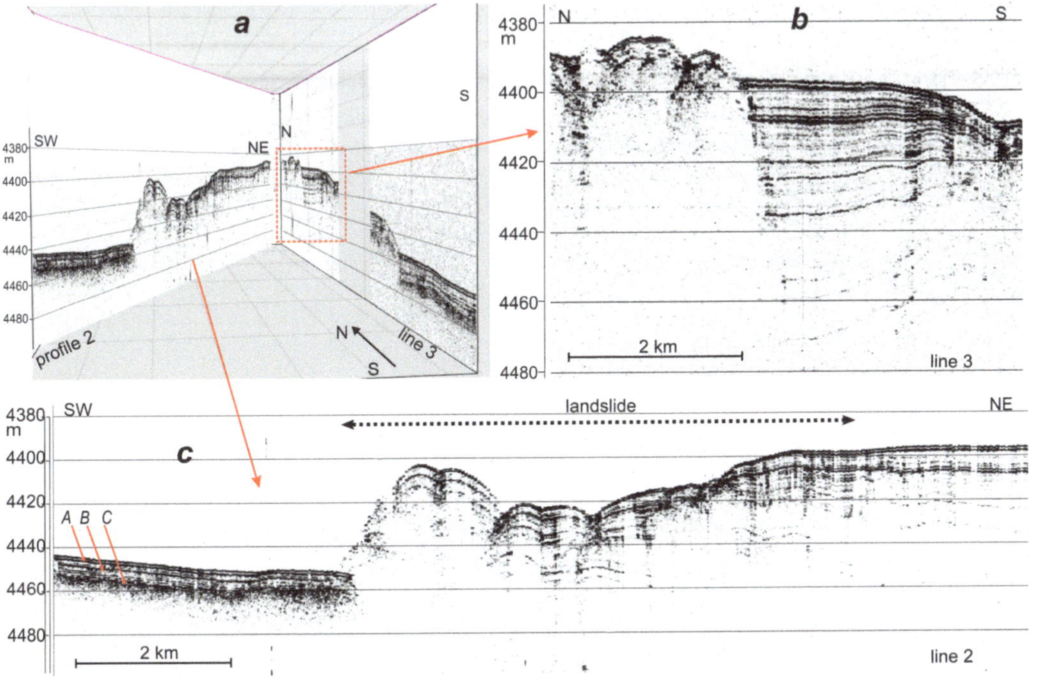

Figure 6. Fragments of seismic profiles 2 and 3 through the landslide bodies in the north of the polygon. (**a**) common view; (**b**,**c**) view in details. The position of the profiles is shown in Figure 1b. Reflectors A, B, and C in (**c**) mark possibly the ash interlayers.

Seismo-stratigraphic unit 1 is located between the bottom and reflector A and has a thickness of about 2–3 m. It is traceable in all seismic sections. reflector A is a continuous, high-amplitude, relatively smooth boundary that almost conformally repeats the bottom topography. Then, seismo-stratigraphic unit 2 follows, the bottom of which is the boundary B. It is characterized by acoustic facies with parallel, high-amplitude discontinuous reflectors. Reflector B stands out at a depth of 7–9 m; it has a lower amplitude and is more discontinuous than reflector A. Seismo-stratigraphic unit 3 has seismic facies with parallel, high-amplitude continuous to discontinuous reflectors. The lower boundary C at a depth of 11−15 m stands out in some places, more often being a series of close intermittent high-amplitude reflections. Below seismo-stratigraphic unit 3, no layering can be traced in our seismic sections.

These reflectors mark the geological boundaries of sediment layers with different physical properties of sediments, primarily density, which is usually due to changes in their lithology. For the lithological and stratigraphic interpretation of seismic profiles, the results from deep ocean drilling sites 758 and 1443 [15,38] were used. We understand that our sections did not cross the drilling sites. Nevertheless, we are trying to give a lithological characteristic to seismo-stratigraphic units. The sites are close to each other and have a similar section. For analysis, we studied the top of section of Site 758. The upper part (0–25 m) of it consists of alternating clay-rich and clay-poor layers of nannofossil ooze with several thick and thin ash layers. Deeper than 25 m of nannofossil ooze with clay was drilled. At a depth of 2 m, the first ash layer of up to 10 cm thickness was identified in the core section. We suppose that the high-amplitude reflector A in the seismic section is correlated with this ash layer (correlation lines are shown by a dotted line between Figure 4b,c), and seismo-stratigraphic unit 1 represents nannofossil ooze.

At a depth of 8 m, an ash layer up to 23 cm thick was identified in the site core. Possibly, reflector B is associated with the appearance of this relatively thick ash horizon in the sedimentary section, and seismo-stratigraphic unit 2 is nannofossil ooze with clay and frequent interbedding of thin layers of ash. In this interval, a change in physical properties is also noted: a decrease in $CaCO_3$ content and a slight increase in density. One of the largest layers of ash for this section is noted at a depth of 11 m and has a thickness of 13 cm. Perhaps this ash layer is the C boundary. The seismo-stratigraphic unit 3 also represents clayey nannofossil ooze. The sediments are more bioturbated and contain less ash layers. Furthermore, this core interval is characterized also by a change in physical properties: a decrease in $CaCO_3$ content and a slight increase in density. Thus, there is a rather high degree of correlation between seismic reflectors and the lithology of the upper part of the sedimentary section. According to core data, we suppose that seismo-stratigraphic units 1, 2, and 3 can be placed in the Holocene–late Pleistocene age.

The northeastern part of the Indian Ocean south of Hindustan is characterized by high tectonic activity manifested by intense crustal fracture deformation and the highest oceanic intraplate seismicity [28–30]. This zone of intense intraplate deformation also captures the northern segment of the Ninetyeast Ridge [15]. Here, strong aftershock events of two mega-earthquakes of a magnitude of around 9 continued in the adjacent Warton Basin [31]. It is believed that the high seismicity of the NER may be related to faults extending down to the bottom surface [20,21]. Our seismic profiles near the seamount show some signs of young, possibly modern neotectonic fault activity in the uppermost part of the sedimentary sequence (up to 50–80 m) (Figure 5). Considering the distortion scale, the faults are relatively gentle normal faults and have a small displacement amplitude. The data obtained do not allow us to trace them in depth.

Further evidence of modern tectonic activity on the seismic profile are the acoustically opaque zones with a chaotic internal sedimentary structure and an uneven bed (Figure 6), which we interpret as landslide sedimentary bodies. The landslide is located in the north-western part of the study area. The source area was observed as chaotic, discontinuous, low-amplitude reflectors in the reflection seismic data were collected, with some continuous stratigraphy visible. The observed pattern is in good agreement with the classical landslide structures on the slope of marine tensional depressions and compressional elevations [39]. Reflectors A, B, and C were determined in the southwest part of the profile and have sharp contact with the frontal deformation landslide zone. At first sight, the vertical boundary, at which the reflectors abruptly cut off, may appear as a steep fault (Figure 6b); however, this is due to a strong distortion of the recording scale (horizontal scale/vertical scale = ~30). On the profile transformed to a more realistic view, which is not given in the paper, this boundary is quite flat and characteristic of the landslide edge.

2.3. *Magnetic Survey*

2.3.1. Data and Methods

The hydromagnetic survey in cruise #42 of R/V *Akademik Boris Petrov* was carried out with the MPMG-4 magnetometer (IO RAS production). A map of the anomalous magnetic field (AMF) was synthesized based on the data of this cruise collected at the studied area and the results of two surveys from the National Centers for Environmental Information, NCEI database [40] (Figure 7a). The RMS error in the survey is estimated from the 20 profile intersection points to be 6.3 nT, which is just over 1% of the AMF amplitude over the seamount. This is a good result for open oceanic areas. To estimate the depth to the center of magnetic masses, we apply the generally recognized method [41], which also uses a radially averaged, frequency-normalized spectrum of magnetic anomalies.

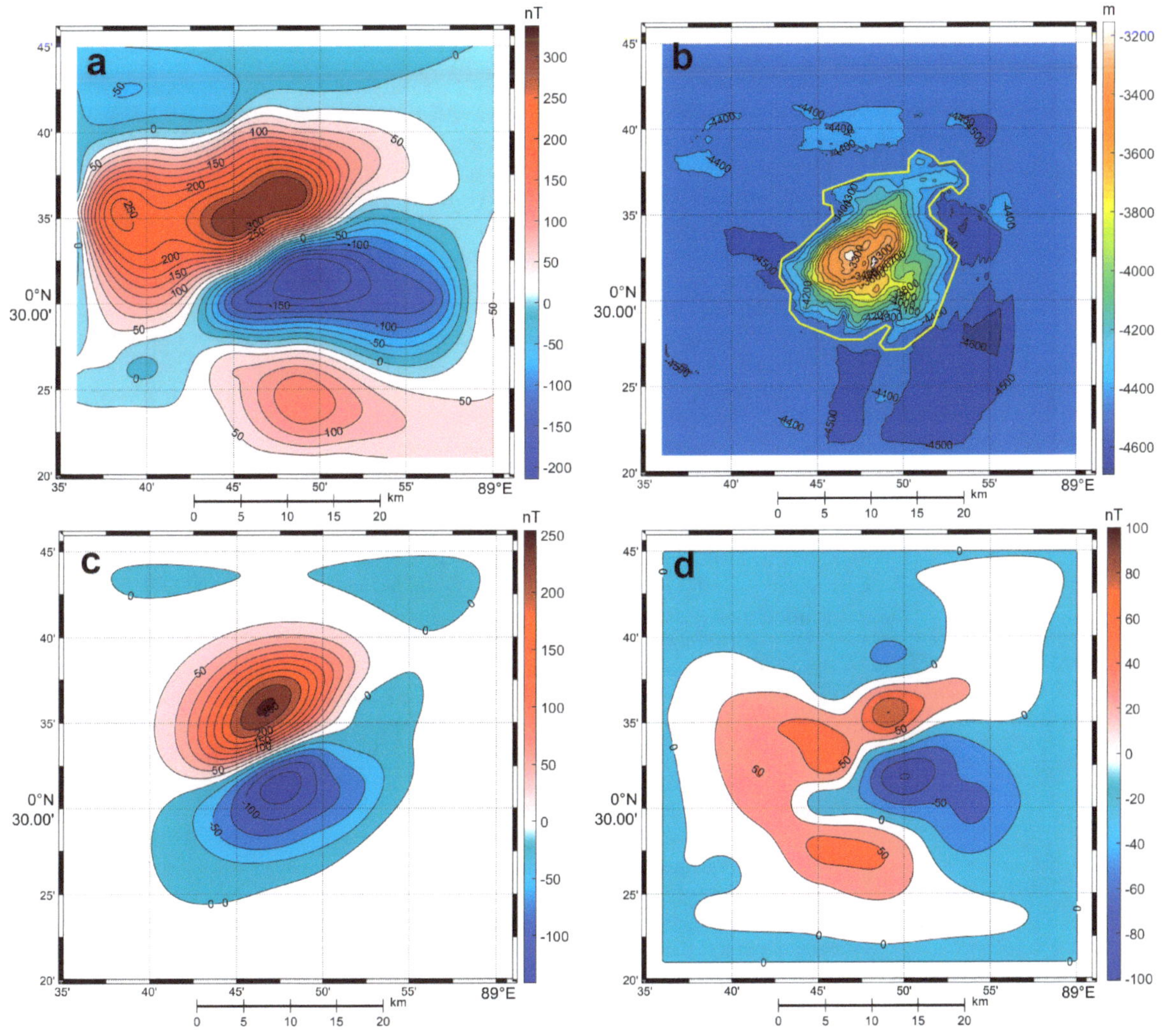

Figure 7. Anomalous magnetic field (AMF) above the seamount and testing of the hypothesis of homogeneous magnetization of this structure: (**a**) measured field synthesized from geomagnetic surveys, in nT; (**b**) seamount topography and contour (yellow bold line) limiting the position of the homogeneous source in the horizontal plane, in m; (**c**) model AMF with a homogeneously magnetized seamount, in nT; and (**d**) residual field (measured−model), in nT.

2.3.2. Results

An intense (>540 nT) alternating magnetic anomaly, extending from SW to NE, is observed above the studied seamount (Figure 7a). The zone of the largest AMF gradients is confined to the seamount top. This gradient zone of NE extension (azimuth 60–65°) separates rather smooth positive and negative parts of the anomaly. Its positive branch is located above the northern slope of the seamount and continues into the adjacent deep-water basin. It has two local maxima of approximately the same amplitude +305 nT and +280 nT to the north of the oval top. This indicates a possible inhomogeneity of the magnetization of the seamount. The negative branch of the anomaly with one minimum ~−240 nT is located above the southern slope of the seamount as well as in the adjacent deep-water basin. It is important to note that the main part of the magnetic anomaly (>150 nT module, i.e., about 0.5 of the maximum amplitude) covers an area much larger than the seamount foot along the 4500 m isobath. This may indicate that the magnetic field in this section of the NER reflects not only the actual relief of the ocean floor, but also deeper sources. In addition, the area of maximum AMF amplitudes is to the north of the seamount, which is a characteristic of normally magnetized objects in the Southern Hemisphere.

A three-dimensional interpretation of the magnetic anomalies over the seamount was performed and includes an analysis of the morphology of AMF and its gradients, in order to highlight the structural boundaries and features of the spatial distribution of the anomaly sources [42–45]. The key issue for this object located practically on the equator was to determine the direction of its magnetization, since the correct analysis of the AMF morphology and its gradients requires a reduction in the anomaly to the pole. This problem was solved by successive application of the methods of the area analysis of AMF—the Helbig magnetic moment method [46] and the method of interactive selection of the equivalent source geometry, with a calculation of the average magnetization vector by the least squares method. Both independent methods yielded results that are close in inclination, -41° and -45°, but quite different in declination, -62° and -27°, respectively. This difference is explained by the fact that the magnetic moment method is more sensitive to edge effects and misaligned anomalies [47], which are obviously present in our data (see Figure 7a). The analysis of the morphostructure of the analytical signal suggests that structural lineaments orthogonal to its strike, possibly limiting faults, adjoin the southwestern and northeastern edges of the main magnetic source. The morphostructure of the analytical signal shows evidence of two centers in the deep structure of the object confined to its flanks, which confirms the modeling results.

To estimate the magnitude and direction of the average magnetization of the seamount, we also used the simplest model of a uniformly magnetized seamount; the results of this fitting are shown in Figure 7b–d. These results clearly show that the magnetic structure of our object is significantly more complex. The residual anomalies clearly demonstrate the presence of at least one more intense source of anomalies in the NE part of the structure. Therefore, we complicated the model and divided the volume of the seamount, considering the morphostructural analysis performed, into four parts: two in the apical part, while the other two are confined to the NE and NW slopes. However, in the process of solving the inverse problem for four uniformly magnetized bodies, it turned out that the calculated magnetization directions of the central and peripheral bodies practically coincide in pairs. Therefore, we left the segmentation into the central and peripheral parts of the seamount as the main one. Further interpretation was performed under the assumption that the magnetization direction changes insignificantly within the given blocks (the geological–structural inverse problem, according to [48]).

Before the vector three-dimensional interpretation, we also estimated the depth to the lower edge of the sources of magnetic anomalies. During the selection of the simplest models, the depth of the base of the model blocks was successively changed with a step of 0.5 km in the range from 4.5 to 8 km. The best selection corresponded to the depth of the seamount base at 7 km from the ocean surface (2.5 km deeper than the bottom of the adjacent basin). The estimate obtained by the widely used spectral method [49,50] has the same value: 7.08 km. Therefore, in further constructions, the lower edge of the models was fixed at a depth of 7.1 km from the ocean level (2.6 km below the bottom in the seamount area). It should be noted that the field morphology analysis and model selection were carried out with simultaneous use of field data and their gradients, which significantly increased the stability of the interpretation.

The determination of the average depth to the upper edge and center of magnetic masses showed that the main depth interval, in which the sources of the anomaly are located, is 3.5–7.1 km from sea level. This indicates the presence of the deeper roots of the investigated seamount and indicates its volcanic nature.

The results of solving the inverse problem, i.e., finding the vector inhomogeneous magnetization in the volume of the mountain, are presented in Figure 8. We obtained an almost complete coincidence between the measured (Figure 8a) and model (Figure 8c) fields (correlation coefficient > 0.99). To verify the model, we also used the solution of the direct problem with the found magnetization values, but under the condition of the verticality of the magnetization vector and the direction of the external field - an analog of reduction to the pole (Figure 8d), which is usually performed only by the AMF maps [45]. The direct correlation of the obtained anomaly with the relief (Figure 8b) is obvious, as well as the confinement of the high magnetization values (Figure 8f) to the main volcanic structures of the seamount—its central two-top block—located inside the white dotted line in Figure 8b. The seamount is characterized by magnetization values up to 5.7 A/m with an average value of 3.2 A/m, which are normal for this type of volcanic structure.

The magnetization vector directions obtained by solving the inverse problem were recalculated into the paleolatitudes and age of the formation according to global modern paleo-reconstructions. For each elementary volume in the seamount body, we can determine the direction of this vector, and hence, the paleolatitude of its formation by the found value of the magnetization vector, assuming that the magnetization, such as that of most igneous rocks in the ocean, has a predominantly remanent component. Next, we used the relationship between the paleolatitudes and age for any point on the Earth [51], based on the modern paleotectonic reconstructions, to recalculate the paleolatitudes we found into the ages. The magnetization directions calculated for the volume of the seamount form two clusters with average values of -36° and -16°, respectively, for the central and peripheral blocks. These inclination values correspond to the paleolatitudes of 20° S and 8° S. The results of determining the age of the seamount sections are presented in Figure 8e. The proximity of the boundaries of the same-age areas to the dividing of the seamount set at the first stage of the interpretation is noteworthy; apparently, this approach was successful, which was confirmed by checking the recalculation of the magnetization in the poleward anomalies. In general, the patterns of the age distribution of the seamount structures are as follows: the central part of the structure, including smoother and steeper western and northern flanks, is about 47 Ma. Less regular, gentler eastern and southern flanks with numerous small uplifts and depressions are about 23 Ma. Such a significant difference indicates at least a two-stage volcanic development within the studied seamount.

In Figure 9, which shows the distribution of the magnetization modulus within and under the volume of seamount, we noticed a thick central body extending from SW to NE (the isosurface is given for magnetization values I = 3 A/m), which is located directly below the oval top of the seamount. The ratio of the horizontal axes of this body is close to 2.5, and the strike azimuth completely coincides with the morphological trends of the seamount.

According to the complex of the studied seamount's geological and geophysical parameters and the results of their interpretation, it becomes obvious that the most adequate explanation of the observed phenomena is the process of volcanism development along the NE strike-slip fault. This volcanism, which originated on the NE strike-slip fault and began 47 Ma ago, led to the growth of almost 3.5 km (considering the thickness of the modern sedimentary cover—up to 2 km [52]) of a conical structure, which we observe as the central part of the seamount. As the paleo-volcan grew, the center of the effusion shifted along the fault to the NE, and then volcanism ceased after a short time. The short duration of this phase of volcanism follows from the fact that the studied seamount has a direct polarity magnetization, and in the middle Eocene, chrons of direct magnetic polarity are less prolonged than those of reversed polarity. The most likely straight magnetic polarity chron, during which our object was formed, is 21n, which began 48.88 and ended 47.76 Ma ago [53]. Intense volcanism resulted in the formation of a regularly shaped seamount extending in the NE direction and up to 3.5 km high.

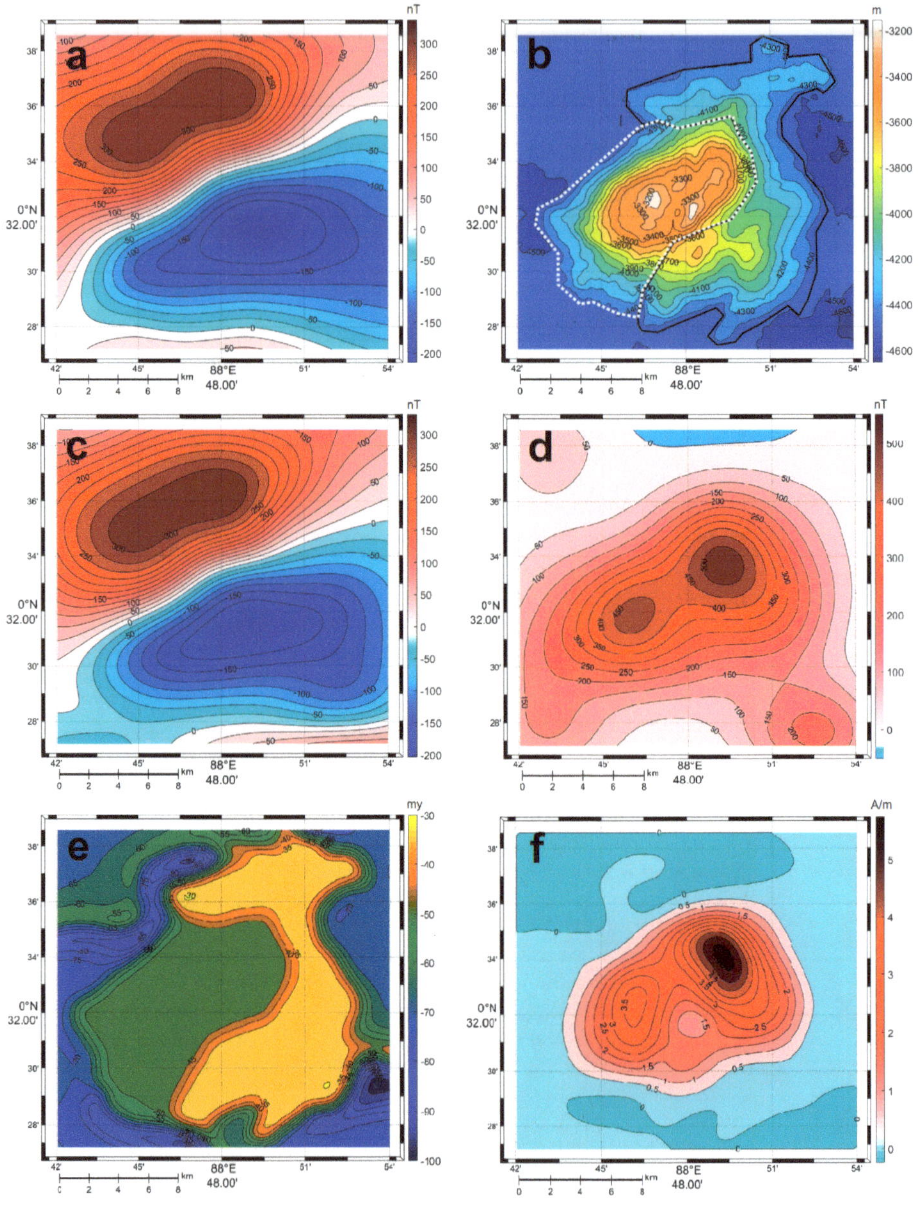

Figure 8. Results of solving the inverse problem − finding the inhomogeneous magnetization in the volume of the seamount: (**a**) the initial AMF directly over the seamount, nT; (**b**) the division of the relief into two main blocks (white dashed and black bold lines), for which the average magnetization directions were previously determined and fixed as constraints; (**c**) the model field from the picked magnetization distribution, nT; (**d**) solution of the direct problem with the selected magnetization distribution, provided that its direction and the direction of the external field are vertical (reduction to the pole), nT; (**e**) age of the seamount fragments, determined by recalculating the found dip angles into the paleolatitude and age using paleo reconstructions, Ma; and (**f**) lateral distribution of magnetization, A/m.

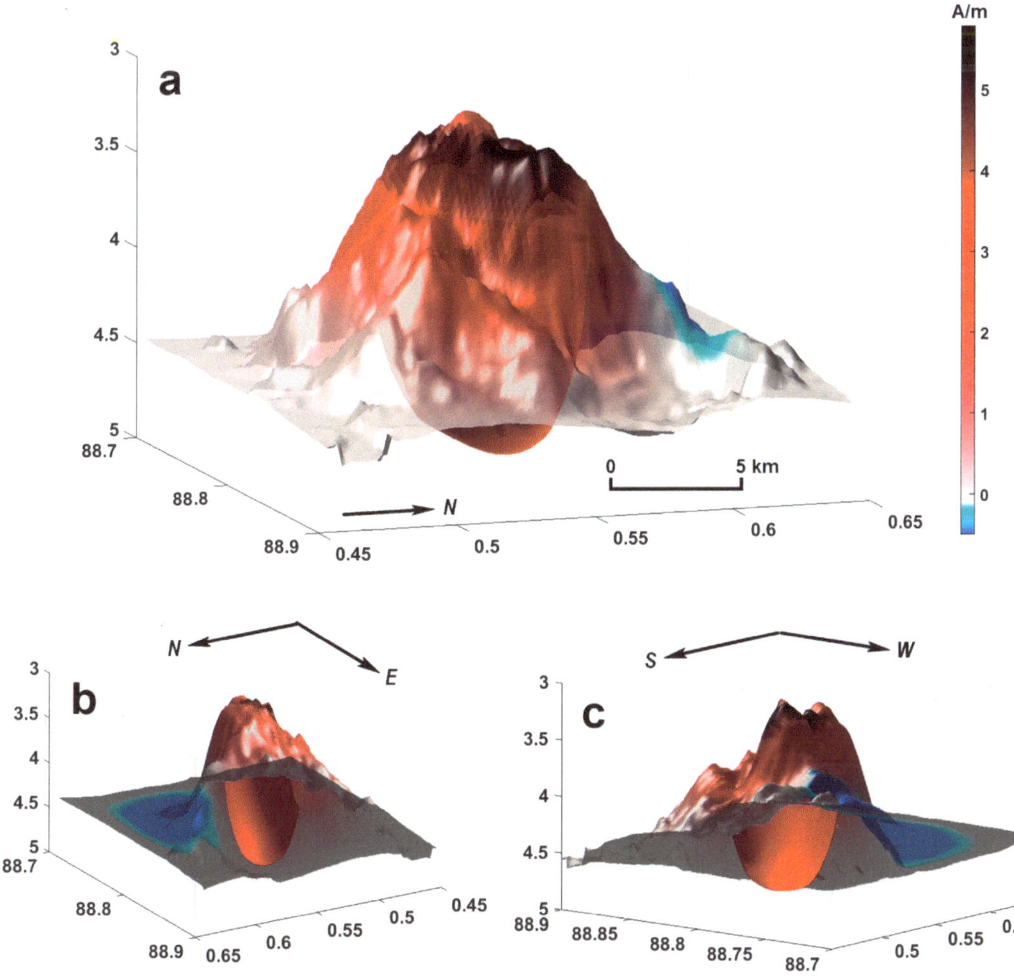

Figure 9. Solution of the inverse problem - finding the inhomogeneous magnetization in the volume of the seamount: (**a**) magnetization, A/m. The isosurface corresponds to 3 A/m. (**b**) The isosurface, bottom view from the NE. (**c**) The isosurface, bottom view from the SW.

Volcanism at the study area resumed about 23 million years ago. The center was confined to the NE summit, but for a number of reasons (a topic beyond the scope of this article), the main effusion occurred in the eastern and SE segments of the structure, and precisely on the seamount slopes, not on its summit. As a result, a characteristic morphology of the structure and a peculiar magnetic structure of the anomaly sources were formed.

3. Discussion

Seamounts are widespread in the Eastern Indian Ocean [2–7]. Many seamounts are located on the Ninetyeast Ridge [19–21]. However, their age and origin are still largely undetermined, as no samples of their bedrock were recovered. The prominent seamount at the base of the NER in its ancient northern segment was investigated by an integrated multidisciplinary detailed geophysical survey in cruise #42 of R/V *Akademik Boris Petrov*. The detailed maps of the topography from the collected multibeam data and the anomalous magnetic field of the surveyed area were constructed the first time. As a result of the bathymetric survey, the morphology of the seamount was significantly clarified. It was found that the bottom relief here has several clear morphological trends of SW-NE, SE-NW, and SSW-NNE strike, which most likely mark the faults. The results obtained are consistent with existing ideas about the tectonics of the northern segment of the NER and add new details to them. Furthermore, the results allow us to make assumptions about the role of the tectonic factor in the formation of the seamount and the possible timing of the volcanism that formed it. The assumption of two stages of fault-related magmatic activity is fundamentally new.

The formation of the main massif of the studied seamount appears to be related to magmatism along the SW-NE strike-slip fault. Two small ridges on the mountain vault are probably associated with the manifestations of magmatic activity along the SSW-NNE strike-slip faults. The strike of the distinguished faults does not coincide with the N-S (transform fracture zones) and E-W (faults in the paleo spreading center) strike of spreading fault structures in the Eastern Indian Ocean [8,9], i.e., they have a different nature and formation time (apparently, more recent). Based on the geomorphological analysis of rare echosounding profiles, some influence of fault tectonics on the formation of the modern bottom topography of the northern segment of the NER [21], located immediately north of the polygon, was previously admitted. The volcanic nature of the ridge itself and its formation as a result of a massive outpouring of basalts in shallow/subaerial conditions is confidently supported by the results of petrological, lithological, and biostratigraphic studies of deep-water drilling cores [14,15,54]. However, this does not exclude a younger subsequent stage of volcanism, which could partly be controlled by fault tectonics.

The flat summit surfaces of the seamounts on the NER vault for the section located immediately north of the polygon are explained by possible erosion activity in subaerial conditions [18]. The elevated summit of the seamount we investigated cannot be related to such a factor. It lies much deeper than it could if submerged after formation from the ocean surface, both according to the theoretical and actual graphs of aseismic ridge submergence, including the NER [54]. Apparently, its morphology with a relatively elevated top is caused by endogenous factors related to the peculiarities of the NER volcanism, and probably, by alignment as a result of pelagic sedimentation and subsequent redistribution of sediments by local exogenous factors.

In the northeastern part of the Indian Ocean, traces of volcanic activity of the Indonesian island arc are widespread in the quaternary sediments. The products of volcanic activity (ash, tephra, and lapilli) in the form of 5–40 cm-thick interlayers were found in many deep-water sites [14,15]. Using tephrochronology, geochemical, and mineralogical analyses, it was established that the ash layers in the upper part of the core section of Site 758 are the result of four stages of the eruption of the Toba volcano on Sumatra Island [55]. Its last eruption (75,000 BP), when volcanic ash covered an area of ~4 million km^2 all the way to the east coast of Africa, was one of the largest in the last 25 million years. An analysis of our seismic data, performed with stratigraphic reference of deep-water drilling data, revealed that the three main reflectors A, B, and C (Figure 4) are possibly associated with the presence of ash interlayers. Their wide distribution within the study area may be associated with the eruptions of the Toba volcano on Sumatra Island.

Furthermore, the seismic data obtained significantly complements the ideas about high neotectonic, up to modern, activity in the northern part of the Ninetyeast Ridge. They confirmed the existence of faults in the uppermost part of the sedimentary cover up to the bottom surface, with which, as suggested by [20,21] high regional intraplate seismicity, they may be associated. Further evidence of modern tectonic activity may be the discovery of a sedimentary body with a chaotic internal structure, which appears to be of a landslide nature.

The small low depression is in the seafloor in the central part of profile 6 (Figure 4). Reflectors A, B, and C bend the immediate in its vicinity and then they abruptly break off, apparently on a fault scarp. Existence of a disturbed zone with a chaotic acoustic structure and no correlation between reflectors B and C is related apparently to tectonic deformation. However, an increased thickness of sediments above reflector A may indicate some activity of any small gravity flow that created a low channel. There is some alternative to its genesis (paleochannel or fault zone) due to the scarce penetration.

The magnetic survey over the seamount registered an intense magnetic anomaly, which is much larger in area than the base of the seamount. It was established that the main source of this anomaly is a deep linear geological object of SW-NE strike, located under the basalt layer of the oceanic crust. Apparently, the deep roots of the seamount are associated with the manifestation of magmatic activity in the zone of one of the faults of such a strike, previously identified in the northern segment of the NER by morphological features [20,21]. It is necessary to explain why the two identified positive local magnetic anomalies in the polygon are slightly offset relative to the ridges in the bottom topography at the top of the seamount. Apparently, the ascending magma flow had a lateral horizontal deflection from the deep source, which creates these anomalies, and magma migrated upward along the already existing weakened fracture zones in the ridge body, which was potentially somewhat distanced from it. A similar setting was previously observed in the formation of relatively young volcanoes to the south on the NER vault during horizontal magma flow [19].

The results of the interactive selection of the equivalent source of magnetic anomalies clarified our preliminary assumptions, which were made on the basis of the geomorphological analysis, about the two-stage formation of the seamount. The obtained values suggest different paleolatitudes of the formation of the isolated segments of the seamount in the two-stage cycle of its formation. According to our calculations of the paleolatitude and age according to the magnetization vector of the seamount, its southwestern part was formed at the first stage approximately 47 million years ago, and later the northeastern part approximately 23 million years ago. At the same time, apparently, two small volcanic structures on its top were also formed. Initially, in our opinion, after the formation of the main volcanic massif of the ridge here in the next phase of secondary outpourings, the more ancient western part of the seamount was formed. Then, during the subsequent phase of magmatic activity, it expanded in the eastern direction. At the same time, the source of magmas, judging by the close values of the estimates of the residual magnetization value, remained the same. As a result, a linear (aspect ratio ~3:1) single object of the northeastern strike apparently formed by magmatism in the zone of one of the faults of the same strike, and was previously identified by different researchers in the northern segment of the NER from the analysis of its morphology [13,20,21].

The explored seamount at 0.5° N lies on the boundary of two segments of the NER that differ sharply in their depth structure. The segment to the south of 0.5° N is characterized by a low effective elastic lithospheric thickness (Te) 0–5 km, whereas the segment to the north is much higher in thickness 10–20 km [16,56]. The formation of this boundary is associated with the second phase of spreading in the multistage late Cretaceous–Cenozoic evolution of the Eastern Indian Ocean and the first major reorganization of plate kinematics about 53 Ma ago during the first soft continental collision of the Indian continent with the island arc of the South Asian subduction zone [8]. The interaction of the Kerguelen mantle plume with the paleospreading ridge potentially created a thermal anomaly that disrupted the oceanic lithosphere and excessive magmatism [12,13]. The timing of the first phase of the formation of the seamount we studied, 47 Ma, closely follows this dramatic event. The third phase of spreading began in the Middle Eocene ~42 Ma, when the second major reorganization of the plates in the Eastern Indian Ocean occurred due to the hard collision of the Indian Plate with the Eurasian Plate - the continental collision of Hindustan with Asia [8], which continues to the present day. The second stage in the formation of the seamount we studied, 23 Ma, seemed to be related to the reorganization of the fault when the stress created by the collision in the northward drifting Indian lithospheric plate reached critical values.

The mechanism of the formation of the studied seamount may be identical to the previously proposed underplating mechanism. It explained the nature of two relatively young volcanoes to the south on the vault of the central segment of the NER near 17° S, which formed 6 Ma later than the main structural phase of the ridge [19]. The aforementioned authors associated the latter phase of volcanism with the relaxation of tensile stresses in the attenuated NE-SW strike-slip lineaments and magma uplift along the small pool apart structures that formed. These stresses could create a regional tectonic and/or heat source beneath the ridge due to an anomalous hot spot regime. They believed that the weakened NE-SW and NW-SE strike-slanted lineaments, obliquely oriented relative to the main regional meridional tectonic structures, were potentially formed by left-lateral strike-slip faults along a giant 90° E fracture zone [21]. The tectonic faults identified by the polygon survey are in good agreement with these directions. A similar two-stage volcanism in this region was also established for the neighboring large intraplate Afanasy Nikitin uplift based on the determination of the absolute age of the dredged basalts there [56]. Initially, the main plateau was formed about 80–73 Ma ago. Then, 6–13 Ma later, a 1200 m-high seamount of the same name was formed on the plateau as a result of magma outpouring from the chamber below the uplift, which drifted together with the lithospheric plate. All of the above is consistent with our assumption of two stages of formation of the studied seamount at the base of the western slope of the NER as a result of magmatic activity apparently associated with faults.

4. Conclusions

(1) The studied seamount appears to consist of two different-age fragments formed about 47 Ma and 23 Ma, respectively, 31 and 55 Ma after the main Ninetyeast Ridge massif. In the first stage, the main western part of the seamount was formed. Then, during the exploratory phase of magmatic activity, the volcanic structure expanded eastward;

(2) The explored seamount located at latitude 0.5° N and the volcano of the late Miocene age located to the south at latitude 10° S [16] are in the zone of one of the giant meridional faults that define the NER fault structure [8]. It is likely that the mechanism of formation of these two volcanic structures is somewhat the same, although the timing of formation differs significantly. As for the two volcanoes on the NER dome near 17° S [19], these appeared to be related to the underplating magmatism of the reactivation of this fault when the critical intraplate compression during continental collision was reached;

(3) The morphology of the studied seamount suggests that it was formed as a result of magma outpourings in the SW-NE strike-slip fault after the formation of the main NER massif in the zone of a giant meridional fault [24]. Presumably, these SW-NE strike-slip faults are second-order structures that apparently formed in the zone of the ancient meridional transform fault [12]. The process was complicated by the fact that shear deformations occurred along the fault during the relaxation of excess regional compression stress with a NW-SE vector in the body of the Indo-Australian lithospheric plate due to continental collision at its northern boundary. These shifts potentially created the local tensile fault zones necessary for this, along which the rise of magmatic material took place;

(4) The results of comprehensive studies performed in cruise #42 of the R/V *Akademik Boris Petrov* suggest rather long cyclic magmatic activity in the northern segment of the NER after 47 Ma already after the formation of its main massif at about 80 Ma.

In conclusion, we note, as well as other researchers [19,56], that the question of secondary magmatism, or its last stage in the Ninetyeast Ridge, remains open to discussion, as dredges of seamounts within it are needed to obtain a more definitive answer.

Author Contributions: Conceptualization, O.L.; methodology, O.L., A.I., I.V. and N.T.; software, A.I. and I.V.; validation, V.Y., O.L. and A.I.; formal analysis, V.Y. and O.L.; investigation, O.L., A.I., V.Y., I.V., Y.M., N.T. and V.Y.; resources, V.Y.; data curation, O.L. and V.Y.; writing—original draft preparation, O.L., A.I., N.T. and Y.M.; writing—review and editing, V.Y. and O.L.; visualization, Y.M., A.I., I.V. and, N.T.; supervision, V.Y. and O.L.; project administration, O.L. and V.Y.; funding acquisition, V.Y. All authors have read and agreed to the published version of the manuscript.

Funding: This work was performed within the framework of IO RAS No. FMWE-2021-0005 and GIN RAS No. FMMG-2023-0005. The APC was partially funded by Instituto Potosino de Investigación Científica y Tecnológica, A.C. and by grant CONACYT project A1-S-29604 to Vsevolod Yutsis.

Institutional Review Board Statement: Not applicable.

Informed Consent Statement: Not applicable.

Data Availability Statement: The original data are property of Shirshov Institute of Oceanology, it could be available to third parts after consideration of official request. The bathymetry data are available at the GEBCO_2019 Grid, https://www.gebco.net/data_and_products/gridded_bathymetry_data/ (accessed on 10 December 2022). Marine geomagnetic data are available at NCEI database https://www.ngdc.noaa.gov/mgg/trk/trackline (accessed on 15 December 2022).

Acknowledgments: This work was performed within the framework of IO RAS No. FMWE-2021-0005 and GIN RAS No. FMMG-2023-0005. The observations described have generally been made in cruise #42 of the R/V *Akademik Boris Petrov*. The authors are grateful for the support provided by colleagues who participated in the geophysical research, as well as the captain and crew of the vessel. The authors thank the anonymous reviewers for their valuable observations, which greatly improved the work. The authors are grateful to Graham Matthew Tippett for editing the English version of manuscript.

Conflicts of Interest: The authors declare no conflict of interest.

References

1. Koppers, A.A.P.; Watts, A.B. Intraplate seamounts as a window into deep earth processes. *Oceanography* **2010**, *23*, 42–57. [CrossRef]
2. Hedervari, P.A. Possible submarine volcano near the central part of Ninety-East Ridge, Indian Ocean. *J. Volcanol. Geotherm. Res.* **1982**, *13*, 199–211. [CrossRef]
3. Mukhopadhyay, R.; Batiza, R. Basinal Seamounts and Seamount Chains of the Central Indian Ocean: Probable Near-Axis Origin from a Fast-Spreading Ridge. *Mar. Geoph. Res.* **1994**, *16*, 303–314. [CrossRef]
4. Mukhopadhyay, R. Post-Cretaceous intraplate volcanism in the Central Indian Basin. *Mar. Geol.* **1998**, *151*, 135–142. [CrossRef]
5. Das, P.; Iyer, S.D.; Kodagali, V.N.; Krishna, K.S. Distribution and Origin of Seamounts in the Central Indian Ocean Basin. *Mar. Geod.* **2005**, *28*, 259–269. [CrossRef]
6. Iyer, S.D.; Das, P.; Kalangutkar, N.G. Seamounts—Windows of opportunities and the Indian scenario. *Curr. Sci.* **2012**, *102*, 1382–1391.
7. Homrighausen, S.; Hoernle, K.; Wartho, J.-A.; Hauff, F.; Werner, R. Do the 85° E Ridge and Conrad Rise form a hotspot track crossing the Indian Ocean? *Lithos* **2021**, *398–399*, 106234. [CrossRef]
8. Royer, J.-Y.; Peirce, J.W.; Weissel, J.K. Tectonic constraints on the hot-spot formation of Ninetyeast Ridge. *Proc. ODP Sci. Res.* **1991**, *121*, 763–776. [CrossRef]
9. Krishna, K.S.; Abraham, H.; Sager, W.W. Tectonics of the Ninetyeast Ridge derived from spreading records in adjacent ocean basins and age constraints of the ridge. *J. Geophys. Res.* **2012**, *117*, B04101. [CrossRef]
10. Nobre Silva, G.; Weis, D.; Scoates, J.S.; Barling, J. The Ninetyeast Ridge and its relation to the Kerguelen, Amsterdam and St. Paul Hotspots in the Indian ocean. *J. Petrol.* **2013**, *54*, 1177–1210. [CrossRef]
11. Frey, F.A.; Silva, I.G.N.; Huang, S.; Pringle, M.S.; Meleney, P.R.; Weis, D. Depleted components in the source of hotspot magmas: Evidence from the Ninetyeast Ridge (Kerguelen). *Earth Planet. Sci. Let.* **2015**, *426*, 293–304. [CrossRef]
12. Sushchevskaya, N.M.; Levchenko, O.V.; Dubinin, E.P.; Belyatsky, B.V. Ninetyeast Ridge: Magmatism and geodynamics. *Geochem. Int.* **2016**, *54*, 237–256. [CrossRef]
13. Levchenko, O.V.; Sushchevskaya, N.M.; Marinova, Y.G. The Nature and Evolution of the Ninetyeast Ridge: A Key Tectonic and Magmatic Feature of the East Indian Ocean. *Geotectonics* **2021**, *55*, 194–218. [CrossRef]
14. Von der Borch, C.C.; Christopher, C.; Sclater, J.G.; Veevers, J.J.; Hekinian, R.; Thompson, R.W.; Pimm, A.C.; McGowran, B.; Gartner, S., Jr.; Johnson, D.A. *Initial Reports of the Deep-Sea Drilling Project*; US Government Printing Office: Washington, DC, USA, 1974; Volume 22, p. 890.
15. Weissel, J.; Peirce, J.; Taylor, E.; Alt, J. *Proceedings of the ODP: Scientific Results*; Ocean Drilling Program: College Station, TX, USA, 1991; Volume 121, p. 990.
16. Kumar, R.T.; Windley, B.F. Spatial variations of effective elastic thickness over the Ninetyeast Ridge and implications for its structure and tectonic evolution. *Tectonophysics* **2013**, *608*, 847–856. [CrossRef]
17. Sreejith, K.M.; Krishna, K.S. Magma production rate along the Ninetyeast Ridge and its relationship to Indian Plate motion and Kerguelen Hot Spot activity. *Geophys. Rev. Lett.* **2015**, *42*, 1105–1112. [CrossRef]
18. Evsyukov, Y.D. Morphology of the Ninetyeast Ridge north of the equator. *Oceanology* **2003**, *43*, 270–275.
19. Kopf, A.; Klaeschen, D.; Weinrebe, W.; Flueh, E.R.; Grevemeyer, I. Geophysical evidence for late stage magmatism at the central Ninetyeast ridge, Eastern Indian Ocean. *Mar. Geophys. Res.* **2001**, *22*, 225–234. [CrossRef]

20. Sager, W.W.; Bull, J.M.; Krishna, K.S. Active faulting on the Ninetyeast Ridge and its relation to deformation of the Indo-Australian plate. *J. Geophys. Res.* **2013**, *118*, 13755–13782. [CrossRef]
21. Levchenko, O.V.; Sborshchikov, I.M.; Marinova, Y.G. Tectonics of the Ninetyeast Ridge. *Oceanology* **2014**, *54*, 231–244. [CrossRef]
22. Kashintsev, G.L.; Levchenko, O.V. Miocene volcanism of the Ninetyeast Ridge. *Dokl. Earth Sci.* **2009**, *428*, 1054–1057. [CrossRef]
23. Bezrukov, P.I.; Neprochnov, Y.P. (Eds.) *Geology and Geophysics of the Sea Floor in the Eastern Indian Ocean*; Nauka: Moskow, Russia, 1981; 255p. (In Russian)
24. Kashintsev, G.L. Aspects of magmatism of the Ninetyeast Ridge. *Oceanology* **2001**, *41*, 413–418.
25. Levchenko, O.V.; Shapovalov, S.M. Return of Russian Oceanographers to the Indian Ocean: Multidisciplinary Study during the 42nd Cruise of the Research Vessel Akademik Boris Petrov. *Oceanology* **2019**, *59*, 164–167. [CrossRef]
26. GEBCO (General Bahtymetric Chart of the Oceans). The GEBCO_2019 Grid. Available online: https://www.gebco.net/data_and_products/gridded_bathymetry_data/ (accessed on 10 December 2022).
27. Multibeam Report for RR1510. Ship Name: Roger Revelle: Chief Scientist: None: Source Organization: UNOLS R2R: Start Date: 8 July 2015: End Date: 2 August 2015. Available online: https://www.ngdc.noaa.gov/ships/roger_revelle/RR1510_mb.html (accessed on 15 December 2022).
28. Stein, S.; Okal, E.O. Seismicity and tectonics of the Ninetyeast Ridge area: Evidence for internal deformation of the Indian plate. *J. Geophys. Res.* **1978**, *83*, 2233–2246. [CrossRef]
29. Petroy, D.E.; Wiens, D.A. Historical seicmicity and implications for diffuse plate convergence in the northeast Indian Ocean. *J. Geophys. Res.* **1989**, *94*, 12301–12319. [CrossRef]
30. Bastia, R.; Radhakrishna, M.; Das, S.; Kale, A.S.; Catuneanu, O. Delineation of the 85° E ridge and its structure in the Mahanadi Offshore Basin, Eastern Continental Margin of India (ECMI), from seismic reflection imaging. *Mar. Pet. Geol.* **2010**, *27*, 1841–1848. [CrossRef]
31. Satriano, C.; Kiraly, E.; Pascal Bernard, P.; Vilotte, J.-P. The 2012 Mw 8.6 Sumatra earthquake: Evidence of westward sequential seismic ruptures associated to the reactivation of a N-S ocean fabric. *Geophys. Res. Lett.* **2012**, *39*, L15302. [CrossRef]
32. Caress, D.W.; Chayes, D.N. MB-System Version 5.5.2252. Open Source Software Distributed from the MBARI and L-DEO Web Sites, 2000–2015. 2015. Available online: https://www.mbari.org/products/research-software/mb-system/ (accessed on 20 November 2022).
33. GEODAS. Available online: https://www.ncei.noaa.gov/maps/bathymetry/ (accessed on 20 November 2022).
34. Conrad, O.; Bechtel, B.; Bock, M.; Dietrich, H.; Fischer, E.; Gerlitz, L.; Wehberg, J.; Wichmann, V.; Böhner, J. System for Automated Geoscientific Analyses (SAGA) v. 2.1.4. *Geosci. Model Dev.* **2015**, *8*, 1991–2007. [CrossRef]
35. Catuneanu, O. *Principles of Sequence Stratigraphy*, 2nd ed.; Elsevier: Amsterdam, The Netherlands, 2022; 375p.
36. Mitchum, R.M. Seismic stratigraphy and global changes of sea level, Part 11: Glossary of terms used in seismic stratigraphy. In *Seismic Stratigraphy—Applications to Hydrocarbon Exploration*; Payton, C.E., Ed.; AAPG Memoir 26: Tulsa, OK, USA, 1977; pp. 205–212.
37. Vail, P.R.; Mitchum, R.M.; Todd, R.G.; Widmier, J.M.; Thompsom, S.; Sangree, J.B.; Bubb, J.N.; Hailelid, W.G. Seismic stratigraphy and global changes in sea level. In *Seismic Stratigraphy—Applications to Hydrocarbon Exploration*; Payton, C.E., Ed.; AAPG Memoir 26: Tulsa, OK, USA, 1977; pp. 49–212.
38. Clemens, S.C.; Kuhnt, W.; LeVay, L.J.; Anand, P.; Ando, T.; Bartol, M.; Bolton, C.T.; Ding, X.; Gariboldi, K.; Giosan, L.; et al. Indian Monsoon Rainfall. In Proceedings of the International Ocean Discovery Program, College Station, TX, USA, 29 November 2014–29 January 2015; International Ocean Discovery Program: College Station, TX, USA, 2016; Volume 353. [CrossRef]
39. Lewis, K.W. Slumping on a continental slope inclined at 1°–4°. *Sedimentology* **1971**, *16*, 97–110. [CrossRef]
40. National Centers for Environmental Information, NCEI Database. NGDC. Available online: https://www.ngdc.noaa.gov/mgg/trk/trackline (accessed on 15 December 2022).
41. Okubo, Y.; Matsunaga, T. Curie point depth in northeast Japan and its correlation with regional thermal structure and seismicity. *J. Geophys. Res. Solid Earth* **1994**, *99*, 22363–22371. [CrossRef]
42. Thébault, E. Magnetic Anomalies, Interpretation. In *Encyclopedia of Solid Earth Geophysics. Encyclopedia of Earth Sciences Series*; Gupta, H.K., Ed.; Springer: Dordrecht, The Netherlands, 2011. [CrossRef]
43. Vine, F.J.; Matthew, D.H. Magnetic anomalies over oceanic ridges. *Nature* **1963**, *199*, 947–949. [CrossRef]
44. Hulot, G.; Olsen, N.; Sabaka, T.J. The Present Field, Chap. 6, Vol. 5. In *Geomagnetism*; Kono, M., Ed.; Treatise on Geophysics; Elsevier: Amsterdam, The Netherlands, 2007; pp. 33–72.
45. Baranov, V. *Potential Fields and Their Transformations in Applied Geophysics*; Borntraeger: Berlin, Germany, 1975; ISBN 3443130089.
46. Helbig, K. Some integrals of magnetic anomalies and their relationship to the parameters of disturbing body. *Z. Geophys.* **1962**, *29*, 83–97.
47. Caratori Tontini, F.; Pedersen, L.B. Interpreting magnetic data by integral moments. *Geophys. J. Int.* **2008**, *174*, 815–824. [CrossRef]
48. Valyashko, G.M.; Ivanenko, A.N.; Czerniawski, G.E.; Lukyanov, S.V. Interpretation Procedure of Marine Magnetic Data: Topical Problems. In *Anomalous Magnetic Field of the World Ocean*; Gorodnitsky, A.M., Ed.; CRC Press: Boca Raton, FL, USA, 1995; pp. 21–66.
49. Spector, A.; Grant, F.S. Statistical models for interpreting aeromagnetic data. *Geophysics* **1970**, *35*, 293–302. [CrossRef]
50. Kelemework, Y.; Fedi, M.; Milano, M. A review of spectral analysis of magnetic data for depth estimation. *Geophysics* **2021**, *86*, J33–J58. [CrossRef]

51. Van Hinsbergen, D.J.J.; de Groot, L.V.; van Schaik, S.J.; Spakman, W.; Bijl, P.K.; Sluijs, A.; Langereis, C.G.; Brinkhuis, H. A Paleolatitude Calculator for Paleoclimate Studies (model version 2.1). *PLoS ONE* **2015**, *10*, e0126946. [CrossRef]
52. Levchenko, O.V.; Milanovsky, V.E.; Popov, A.A. Sediment thickness map and tectonics of the Distal Bengal Fan. *Oceanology* **1993**, *33*, 269–275.
53. Ogg, J. Geomagnetic Polarity Time Scale. In *Geologic Time Scale*, 1st ed.; Elsevier: Amsterdam, The Netherlands, 2022; pp. 159–192. [CrossRef]
54. Coffin, M.F. Emplacement and subsidence of Indian Ocean Plateaus and submarine ridges. In *Synthesis of Results from Scientific Drilling in the Indian Ocean*; Monograph 70; American Geophysical Union: Washington, DC, USA, 1992; pp. 115–125.
55. Ninkovich, D. Distribution, age and chemical composition of tephra layers in deep-sea sediments of western Indonesia. *J. Volcanol. Geotherm. Res.* **1979**, *5*, 67–86. [CrossRef]
56. Krishna, K.S.; Bull, J.M.; Ishizuka, O.; Scrutton, R.A.; Jaishankar, S.; Banakar, V.K. Growth of the Afanasy Nikitin seamount and its relationship with the 85° E Ridge, northeastern Indian Ocean. *J. Earth Syst. Sci.* **2014**, *123*, 33–47. [CrossRef]

Journal of
Marine Science and Engineering

MDPI

Article

Numerical Simulation-Based Analysis of Seafloor Hydrothermal Plumes: A Case Study of the Wocan-1 Hydrothermal Field, Carlsberg Ridge, Northwest Indian Ocean

Kanghao Wang [1,2,†], Xiqiu Han [1,2,*,†], Yejian Wang [2], Yiyang Cai [1,2], Zhongyan Qiu [2] and Xiaoquan Zheng [1,2]

1 School of Oceanography, Shanghai Jiao Tong University, Shanghai 200240, China; wangkanghao@sjtu.edu.cn (K.W.); caiyiyang@sio.org.cn (Y.C.); xiaoquanzheng@sjtu.edu.cn (X.Z.)
2 Key Laboratory of Submarine Geosciences & Second Institute of Oceanography, Ministry of Natural Resources, Hangzhou 310012, China; yjwang@sio.org.cn (Y.W.); zhongyan_qiu@126.com (Z.Q.)
* Correspondence: xqhan@sio.org.cn
† These authors contributed equally to this work.

Citation: Wang, K.; Han, X.; Wang, Y.; Cai, Y.; Qiu, Z.; Zheng, X. Numerical Simulation-Based Analysis of Seafloor Hydrothermal Plumes: A Case Study of the Wocan-1 Hydrothermal Field, Carlsberg Ridge, Northwest Indian Ocean. *J. Mar. Sci. Eng.* **2023**, *11*, 1070. https://doi.org/10.3390/jmse11051070

Academic Editors: George Kontakiotis, Assimina Antonarakou and Dmitry A. Ruban

Received: 24 April 2023
Revised: 8 May 2023
Accepted: 8 May 2023
Published: 18 May 2023

Abstract: Understanding the dynamics of deep-sea hydrothermal plumes and the depositional pattern of hydrothermal particles is essential for tracking the submarine hydrothermal venting site, prospecting polymetallic sulfide resources, as well as deciphering biogeochemistry cycling of marine elements. In this paper, a numerical model of the deep-sea hydrothermal plume is established based on the topography and long-term current monitoring data of the Wocan-1 hydrothermal field (WHF-1), Carlsberg Ridge, Northwest Indian Ocean. The model allows for a reconstruction of the hydrothermal plume in terms of its structure, velocity field, and temperature field. The relationships between the maximum height of the rising plume and the background current velocity, and between the height of the neutral-buoyancy layer and the background current velocity are established, respectively. The transport patterns of the hydrothermal particles and their controlling factors are revealed. Using hydrothermal particles with a density of ~5000 kg/m^3 (i.e., pyrite grains) as an example, it is found that pyrite larger than 1 mm can only be found near the venting site. Those in the size 0.3–0.5 mm can only be found within 137–240 m from the venting site, while those smaller than 0.2 mm can be transported over long distances of more than 1 km. Using the vertical temperature profiling data of WHF-1 obtained during the Jiaolong submersible diving cruise in March 2017, we reconstruct the past current velocity of 10 cm/s, similar to the current data retrieved from the observational mooring system. Our model and the findings contribute to a better understanding of the hydrothermal system of WHF-1, and provide useful information for tracing the hydrothermal vents, prospecting the submarine polymetallic sulfide resources, designing the long-term observation networks, and relevant studies on element cycling and energy budget.

Keywords: deep-sea hydrothermal plumes; numerical simulation; plume rise height; particle transport; Indian Ocean

1. Introduction

Since the first discovery of submarine hydrothermal vents along the East Pacific Galapagos Rift in 1977 [1], more than 700 hundred active hydrothermal fields have been discovered in the world's oceans [2]. Hydrothermal systems play a significant role in the cycling of energy and mass between the solid earth and the oceans [3–5]. Hydrothermal plumes form when high-temperature vent fluids are expelled into the cold and stratified oceanic water column, which represents a significant dispersal mechanism for chemicals and heat released from the seafloor to the oceans [6]. Metallic sulfide particles form when this hot metal- and H_2S-rich, and highly reduced vent fluid mixes with cold, partially oxygenated ocean bottom water [7]. Detection of hydrothermal anomalies in the water column has been widely used to discover hydrothermal fields and associated polymetallic sulfide deposits [8–12].

Compared with on-site investigation and observations, the numerical simulation method, which can realize the quantitative description of the structure of a hydrothermal plume after the model of the hydrothermal plume flow field is established, has the unique advantage of cost-effectiveness and easy use. Morton et al. [13] first proposed the MTT analytical model to describe the transport and mixing process of buoyant plumes and presented an empirical formula for estimating the maximum rising height of plumes through buoyancy flux and buoyancy frequency. Lavelle et al. [14] were the first to use numerical models to study the transverse variation of plumes generated by line-source hydrothermal vents and the distribution of temperature and salinity anomalies, discovering backflow and vortex structures between the plume cap zone and the plume stem zone. Tao et al. [15] established a three-dimensional unsteady numerical model to study the evolution of plumes and found a correlation between the rising height of plumes and the transverse velocity, suggesting that the ratio of the neutral-buoyancy layer height to the maximum rising height of hydrothermal plumes can be used to estimate the heat flux of hydrothermal plumes. Lou et al. [16] took into consideration the interaction between two vent plumes and concluded that the maximum values of turbulent kinetic energy and turbulent dissipation rate decrease with the increase of the distance between the two vents and proposed an empirical formula for the maximum rising height of double-vent plumes.

The numerical simulation has been applied in the study of submarine hydrothermal plumes. For instance, Jiang et al. [17] conducted a numerical simulation of plumes in the ABE hydrothermal field located in the Eastern Lau Spreading Center of the West Pacific. They found that the maximum values of turbulent kinetic energy and turbulent dissipation rate reach their maximum near the vent, while the maximum value of turbulent viscosity is near the plume top, which indicates the presence of strong turbulent mixing that affects chemical reactions and transport processes. Adams et al. [18] studied the turbulent mixing characteristics of high-temperature hydrothermal vent plumes and estimated the heat flux of the Dante sulfide mound on the Juan de Fuca Ridge in the northwest Pacific. Lou et al. [19,20] studied the transport and sedimentation characteristics of particles driven by hydrothermal vent plumes in the Daxi hydrothermal field in the northwest Indian Ocean and classified particles of different sizes and densities.

However, the existing numerical models were often established under ideal environmental conditions. In reality, a hydrothermal field usually has several to dozens of black smokers with complex terrain [21]. Ocean currents also change temporally in terms of speed and direction. Therefore, it is important to consider the interference of multiple hydrothermal vents and the variability of ocean currents while developing the numerical models for submarine hydrothermal plumes.

The Wocan-1 hydrothermal field (WHF-1) (6°22′ N, 60°31′ E), discovered in 2013 by the Chinese DY28th cruise, is located on an axial volcanic ridge along the Carlsberg Ridge, at a water depth of ~3000 m [22,23]. It covers an area of about 450 m × 400 m (Figure 1) [22], consisting of a high-temperature "Flaming Hill" area and a medium-temperature "Mulberry Forest" area [24]. In the high-temperature Flaming Hill area, approximately 60 black smokers were observed, with the measured temperatures of venting fluid reaching up to 359 °C [22,25]. The Mulberry Forest area is located ~70 m to the northwest of the Flaming Hill area [25]. Four black smoker clusters were observed there, with the temperature ranging from 250 to 290 °C, according to the spherite geothermometer [22]. The hydrothermal plumes of the Wocan-1 are mainly observed at a water depth of 2750–2880 m. Figure 2 shows the section where "Flaming Hill" area and "Mulberry Forest" area are located.

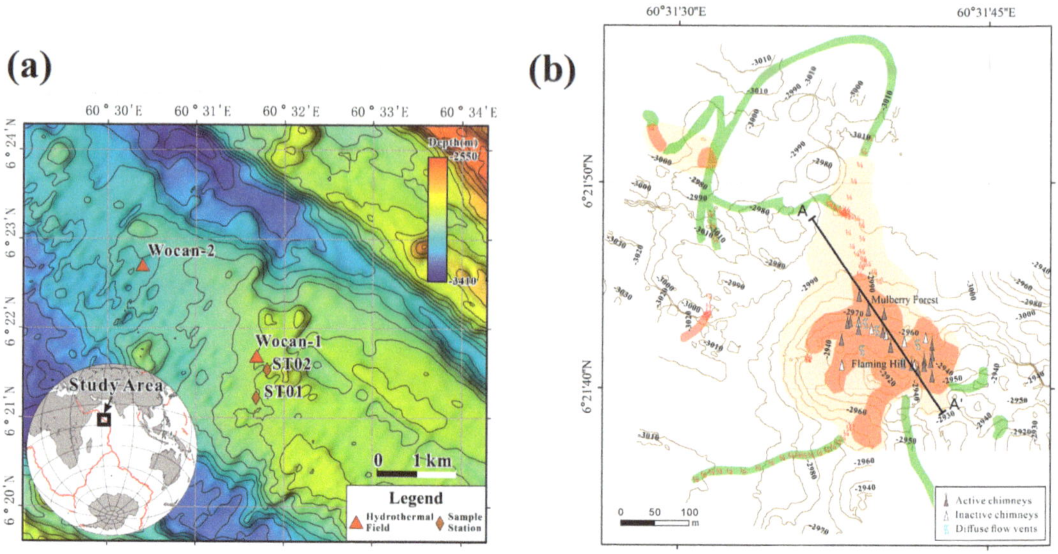

Figure 1. (**a**) Bathymetric map of the study area showing the locations of the Wocan-1 and Wocan-2 hydrothermal fields and the mooring stations for ocean current observation (ST01 and ST02); (**b**) Geological map of the Wocan-1 hydrothermal field, showing the locations of black smokers, adapted from Ref. [22] with permission from Elsevier B.V., 2021.

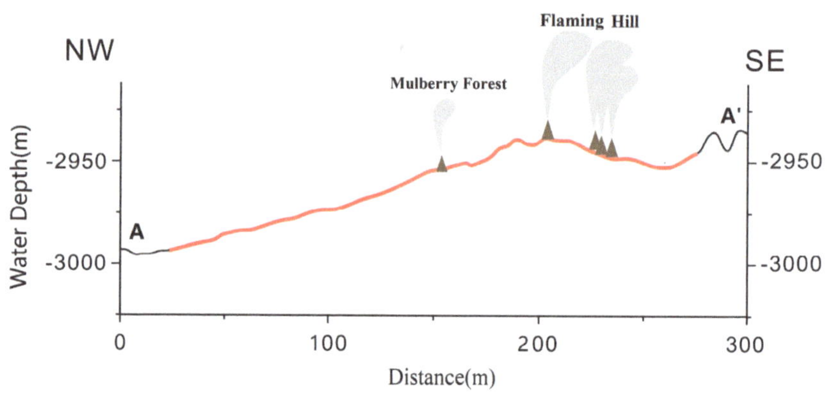

Figure 2. Variation of water depths along the section AA′ crossing WHF-1 as shown in Figure 1.

In this paper, we use a numerical simulation method to construct a three-dimensional model of the hydrothermal plume from WHF-1, with the aim of understanding the dynamics of the hydrothermal plume, the transport law of hydrothermal sulfide particulates, and its controlling factors. The model is built based on the topography of WHF-1, the occurrence and features of black smokers, as well as the measured variability of ocean current velocity. Our results provide theoretical guidance and a scientific basis for the interpretation of hydrothermal anomalies detected in the water column and in the sediments, which could provide important clues on the discovery of unknown active hydrothermal vents.

2. Methods

2.1. Numerical Simulation Method and Model

We built a three-dimensional numerical model based on the actual topography of the WHF-1. In order to enhance calculation accuracy and simulation precision, the fluid domain volume method (VOF) and the separated flow model are employed. The Realizable K-epsilon two-layer turbulence model is adopted as the turbulence model [26,27]. Additionally, the Lagrange multiphase and multiphase interaction are integrated into the model to account for the transport of particles in the plume [28].

A section AA′ crossing the active venting sites of WHF-1 was selected as the geometric model for simulation (Figure 1b). According to the topography of section AA′, it is simplified into a mound with a height of 50 m and a radius of 220 m at the base. To balance computational efficiency and accuracy, the two black smoker clusters found along the section, i.e., the Flaming Hill area and the Mulberry Forest area, are treated as one composite venting site, respectively. The length of the model is set to 2000 m and its height and width are set to 400 m and 200 m, respectively, according to the actual size of WHF-1 and the height of the hydrothermal plume detected in this area. The model is divided via a mesher with a basic size of 1 m. The grid size in the vent area is refined to 0.05 m while that in the rising area of the vent to 0.2 m. Figure 3 illustrates the specific refinement of the whole computational domain grid and the vent area grid. The overall model comprises 7,863,395 squares, including 23,581,668 planes and 8,685,514 points.

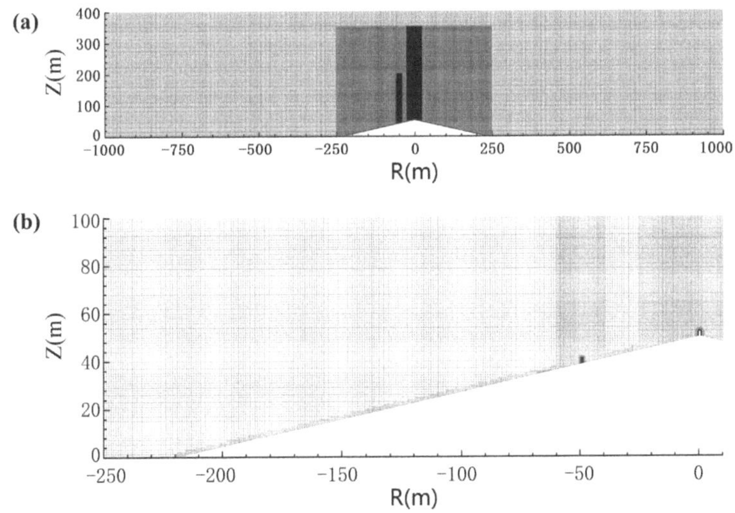

Figure 3. Section of grid model of the Wocan-1 hydrothermal field. (**a**) whole computational domain; (**b**) details of (**a**) showing the part of the mound with the black smokers.

2.2. Parameter Settings of the Numerical Model

The parameters of the numerical model were set based on the measured data of the WHF-1 collected during our various cruises (Tables 1 and 2). Specifically, the data of the background temperature field and density field required by the model were obtained during the Chinese DY57th cruise. According to the CTD profiling, the temperature measured in the water depth of 3000 m was 1.535 °C and 1.907 °C in the water depth of 2600 m, with a temperature gradient of 0.0093 °C/m and a density gradient of 0.00463 kg/m. The occurrence and features of black smokers were obtained based on geological mapping conducted by the Jiaolong submersible during the Chinese DY38th cruise [22]. The temperature of the vent fluids in the Flaming Hill area was 359 °C according to the measurement, while that in the Mulberry Forest area was 250 °C according to the mineralogy of the sulfide chimney [25]. The vent flow rates were calculated through image simulation based on captured images of video [29], which were determined to be 20 cm/s and 15 cm/s for the Flaming Hill area and Mulberry Forest area, respectively. The background ocean current data were obtained from the long-term observation of a mooring system deployed at 1000 m south to the WHF-1 (station 49V-ST01) (Figure 1a) [30], the main direction of the background current being northwest-southeast, which is consistent with the trend of the selected section AA'. The range of current velocity varied from 0.07 to 23 cm/s, dominated by 2 to 10 cm/s, with an average of 6.7 cm/s. Accordingly, four sets of background current velocities: 0 (no background current), 2 cm/s (minimum), 6 cm/s (average), and 10 cm/s (maximum) were used for the comparison study.

Table 1. Basic parameters and boundary conditions of the model used in this study.

Type	Parameter	Setting
computational domain	length width height	2000 m 200 m 400 m
background	temperature gradient density gradient background current direction background current velocity	0.0093 °C/m 0.00463 kg/m^4 NW-SE 2~10 cm/s
boundary condition	bottom top side vent	wall pressure outlet symmetry plane velocity inlet

Table 2. Parameters of hydrothermal vents used in the model based on the observation and measurement of black smokers in the Wocan-1 hydrothermal field.

Venting Area	T_{exit}/°C	T_{bottom}/°C	T_{top}/°C	R/m	v/(cm·s^{-1})
Flaming Hill	360	1.535	1.907	0.4	20
Mulberry Forest	250	1.535	1.907	0.1	15

T_{exit} represents the temperature of hydrothermal fluid, T_{bottom} represents the temperature of background seawater at the bottom, T_{top} represents the temperature of background seawater at the top, R represents the equivalent radius of the venting orifice, v represents the velocity of venting fluid flow.

3. Results

3.1. Plume Structure

As illustrated in Figure 4a, in the absence of a background current, the plume rises and expands to form a cone-shaped plume stem, reaching a maximum height of 315.1 m before forming a plume cap and eventually being transported horizontally. The height of the neutral-buoyancy layer reaches 255 m, which is 80.9% of the former and close to the theoretical value of 76.1%, based on the MTT integral model [12]. When a background current velocity of 2 cm/s is introduced (Figure 4b), the overall structure of the plume is clearly shown and the maximum rise height is 312.9 m. However, there is a presence of background current that results in a slight bending (3°) of the plume axis. If the background current velocity increases to 6 cm/s (Figure 4c), the plume structure is still discernible, but the enhanced current causes more significant bending, with the central axis of the plume bending by 12° and the maximum rise height of the plume reduced considerably to 287.0 m. When the background current velocity is increased to 10 cm/s (Figure 4d), the plume is characterized by turbulent mixing with no distinct rebound current feature at the top. The buoyancy plume structure is disrupted and the maximum rise height is reduced to 134.6 m, which is less than half of the height observed under no current conditions. Overall, increasing the background current velocity would cause the plume structure to bend more, which facilitates the diffusion and mixing of the plume with the ambient water and enhances its horizontal transport capacity.

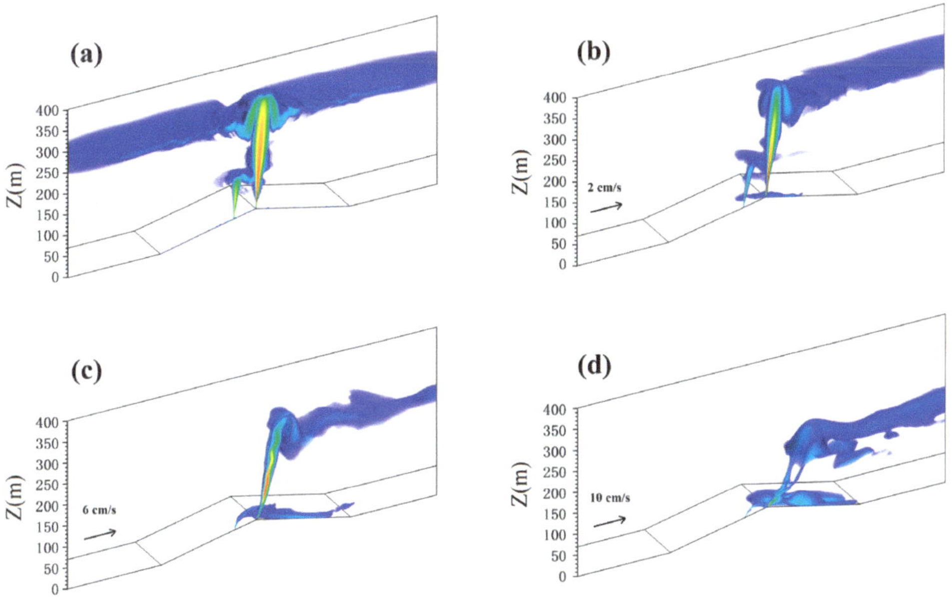

Figure 4. Simulation results showing the three-dimensional structure of the plume. (**a**) no background current; (**b**) background current velocity is 2 cm/s; (**c**) background current velocity is 6 cm/s; (**d**) background current velocity is 10 cm/s.

3.2. Velocity Field of Hydrothermal Plumes

Figure 5 shows the steady-state velocity field of the plume obtained from the numerical model. In the condition of no background current, the rising part of the plume can be divided into three intervals. In the interval of 50–180 m, the plume velocity field is cone-shaped, with the velocity ranging from 1 m/s to 1.62 m/s. The maximum velocity occurs at a height of 70 m. In this interval, the plume and the ambient water undergo continuous entrainment, with similar velocities observed at different heights. In the interval of 180–300 m, the plume velocity field is columnar shaped, with the velocity ranging from 0.33 m/s to 1 m/s. It shows a decreasing trend from the inner part to the outer part. A vortex structure can also be observed between the rising part of the plume and the buoyancy layer. In the interval from the upper part of the plume to the maximum rise height, the buoyancy plume rises and then falls back, and the velocity decreases to zero at the top, coiling and sucking toward the inside of the plume at the boundary and forming a neutral-buoyancy layer with a thickness of 30 m at 60 m away from the center of the plume and eventually diffusing horizontally near the neutral-buoyancy layer. With the background ocean current increasing from 2 cm/s to 6 cm/s, the plume deformed with lower rising height but the plume structure is still clear. When the ocean current increases to 10 cm/s, the plume is well mixed with the ambient water that the structure of the plume is hardly seen.

3.3. Temperature Field of Hydrothermal Plumes

Figure 6 shows that in the absence of background current, the temperature at the centerline of the plume drops from 360 °C at the orifice to 219 °C at 0.5 m above the orifice, 11.4 °C at 20 m, and 2 °C at 86 m, respectively. When the background current velocities increase to 2 cm/s (Figure 6b), 6 cm/s (Figure 6c), and 10 cm/s (Figure 6d), respectively, the corresponding temperature at the axis of the plume (using the curved centerline as the axis) decreases to 2 °C at 89.4 m, 18.1 m, and 10.3 m above the orifice of black smokers, respectively.

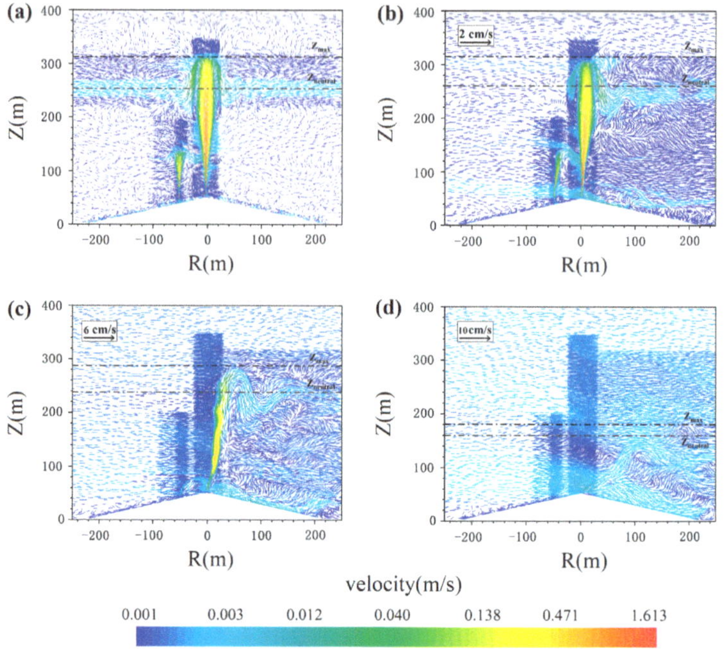

Figure 5. Simulation results show the structure of the velocity field of a stable hydrothermal plume. The dotted lines mark the heights of the maximum rising layers and the neutral-buoyancy layers, respectively. (**a**) no background current; (**b**) background current velocity is 2 cm/s; (**c**) background current velocity is 6 cm/s; (**d**) background current velocity is 10 cm/s.

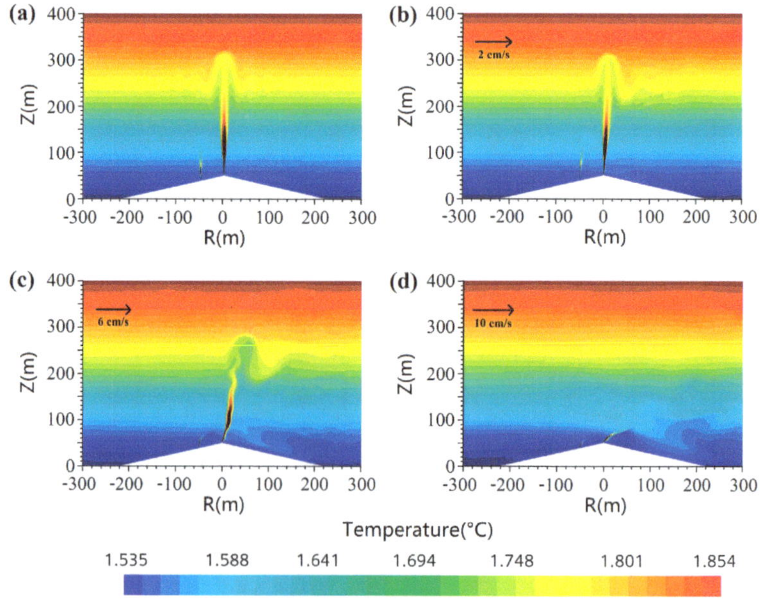

Figure 6. Simulation results showing the temperature field of the plume (**a**) no background current; (**b**) background current velocity is 2 cm/s; (**c**) background current velocity is 6 cm/s; (**d**) background current velocity is 10 cm/s.

4. Discussion

4.1. Relationship between Plume Height and Background Current Velocity

The structure of hydrothermal plumes is mainly affected by the physical properties and velocities of the background ocean current and the fluid ejected from the hydrothermal vents. For a certain hydrothermal field on the seafloor of an open ocean, ocean current velocity is a major variant compared to the other variants such as temperature, salinity and density of the background seawater. Understanding the relationship between the height of a rising plume and the background current velocity is not only important for the design of instrumentation for hydrothermal plume survey in order to detect the hydrothermal anomalies successfully, but also important for the prediction of unknown hydrothermal venting site. Using our numerical model, we extracted the data of the maximum height of the plume and the height of the neutral-buoyancy layer under various background ocean current velocities. Here, the maximum height of the plume is defined as the height at which the vertical velocity of the plume axis is zero, and the height of the neutral-buoyancy layer is the layer where the density of the plume axis first equals that of the environment. Two empirical formulas are obtained via Polynomial Fitting of the results and illustrated in Figure 7:

$$Z_{max} = 315.1 - 1.252v + 0.4v^2 - 0.162v^3, \quad (1)$$

$$Z_{neutral} = 255 - 0.016v + 0.245v^2 - 0.112v^3 \quad (2)$$

where Z_{max} represents the maximum height of a plume, and v is the background current velocity.

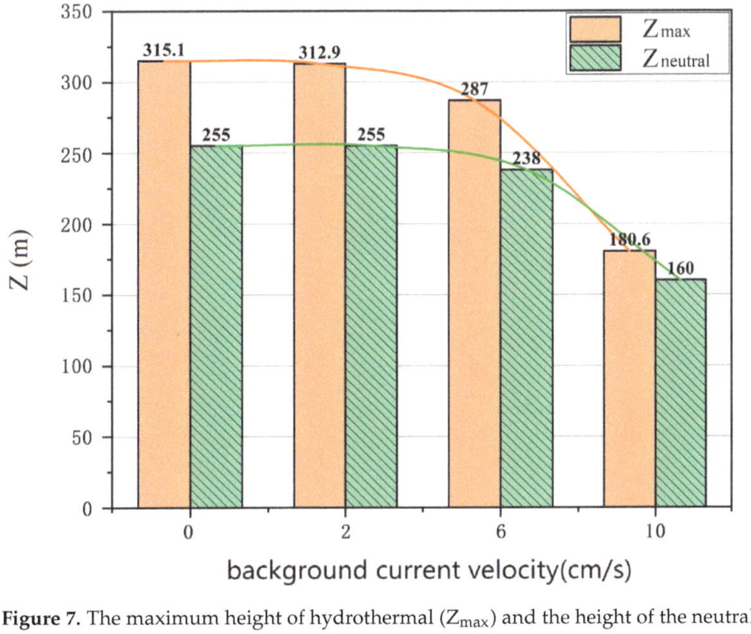

Figure 7. The maximum height of hydrothermal (Z_{max}) and the height of the neutral-bouncy layer ($Z_{neutral}$) under various background current velocities.

Figure 7 shows clearly that with the increase of background current velocity, both the maximum height and the height of the neutral-buoyancy layer of the plume decrease to some extent. However, the overall influence of background current velocity on the height of the neutral-buoyancy layer is smaller than that of the maximum height of the plume, especially when the background current velocity is <6 cm/s. This suggests that the height of the neutral-buoyancy layer is more insensitive to the variation of background ocean current intensity compared to the maximum height of the plume.

4.2. Reconstruction of Background Current Velocity

The temperature anomaly of the hydrothermal plumes is mainly controlled by the extent of mixing of the hydrothermal fluid with the ambient seawater [31]. The temperature gradients between the hydrothermal plume and the ambient seawater decrease dramatically with increasing distance to the venting site. The reported temperature anomalies of hydrothermal plumes detected mostly vary between 0.005 °C and 0.05 °C, and those near the vent area can reach 0.4 °C or higher. Using the temperature anomaly data, it is possible to infer the current velocity of the region by numerical simulation in case no current observation data is available. Here, we use CTD data (38I-CR-CTD-JL125) obtained by the Jiaolong submersible within the Wocan-1 hydrothermal field to conduct an inversion study of the background current velocity and test the robustness of our model. A significant temperature anomaly was observed in the water depth between 2930–2960 m about 20 m from the high-temperature black smoker in Flaming Hill, with the maximum temperature anomaly of 1.65 °C at the water depth of 2945 m. Figure 8 compares the measured temperature profile on site and the temperature profiles extracted from our numerical model at the same location under different ocean current velocities. It is clear that in the absence of a background current, no significant temperature anomaly can be observed. With the increasing background ocean current, the temperature anomaly becomes stronger, the neutral-buoyant layer becomes thinner and lower. Specifically, when the background current is 2 cm/s, a temperature anomaly is observed in the water depth of 2750–2880 m, with the maximum temperature anomaly of 0.15 °C at 2850 m. When the background current velocity increases to 6 cm/s, a stronger temperature anomaly can be observed in the water depth between 2850–2930 m, with the maximum temperature anomaly of 0.46 °C at 2907 m. When the ocean current is 10 cm/s, a strong temperature anomaly is seen in the water depth between 2938–2943 m, with a maximum of 1.36 °C at 2940 m. This case fits the CTD profile very well. The mooring system deployed at 300 m south to the WHF-1 (station 49V-ST02) registered the ocean current velocities vary from 0–24 cm/s, dominated by 3.5–11 cm/s, with a mean value of about 9 cm/s, indicating that the reconstructed background current velocity agrees well with the on-site observation. This also suggests that our numerical model is robust. Note, there is some offset with depth between our observed data and the computational outcome, this could be originated from the different environment between the location of CTD measurements and the location of the temperature field extracted from the 3D numerical model.

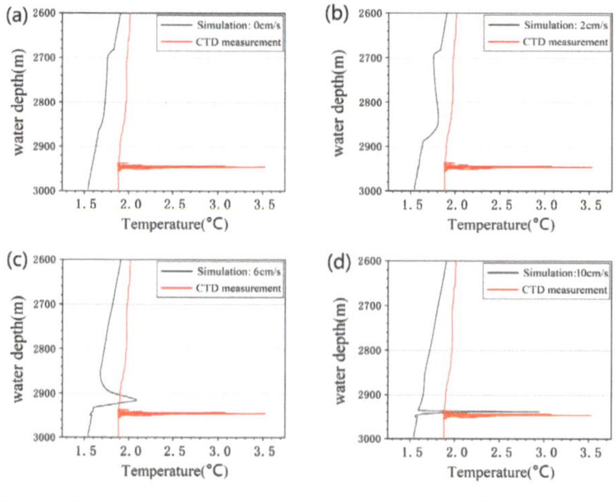

Figure 8. The temperature anomaly obtained by CTD measurement during the DY57 cruise and the modeling results under various background current velocities (**a**) no background current; (**b**) background current velocity is 2 cm/s; (**c**) background current velocity is 6 cm/s; (**d**) background current velocity is 10 cm/s.

4.3. Transport Pattern of Plume Particles and the Controlling Factors

When the high-temperature metal- and H_2S-rich hydrothermal fluid is ejected from the black smoker, the rapid changes in temperature and redox conditions result in the formation of sulfide mineral particles in the plume. The settling of hydrothermal particles laden in the hydrothermal plumes would accumulate to form metalliferous sediments in the near field [32], which can be used as a good indicator for the prospecting and exploration of hydrothermal sulfide deposits [33]. Numerical simulation has the advantage to conduct quantitative analysis of the depositional patterns of hydrothermal particles with different physical properties. The background current velocity is set to 10 cm/s according to the dominant frequency of the current velocity in the study area. Considering the morphology, mineralogy, and geochemical characteristics of WHF-1 [34,35], and the fact that pyrite is the most common hydrothermal sulfide mineral [36], we investigate the transport behavior of pyrite particles of various sizes in the plume. The density of pyrite particles is set to 5000 kg/m^3, and the sizes of pyrite particles are set at 50 μm, 200 μm, 300 μm, 400 μm, 500 μm, and 1000 μm, respectively, based on the results of field surveys.

Figure 9 shows the transport trajectories of pyrite particles in the plume field. Pyrite particles with a size smaller than 200 μm are strongly influenced by the background current and can be transported at a height of 160 m over a distance of more than 1 km. Particle sizes ranging from 300 to 500 μm can rise to 124–161 m but only can migrate a short distance not farther than 240 m. Specifically, a 300 μm size particle can be transported as far as 240 m, a 400 μm particle can be transported to 198 m, and a 500 μm particle can be transported to 137 m. Particles >1000 μm cannot rise and be transported by the plume.

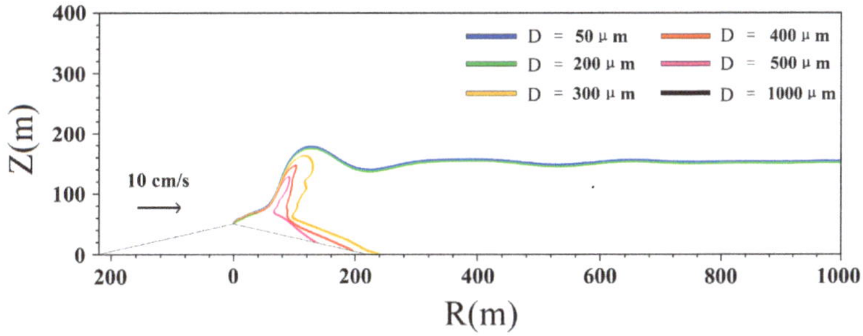

Figure 9. The modeling results showing the transportation behavior of pyrite in different sizes laden in the hydrothermal plume in the Wocan-1 Hydrothermal Field under the background current velocity of 10 cm/s. D = diameter of the hydrothermal particle.

Lou et al. [19] did the numerical modeling for the hydrothermal plume of the Daxi hydrothermal field (~62 km northwest to WHF-1) and investigated the depositional pattern of ferrous disulfide particles (density of 5020 kg/m^3). The difference between the two studies is that the temperature of the black smoker of Daxi is 300 and WHF-1 is 360. In addition, this study also considers the potential interference of topography on the background current and particle deposition. They found that under a background current of 2 cm/s, particles with a size of 50 μm can participate in long-distance transport. Particles with a size of 100 μm will deposit within 760.41 m from the vent, and particles with a size larger than 200 μm cannot be transported. If the background current increase to 5 cm/s, particles with a size of 100 μm can be transported farther to 826.45 m, the particles with a size of 200 μm can be transported as far as 584.40 m from the vent, while the particles with a size larger than 500 μm cannot be transported but suspended above the vent and eventually settled near the venting site.

The deposition of particles under different environmental conditions exhibits distinct patterns: (1) The transportation distance of hydrothermal particles such as pyrite decreases with the increase of particle size under a certain background current velocity. The increase of background current velocity facilitates the particle to transport a longer distance; (2) Under the conditions of this study, i.e., hotter temperature and stronger background current than the study of Lou et al. [19], a wider range of particle sizes can participate in medium- to long-distance transport. Particles within 200 μm are more likely to be transported over a long distance of 1 km, and particles with a size of 500 μm can also be transported and deposited to a region more than 100 m from the vent; (3) The topography of the mound in our model produces stronger turbulent mixing near the bottom, which affects the transport capacity of particles as well, especially for the particles larger than 300 μm. In this regard, Walker et al. [37] also observed that particles along the slope are more active and continuous in the real environment.

5. Conclusions

1. In this study, we build a three-dimensional numerical model of a submarine hydrothermal plume using the field observation data collected in the Wocan-1 Hydrothermal field. The results lead to the following conclusions: The structure of the plume consists of a cone-shaped rising zone, columnar-shaped rising zone, and neutral-buoyancy layer. Background ocean current alters the structure of the plume, causing its bending and increasing of diffusion capacity. The bending angle of the plume axis is positively correlated with the background current velocity. The plume structure would damage when the current velocity would be larger than 10 cm/s.
2. An increase in background current velocity significantly reduces the maximum rise height and a neutral-buoyancy layer height of the plume. Two empirical formulas are established to describe the Wocan-1 hydrothermal plume: the maximum rise height $Z_{max} = 315.1 - 1.252v + 0.4v^2 - 0.162v^3$, and the neutral-buoyancy layer height $Z_{neutral} = 255 - 0.016v + 0.245v^2 - 0.112v^3$.
3. Using the temperature anomaly profile obtained by on-site CTD measurement near the flaming hill of WHF-1, by fitting the temperature profile extracted from our numerical model, the background current velocity of 10 cm/s is reconstructed, which is within the dominant range of ocean current velocities registered by the long-term observational mooring system deployed near the WHF-1. This also attests to the robustness of our model.
4. Under the background current velocity of 10 cm/s, the pyrite grains smaller than 2 mm can be transported over 1 km by the high-temperature black smokers from the flaming hill of WHF-1, those between 3 mm and 5 mm would deposit within 137–240 m from the venting site, while particles larger than 10 mm cannot be transported with the plume and deposit near the venting site.

Overall, this study complements the existing numerical models of hydrothermal plumes by considering the influence of strong background current and complex topography. The 3D structure of the hydrothermal plumes and the depositional patterns of pyrite obtained in this study provide an important basis for relevant studies on the sediments from the near field and far field of the Wocan-1 hydrothermal field. Besides, our model demonstrates the application potential for data interpretation during hydrothermal plume searching and hydrothermal sulfide resource prospecting elsewhere.

Author Contributions: Conceptualization, X.H.; methodology, X.H. and K.W.; modeling, K.W.; investigation, X.H., Y.W., Z.Q., Y.C. and X.Z.; writing—K.W. and X.H.; visualization, K.W.; supervision, X.H.; project administration, X.H. All authors have read and agreed to the published version of the manuscript.

Funding: This study was funded by the National Key Research and Development Program of China (2021YFF0501304), the National Natural Science Foundation of China (No. 91951201, No. 41976076, No. 19151201), and the China Ocean Mineral Resources R&D Association Project (DY135-S2-1), and the Talent Program of Zhejiang Province (grant no. 2018R51003).

Institutional Review Board Statement: Not applicable.

Informed Consent Statement: Not applicable.

Data Availability Statement: Not applicable.

Acknowledgments: We would like to thank Zhiguo He and Yingzhong Lou for their assistance in the technical aspects of modeling. We would also like to thank the editor and the anonymous reviewers for their valuable comments and suggestions.

Conflicts of Interest: The authors declare no conflict of interest.

References

1. Corliss, J.B.; Dymond, J.; Gordon, L.I.; Edmond, J.M.; von Herzen, R.P.; Ballard, R.D.; Green, K.; Williams, D.; Bainbridge, A.; Crane, K.; et al. Submarine thermal springs on the Galapagos Rift. *Science* **1979**, *203*, 1073–1083. [CrossRef] [PubMed]
2. Beaulieu, S.E.; Baker, E.T.; German, C.R.; Maffei, A. An authoritative global database for active submarine hydrothermal vent fields. *Geochem. Geophys. Geosystems* **2013**, *14*, 4892–4905. [CrossRef]
3. Baker, E.T.; Massoth, G.J. Characteristics of hydrothermal plumes from two vent fields on the Juan de Fuca Ridge, northeast Pacific Ocean. *Earth Planet. Sci. Lett.* **1987**, *85*, 59–73. [CrossRef]
4. Bemis, K.; Lowell, R.P.; Farough, A. Diffuse flow: On and around hydrothermal vents at Mid-Ocean Ridges. *Oceanography* **2012**, *25*, 182–191. [CrossRef]
5. Baker, E.T. Hydrothermal plumes. In *Encyclopedia of Marine Geosciences*; Springer: Berlin/Heidelberg, Germany, 2014. [CrossRef]
6. German, C.R.; Seyfried, W.E. Hydrothermal Processes. In *Treatise on Geochemistry*; Elsevier: Amsterdam, The Netherlands, 2014; Volume 8, pp. 191–233.
7. Haymon, R.M. Growth history of hydrothermal black smoker chimneys. *Nature* **1983**, *301*, 695–698. [CrossRef]
8. Gartman, A.; Findlay, A.J. Impacts of hydrothermal plume processes on oceanic metal cycles and transport. *Nat. Geosci.* **2020**, *13*, 396–402. [CrossRef]
9. Sudarikov, S.; Narkevsky, E.; Petrov, V. Identification of two new hydrothermal fields and sulfide deposits on the Mid-Atlantic Ridge as a result of the combined use of exploration methods: Methane detection, water column chemistry, ore sample analysis, and camera surveys. *Minerals* **2021**, *11*, 726. [CrossRef]
10. Zeng, Z.; Wang, X.; Murton, B.J.; Qi, H.; Lehrmann, B.; Li, X.; Shu, Y. Dispersion and intersection of hydrothermal plumes in the Manus back-arc basin, western pacific. *Geofluids* **2020**, *2020*, 4260806. [CrossRef]
11. Kim, J.; Son, S.K.; Kim, D.; Pak, S.J.; Yu, O.H.; Walker, S.L.; Son, J. Discovery of active hydrothermal vent fields along the Central Indian Ridge, 8–12 S. *Geochem. Geophys. Geosystems* **2020**, *21*, e2020GC009058. [CrossRef]
12. De Ronde, C.E.J.; Baker, E.T.; Massoth, G.J.; Lupton, J.E.; Wright, I.C.; Sparks, R.J. Submarine hydrothermal activity along the mid-Kermadec Arc, New Zealand: Large-scale effects on venting. *Geochem. Geophys. Geosystems* **2007**, *8*. [CrossRef]
13. Morton, B.R.; Taylor, G.I.; Turner, J.S. Turbulent gravitational convection from maintained and instantaneous sources. *Proc. R. Soc. London. Ser. A Math. Phys. Sci.* **1956**, *234*, 1–23.
14. Lavelle, J.W. A convection model for hydrothermal plumes in a cross flow. *NOAA* **1994**, *102*, 1–18.
15. Tao, Y.; Rosswog, S.; Brüggen, M. A simulation modeling approach to hydrothermal plumes and its comparison to analytical models. *Ocean. Model.* **2013**, *61*, 68–80. [CrossRef]
16. Lou, Y.; He, Z.; Jiang, H.; Han, X. Numerical simulation of two coalescing turbulent forced plumes in linearly stratified fluids. *Phys. Fluids* **2019**, *31*, 037111.
17. Jiang, H.; Breier, J.A. Physical controls on mixing and transport within rising submarine hydrothermal plumes: A numerical simulation study. *Deep. Sea Res. Part I Oceanogr. Res. Pap.* **2014**, *92*, 41–55. [CrossRef]
18. Adams, I.; Di Iorio, D. Turbulence Properties of a Deep-Sea Hydrothermal Plume in a Time-Variable Cross-Flow: Field and Model Comparisons for Dante in the Main Endeavour Field. *J. Geophys. Res. Ocean.* **2021**, *126*, e2020JC016638. [CrossRef]
19. Lou, Y.; Han, X.; He, Z.; Wang, Y.; Qiu, Z. Numerical modeling of hydrodynamic processes of deep-sea hydrothermal plumes: A case study on Daxi hydrothermal field, Carlsberg Ridge. *Sci. Sin. Tech.* **2020**, *50*, 194–208. (In Chinese)
20. Lou, Y.; He, Z.; Han, X. Transport and deposition patterns of particles laden by rising submarine hydrothermal plumes. *Geophys. Res. Lett.* **2020**, *47*, e2020GL089935. [CrossRef]
21. Jamieson, J.W.; Gartman, A. Defining active, inactive, and extinct seafloor massive sulfide deposits. *Mar. Policy* **2020**, *117*, 103926. [CrossRef]
22. Qiu, Z.; Han, X.; Li, M.; Wang, Y.; Chen, X.; Fan, W.; Zhou, Y.; Cui, R.; Wang, L. The temporal variability of hydrothermal activity of Wocan hydrothermal field, Carlsberg Ridge, northwest Indian Ocean. *Ore Geol. Rev.* **2021**, *132*, 103999. [CrossRef]
23. Wang, Y.; Han, X.; Petersen, S.; Frische, M.; Qiu, Z.; Li, H.; Li, H.; Wu, Z.; Cui, R. Mineralogy and trace element geochemistry of sulfide minerals from the Wocan Hydrothermal Field on the slow-spreading Carlsberg Ridge, Indian Ocean. *Ore Geol. Rev.* **2017**, *84*, 1–19. [CrossRef]
24. Xie, Q.; Han, X.; Wei, M.; Qiu, Z.; Dong, C.; Wu, Y.; Wu, X.; Yu, J. Characteristics and evolution of bacterial communities in the Wocan hydrothermal plume-influenced zone, Carlsberg Ridge, northwestern Indian Ocean. *Acta Microbiol. Sin.* **2022**, *62*, 1974–1985. (In Chinese)
25. Cai, Y. Characteristics of Hydrothermal Activity and Sulfide Minerals of Wocan-1 Hydrothermal Field, Carlsberg Ridge. Master's Thesis, Second Institute of Oceanography, MNR, Zhejiang, China, 2020. (In Chinese)

26. Rodi, W. Experience with Two-Layer Models Combining the k-e Model with A One-Equation Model Near the Wall. In Proceedings of the 29th Aerospace Sciences Meeting, Reno, NV, USA, 7–10 January 1991.
27. Shih, T.H.; Liou, W.W.; Shabbir, A.; Yang, Z.; Zhu, J. A new k-ϵ eddy viscosity model for high reynolds number turbulent flows. *Comput. Fluids* **1995**, *24*, 227–238. [CrossRef]
28. Crowe, C.T.; Schwarzkopf, J.D.; Sommerfeld, M.; Tsuji, Y. *Multiphase Flows with Droplets and Particles*; CRC Press: Boca Raton, FL, USA, 2011.
29. Chen, Y.; Lou, Y.; He, Z.; Wang, Y. Flow fields and outputs of the hydrothermal plume at the Wocan-1 field based on an in-situ video. *Sci. Sin. Tech.* **2022**, *52*, 1705–1715. (In Chinese) [CrossRef]
30. Han, X.; Wang, Y.; Wu, Z. *Survey Report for the Second Stage of DY 57*; Second Institute of Oceanography, MNR: Zhejiang, China, 2019. (In Chinese)
31. German, C.R.; Briem, J.; Chin, C.; Danielsen, M.; Holland, S.; James, R.; Jónsdóttir, A.; Ludford, E.; Moser, C.; Ólafsson, J.; et al. Hydrothermal activity on the Reykjanes Ridge: The Steinahóll vent-field at 63 06′ N. *Earth Planet. Sci. Lett.* **1994**, *121*, 647–654. [CrossRef]
32. Baker, E.T.; Lavelle, J.W. The effect of particle size on the light attenuation coefficient of natural suspensions. *J. Geophys. Res. Ocean.* **1984**, *89*, 8197–8203. [CrossRef]
33. Feely, R.A.; Massoth, G.J.; Trefry, J.H.; Baker, E.T.; Paulson, A.J.; Lebon, G.T. Composition and sedimentation of hydrothermal plume particles from North Cleft segment, Juan de Fuca Ridge. *J. Geophys. Res. Solid Earth* **1994**, *99*, 4985–5006. [CrossRef]
34. Popoola, S.O.; Han, X.; Wang, Y.; Qiu, Z.; Ye, Y. Geochemical investigations of Fe-Si-Mn oxyhydroxides deposits in wocan hydrothermal field on the slow-spreading carlsberg ridge, Indian Ocean: Constraints on their types and origin. *Minerals* **2018**, *9*, 19. [CrossRef]
35. Popoola, S.O.; Han, X.; Wang, Y.; Qiu, Z.; Ye, Y.; Cai, Y. Mineralogical and geochemical signatures of metalliferous sediments in Wocan-1 and Wocan-2 hydrothermal sites on the Carlsberg Ridge, Indian Ocean. *Minerals* **2019**, *9*, 26. [CrossRef]
36. Breier, J.A.; Toner, B.M.; Fakra, S.C.; Marcus, M.A.; White, S.N.; Thurnherr, A.M.; German, C.R. Sulfur, sulfides, oxides and organic matter aggregated in submarine hydrothermal plumes at 9°50′ N East Pacific rise. *Geochim. Et Cosmochim. Acta* **2012**, *88*, 216–236. [CrossRef]
37. Walker, S.L.; Baker, E.T.; Resing, J.A.; Chadwick Jr, W.W.; Lebon, G.T.; Lupton, J.E.; Merle, S.G. Eruption-fed particle plumes and volcaniclastic deposits at a submarine volcano: NW Rota-1, Mariana Arc. *J. Geophys. Res. Solid Earth* **2008**, *113*. [CrossRef]

Journal of
Marine Science and Engineering

MDPI

Article

Sedimentological, Diagenetic, and Sequence Stratigraphic Controls on the Shallow to Marginal Marine Carbonates of the Middle Jurassic Samana Suk Formation, North Pakistan

Shazia Qamar [1,2], Mumtaz Muhammad Shah [2,*], Hammad Tariq Janjuhah [3], George Kontakiotis [4,*], Amir Shahzad [1] and Evangelia Besiou [4]

1 Institute of Geology, University of Azad Jammu and Kashmir, Muzaffarabad 13101, Pakistan; shazia.qamar@ajku.edu.pk (S.Q.); amirshahzadgeo@gmail.com (A.S.)
2 Department of Earth Sciences, Quaid-i-Azam University, Islamabad 45320, Pakistan
3 Department of Geology, Shaheed Benazir Bhutto University, Sheringal 18050, Pakistan; hammad@sbbu.edu.pk
4 Department of Historical Geology-Paleontology, School of Earth Sciences, Faculty of Geology and Geoenvironment, National and Kapodistrian University of Athens, Panepistimiopolis, Zografou, 15784 Athens, Greece; wwweua@hotmail.com
* Correspondence: mshah@qau.edu.pk (M.M.S.); gkontak@geol.uoa.gr (G.K.)

Abstract: This study presents a thorough analysis of the sedimentology, diagenesis, and sequence stratigraphy of the Middle Jurassic Samana Suk Formation in the Hazara Basin of northern Pakistan. Focusing on two sections, namely Mera Rehmat and Por Miana, the research aims to unravel the complex geological processes within the formation. The examination of microfacies reveals nine distinct depositional textures, ranging from mudstone to wackestone, packstone, and grainstone, indicating various inner ramp environments such as open marine, lagoon, and coastal settings. Petrographic investigations shed light on diagenetic processes, including micritization, cementation, dissolution, compaction, neomorphism, and dolomitization. Six cementation types are identified, and the dolomitization patterns vary, providing insights into lagoonal environments and mudstone replacement. Sequence stratigraphic analysis uncovers intriguing patterns within the Samana Suk Formation. The high-stand system tract is characterized by mudstones, pelloidal grainstones, and dolomitized mudstones, indicating periods of high sea level. In contrast, the transgressive system tract displays ooidal grainstones, pelloidal packstones, and pel-bioclastic grainstones, representing transgression and inundation of previously exposed areas. A significant finding is the impact of diagenesis on reservoir quality parameters, specifically porosity, and permeability. Diagenetic processes, cementation types, and dolomitization patterns have significantly altered the pore network, highlighting the importance of considering diagenesis in assessing the Samana Suk Formation as a hydrocarbon reservoir. This research provides a comprehensive understanding of the sedimentology, diagenesis, and sequence stratigraphy of the Middle Jurassic Samana Suk Formation. The findings contribute to our knowledge of similar carbonate reservoirs globally, enhancing the exploration and development of hydrocarbon resources in comparable depositional environments.

Keywords: diagenesis; sequence stratigraphy; Samana Suk formation; microfacies; depositional environments; shallow marine carbonates; dolomitization stages; high-stand system track; transgression; reservoir quality

Citation: Qamar, S.; Shah, M.M.; Janjuhah, H.T.; Kontakiotis, G.; Shahzad, A.; Besiou, E. Sedimentological, Diagenetic, and Sequence Stratigraphic Controls on the Shallow to Marginal Marine Carbonates of the Middle Jurassic Samana Suk Formation, North Pakistan. *J. Mar. Sci. Eng.* **2023**, *11*, 1230. https://doi.org/10.3390/jmse11061230

Academic Editor: Markes E. Johnson

Received: 5 May 2023
Revised: 8 June 2023
Accepted: 12 June 2023
Published: 15 June 2023

1. Introduction

Reservoir characteristics are primarily controlled by depositional and diagenetic processes [1–3]. The geometry of the carbonate platform governs the lateral arrangement of depositional facies and the resulting primary porosity [1]. Diagenetic alterations bring about changes in the initial features of limestone deposition, leading to a redistribution of porosity. Only by researching the diagenetic history of carbonates can successive alterations and pore rearrangements of carbonate reservoirs be identified [4–7]. Although the application of sequence stratigraphy enables the anticipation of facies arrangements [8–11], there is insufficient information available regarding the depositional patterns of primary porosity and permeability within the formation [9–11]. The geometry, sorting, and particle size of sediments are the key factors in depositional reservoir quality. The distribution of mudstones and other fine-grained deposits that may act as seals, baffles, and barriers for fluid flow within reservoir rocks and as petroleum source rocks may be predicted using sequence stratigraphy [9–11]. Spatial trends in diagenesis and their influence on reservoir porosity-permeability patterns can be used to relate indirectly to relative sea-level change and the stratigraphic sequence framework [12,13]. As a result, sequence stratigraphy could be used to forecast intraformational reservoir quality. While a sequence stratigraphic framework can provide insights into facies distribution and the deposition of porosity and permeability in sedimentary successions, particularly in deltaic, coastal, and shallow marine environments, it is not applicable for inferring facies distribution and the depositional characteristics of porosity and permeability [10]. Diagenesis can be linked to sequence stratigraphy since most of the constraints on early diagenetic processes are also related to relative sea-level changes (e.g., pore water contents and flow, duration of subaerial exposure) [14–17]. As a result, integrating diagenesis with sequence stratigraphy will result in a powerful tool for predicting the geographical and temporal distribution and development of quality in clastic reservoirs [18,19]. The geographic distribution of diagenetic characteristics in different types of sedimentary successions has been extensively studied, and it has been shown that this distribution is best understood when it is connected to a stratigraphic model [14,20,21]. Carbonate successions of the Middle Jurassic, represented by the Samana Suk Formation, are located in the foreland basin of Pakistan's Lesser Himalayas, which has been significantly altered by diagenetic processes such as dolomitization and has good outcrop exposures. Pakistan's Indus Basin carbonate sequence is a reservoir. The Pakistani Hazara Fold and Thrust Belt, located in the Indus Basin, is bordered by the Mansehra Precambrian crystalline zone to the north, the Potwar Plateau to the south, the Kashmir Basin to the east, and the Peshawar Basin to the west, with subbasins like Kohat and Potwar. Previously, Shah et al. [22] and Rahim et al. [23] were documented the selectivity of carbonate facies in the Samana Suk Formation dolomitization process in the Por Miana and Mera Rehmat, respectively. However, these investigations were associated with vertical facies shift in one outcrop section, multiphase dolomitization, and other related diagenetic processes in the studied section.

This work examines the distribution and characterization of diagenetic heterogeneities related to sequence stratigraphic models and re-establishes the facies analysis and deposition architecture of the Samana Suk Formation. Moreover, it computes reservoir property controls and explains the Samana Suk Formation's diagenetic history and its influence on reservoir quality evolution. This investigation will assist in modeling Jurassic play exploration in the nearby Indus Basin. The distribution and characterization of diagenetic heterogeneities within the Middle Jurassic Samana Suk Formation are of crucial importance for understanding the reservoir quality evolution and modeling Jurassic play exploration in the Indus Basin. Although previous studies have documented the selective dolomitization process and related diagenetic processes in specific sections, there is a need to establish a comprehensive understanding of the facies analysis, deposition architecture, and diagenetic history of the formation. This research aims to fill these gaps by investigating the distribution and characterization of diagenetic heterogeneities within the Samana Suk Formation using a sequence stratigraphic framework. By integrating sequence stratigraphy and diagenesis, this study seeks to provide insights into the controls on reservoir properties and the evolution of reservoir quality. The novelty of this research lies in its comprehensive approach, combining facies analysis, deposition architecture, and diagenetic history. It goes beyond previous studies that focused on vertical facies shifts in isolated sections and multiphase dolomitization. By examining the diagenetic heterogeneities within the context of sequence stratigraphic models, this study will provide a more holistic understanding of the Samana Suk Formation and its reservoir characteristics.

2. Geological Setting

The study area is in Pakistan's Lesser Himalayan Hill Ranges, which includes the Hazara region (Figure 1). These mountain ranges are part of the large Himalayan orogenic belt's foreland region. The study area is located on the hanging wall block of the Main Boundary Thrust (MBT). It formed as a result of the Cenozoic collision of the Indian and Eurasian plates [24–30]. The Indian Plate evolved during the Early Jurassic period when it was part of the southern supercontinent (i.e., Gondwanaland) that began sliding northward from neighboring Australia and Antarctica plates. Around 160 million years ago, the Indian Plate moved around 9000 km before colliding with the Eurasian Plate [31,32]. Initially, the supercontinent Pangaea was divided into Laurasia and Gondwanaland. Gondwanaland was later divided into West and East Gondwana during the Early to Middle Jurassic. During this period, this region was a part of the Indian Plate's shelf edge, where thick depositions of platform sediments accumulated [33–35]. The thick shallow water carbonate platform deposits of the Samana Suk Formation, comprising oolitic, peloidal, fossiliferous, and micritic facies, were formed throughout the Toarcian to Callovian periods along the shelf margin of Neo-Tethys [34,36–39]. This formation contains most of the examined dolomites, whereas the upper and lower formations are largely undolomitized. This carbonate sequence is about 200 m thick, with lateral thickness increasing from the northwest to the southwest. Because of the restricted environment that occurred after the deposition of these shelf carbonates, as well as the deposition of pyrite-rich and belemnite-bearing black shale and siltstone during the Cretaceous period, the high-energy carbonates exhibit an unconformable contact with the Cretaceous shale (Chichali Formation) on the northern margin of the Indian plate.

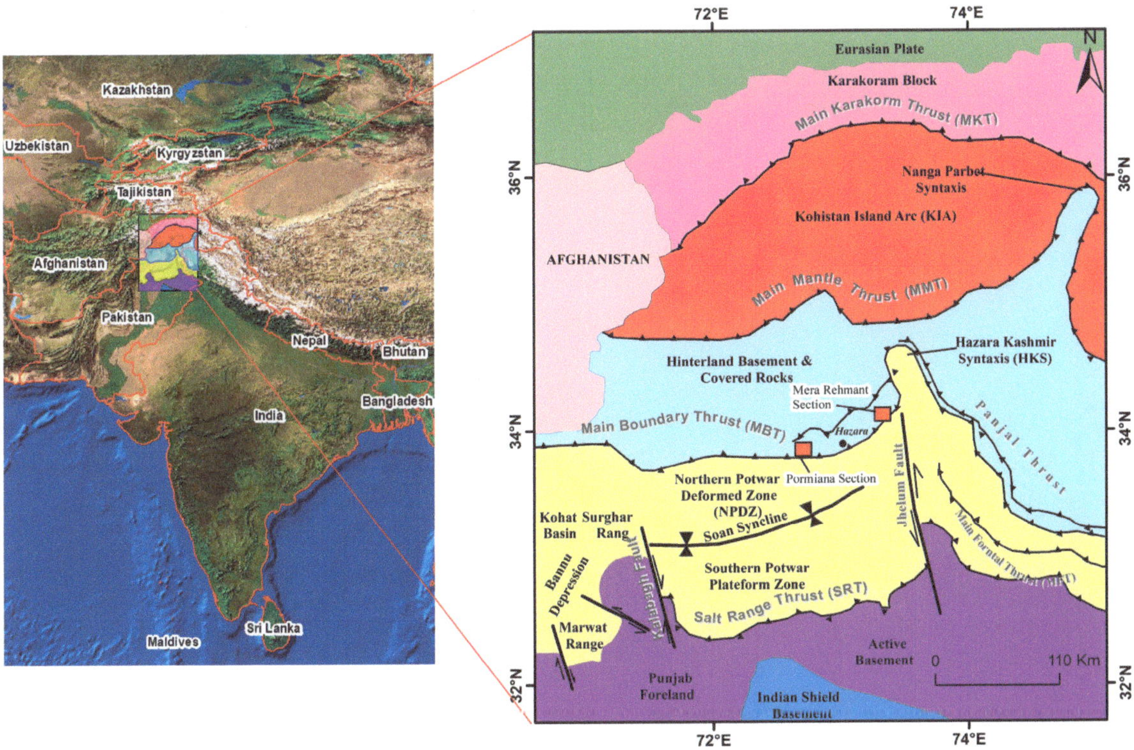

Figure 1. Regional tectonic map of the studied section (adapted from [40]). The rectangles show the study locations.

The Indian Plate separated from Gondwanaland and began a rapid northward migration from Madagascar during the Upper Cretaceous (Cenomanian). During this period, the Indian plate experienced dome uplift as a result of its passage through the Ninety-East Keregulen hotspot, which led the Rajmahal traps to erupt [41]. The domal uplift caused detritus to be eroded and spread over the basin, resulting in the deposition of Lumshiwal Sandstone. Following the formation of the Chichali and Lumshiwal Formations. The rapid northward movement resulted in the depth of the Indian Plate's NW section and the formation of homoclinal ramps, which resulted in the deposition of micritic carbonates as the Kawagarh Formation [36,42]. The Cretaceous–Paleogene boundary denotes an unconformable contact between the overlying Cenozoic succession and the Mesozoic succession. Meanwhile, the northern half of the Indian plate underwent compression-related elevating as a result of its collision with the Kohistan-Ladakh Island arc. During this epoch, ophiolites were deposited in the northern section of the Indian Plate, such as the Muslim Bagh Waziristan-Dargai ophiolites in Pakistan [43]. The Hangu Formation, which consists of laterite, fireclay, and sandstone, was formed during this time period and represented subaerial exposure. The overlying Lokhart Limestone has lagoonal facies, while the Patala Formation has alternating shale and limestone [44]. The highest strata of the Patala Formation are composed of paleosols that reflect a Paleocene-Eocene unconformity. Later, a collision between India and the Kohistan Island Arc (KIA) resulted in sea-level regression during the Early Eocene, and thus a shallow and marginal marine and evaporitic environment prevailed in the Hill ranges [44,45]. Early Eocene Margalla Hill limestone, which mainly consists of nodular limestone [46], followed by the deposition of thinly bedded limestone of Chorgali Formation and later on Kuldana Formation (shales, marls, and limestone layers [47] and subsequent Himalayan uplift resulting in the deposition of Murree Formation [48,49]. The stratigraphic column of Hazara Basin is shown in Table 1.

Table 1. Regional Stratigraphic Column of Hazara Basin, Northwest Himalayas, Pakistan.

Age			Formation	Description
Cenozoic	Miocene		Murree Unconformity	Sandstone, siltstone, clay
	Eocene		Kuldana	Shale, gypsum with interbeds of limestone
			Chorgali	Limestone with interlayers of shale/marl
			Margala-Hill Limestone	Nodular limestone with interbedded shale/marl
	Paleocene	Late	Patala	Marly shale with few thin limestone beds
			Lockhart Limestone	Nodular limestone with occasional marl/shale layers
		Early	Hangu Unconformity	Siltstone, sandstone, shale, bituminous shale
Mesozoic	Cretaceous	Late	Kawagarh	Micritic limestone with shale interbeds
		Early	Lumshiwal	Sand, siltstone with shale interlayer
			Chichali	Gluconitic shale, sandstone
	Jurassic		Samana Suk	Limestone with intra-formational conglomerate
			Datta Unconformity	Calcareous sandstone with fire clay and shale

Table 1. *Cont.*

Age		Formation	Description
Paleozoic	Cambrian	Abbottabad Unconformity	Dolomites with sandstone, shale, and conglomerate
	Pre-Cambrian	Hazara	Shale, phyllite, and shale with minor limestone and graphite

The Indian and Eurasian plates collided, resulting in the construction of an S-W-directed thrust system in the Indian plate. The major thrust systems are the Main Karakoram Thrust (MKT), the Main Mantle Thrust (MMT), the Main Boundary Thrust (MBT), and the Salt Range Thrust (SRT). The north-south compressional regime is represented by the presence of east-west trending major structures (i.e., MBT) and other local faults in the study area and its surroundings [25,26,29].

3. Material and Methods

A total of 182 thin sections were prepared from samples taken from two localities, i.e., Mera Rehmat (73°20′29″ E; 34°14′33″ N) and Por Miana (72°43′9″ E; 33°54′53″ N), both in the upper Indus basin, to develop a depositional model and explain the diagenetic history of the Samana Suk Formation. Thin-section samples were obtained every 0.5 m. Thirty-thin sections were treated with Alizarin Red-S and potassium ferrocyanide, as described by Dickson [50], and 100 were impregnated with blue-dyed epoxy resin. Thin section descriptions include the texture, size, type of allochems and diagenetic characteristics, the sequence stratigraphic effects, and the relationship between the last two. The Dunham [51] and Embry and Klovan [52] limestone classifications were used to classify the rocks. The interpretation of facies is based on rock characteristics and the diagenetic comparison of successions with well-established models [53–55]. According to Flügel and Munnecke [55], special attention was put on identifying standard ramp microfacies types (RMF). Plumley et al. [56] classification was used to evaluate the energy index of sedimentary facies. Overall, the texture, grain type and size, sorting, bedding style, primary sedimentary structures of the rocks, facies, interpretation of depositional setting, appearance according to lithofacies, erosive surfaces, sequence boundaries, and the impact of sequence stratigraphic events on diagnostic outcomes were described in the study thin sections. Two hundred and twenty-eight thin-sections in total were prepared and examined under a polarizing microscope to integrate lithological and diagenetic data for facies characterization, and to construct a conceptual 3-D depositional model.

4. Results

4.1. Field Observations

In the field observations, it was found that limestone and dolomite are recognizable based on their color contrast, hardness, and behavior toward acid treatment. In both sections, the limestone was mostly light to dark gray in color, thin to thickly bedded, and fine to coarsely grained (Figure 2a). The limestone was micritic in the lower part of the Mera Rehmat section and mostly dolomitized in the upper part, while it was mostly dolomitized in the Por Miana section. Moreover, it was massive, hard, and compact, with calcite veins and minor cross-bedding and fractures. Dolomites are light brown to yellowish in color. Detailed observations showed dolomitization occurred in the studied sections as (i) bedding parallel units (Figure 2b) and (ii) patchy dolomites (Figure 2c). Field observations indicate that bedding parallel or stratiform dolomites are restricted to the basal part of the studied formation, while patchy dolomite is distributed unevenly in the basal and middle parts of the section. Stylolites are abundant in the outcrops and are found both in dolomite and limestone. They are mostly filled with dark brown, patchy dolomite and are distributed at low or high angles or parallel to major bedding planes (Figure 2d). The pore-filling white saddle dolomite is present in some places at the top of the unit. Dolomitization and dedolomitization were also observed on the outcrop (Figure 2e). Various hard ground surfaces are also marked in the Mera Rehmat section, which indicates sea level fluctuations (Figure 2f).

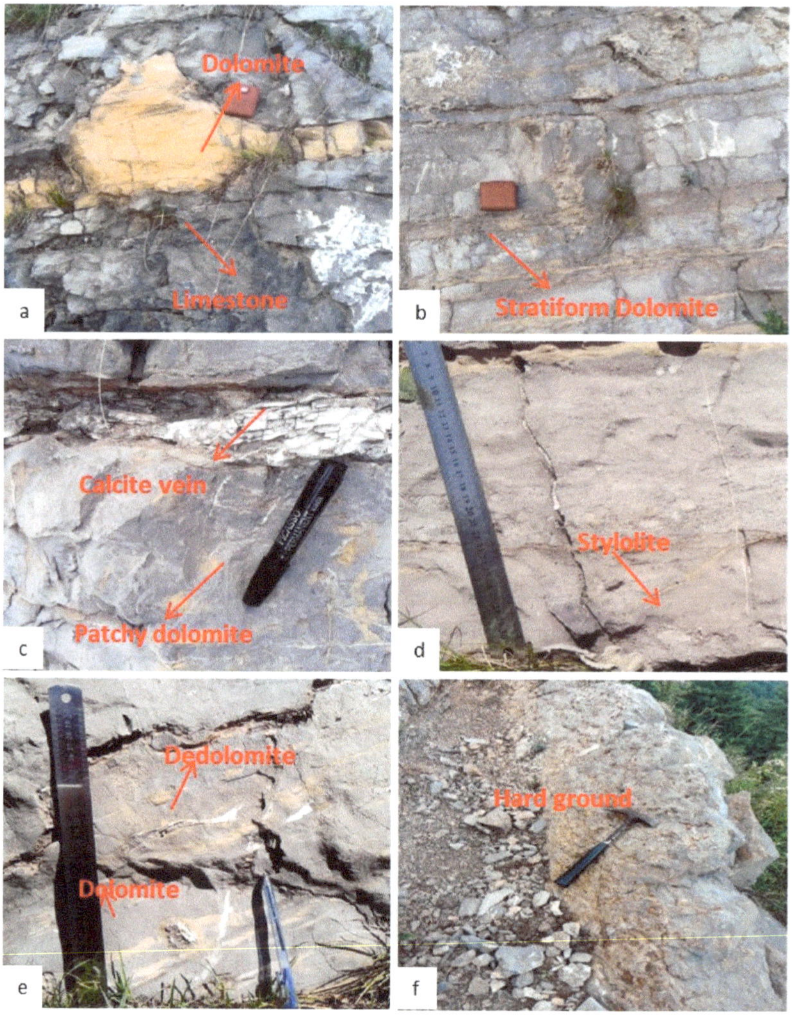

Figure 2. Field photographs showing: (**a**) grey limestone and yellow dolomite, (**b**) bedding parallel dolomite, (**c**) patchy dolomite with a thin calcite vein, (**d**) bedding parallel stylolite, (**e**) dolomite and Dedolomite, and (**f**) erosive surface.

4.2. Microfacies Analysis

In the examined sections, nine microfacies types were recognized in the Middle Jurassic rocks.

4.2.1. Mudstone (MF-1)

In the Mera Rehmat region, this microfacies represents 20% of the overall thickness, whereas, in Por Miana, it comprises 10% of the entire thickness. In the outcrop, this facies consists of yellow sandy limestone that is faulted and fractured and contains parallel bedding stylolites with yellow dolomite patches (Figure 2). This microfacies is more common in the middle to upper Por Miana section, but it occurs at irregular intervals in the middle to the upper half of the Mera Rehmat section. This microfacies consists of 85% matrix, 5–10% dolomite rhombs, 4% twin calcite, and 1% peloids, whereas diagenetic features include twinned calcite, burrowed dolomite and stylolites, respectively (Figure 3a,b). This facies is mostly micritic in both sections (Figure 3a), with only minor cementation seen in veins.

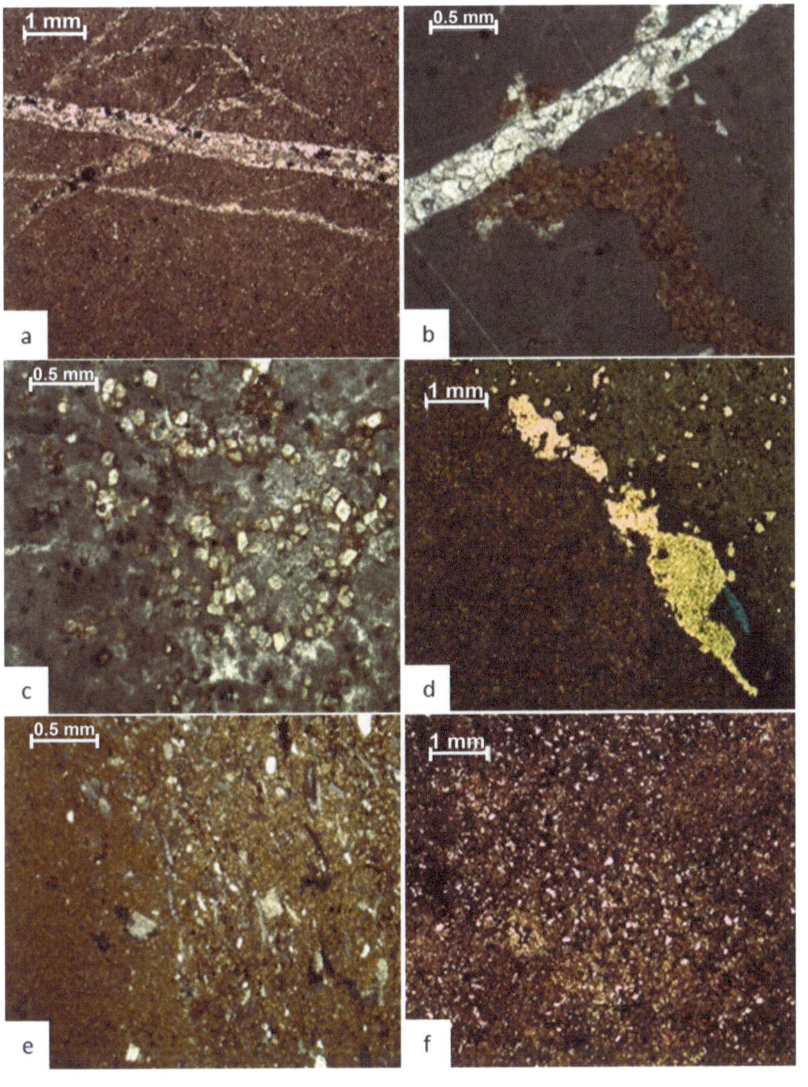

Figure 3. Photomicrographs of microfacies of Samana Suk Formation (**a**,**b**) mudstone microfacies, (**c**,**d**) dolomitized mudstone microfacies, and (**e**,**f**) dolomite.

4.2.2. Dolomitized Mudstone (MF-2)

This facies comprises 6% of the overall thickness in the Mera Rehmat section and is evident in the first, second, and fourth cycles of the Mera Rehmat section, but less than 1% in the Por Miana section. It is composed of light gray limestone with parallel bedding stylolites within dolomite. Microscopic observations reveal that MF-2 contains 60% matrix, of which 20% are sucrosic dolomite rhombs and 5% are Fe oxides, which are scattered as authigenic minerals (Figure 3c). While the blue color of dolomites indicates diagenetic porosity (Figure 3d), Equant cement was observed scattered throughout the matrix. Along with extensive dolomitization, stylolites, parallel bedding, and calcitized euhedral cores, dedolomitization was also observed. In the case of diagenesis, the first stage consists of a matrix, the second of matrix-replacing dolomite, the third of sucrosic dolomite, and the fourth of authigenic Fe oxide.

4.2.3. Dolomite (MF-3)

This microfacies comprises 1% of the overall thickness of the Mera Rehmat section and 5% of the thickness of the Por Miana section. This outcrop is composed of medium-bedded limestone with modest cross-bedding and parallel-bedded dolomites. Flaser bedding, cross-bedding, and minor fossiliferous beds were also observed. This microfacies occurs at the base of Por Miana and in veins containing 4–5% calcite. Matrix, twin and syntaxial calcite, stylolite, pervasive dolomitization, and complete dolomitization are all diagenetic characteristics (Figure 3a–f). This microfacies exhibits dolomitization and dedolomitization in the middle and upper portions of Mera Rehmat, while the base is micritic, cemented, dolomitized, and dedolomitized. In the case of Por Miana, this facies is minor micritic and cemented, as well as dolomitized and sometimes dedolomitized. As a matrix replacement during the diagenetic stage, dolomite rhombs have formed. This facies is common near the base of the Por Miana section, and it repeatedly occurs in thin layers in the Mera Rehmat section.

4.2.4. Ooidal Grainstone (MF-4)

This microfacies comprises 10% of the thickness of the Mera Rehmat section and 3% of the thickness of the Por Miana section. On the outcrop, this facies consists of medium- to thickly bedded, sand-textured, light-gray limestone that is locally brecciated. In the Mera section, ooidal grainstones are present in the lower and middle parts, but in the Por Miana section, they are only found in the top part. This facies comprises 90% ooids, 1% bioclasts, 1% peloids, and 8% cement in thin sections (equant, blocky, twin, and syntaxial). Ooids are deformed and micritized (Figure 4a). These ooids, which are selectively dolomitized, are bound together by cement crystals. In the first step of diagenesis, micritic ooids are produced. In the second stage, calcite veins with equant cement are cut across ooids. The third diagenetic stage is stylolization, and the fourth diagenetic stage is the inflow of authigenic quartz crystals along stylolites. Large euhedral dolomite crystals with cloudy cores and saddle dolomite are very sensitive to telogenetic meteoric alterations during uplift and meteoric water inflow. This microfacies exhibits micritized (Figure 4b), calcitized, stylolites, dolomitized, and dedolomitized behavior at the base of the Mera section, and only micritization and cementation are visible in the middle and upper parts. In the Por Miana area, the upper portion of this facies is micritic, calcitized, and dolomitized.

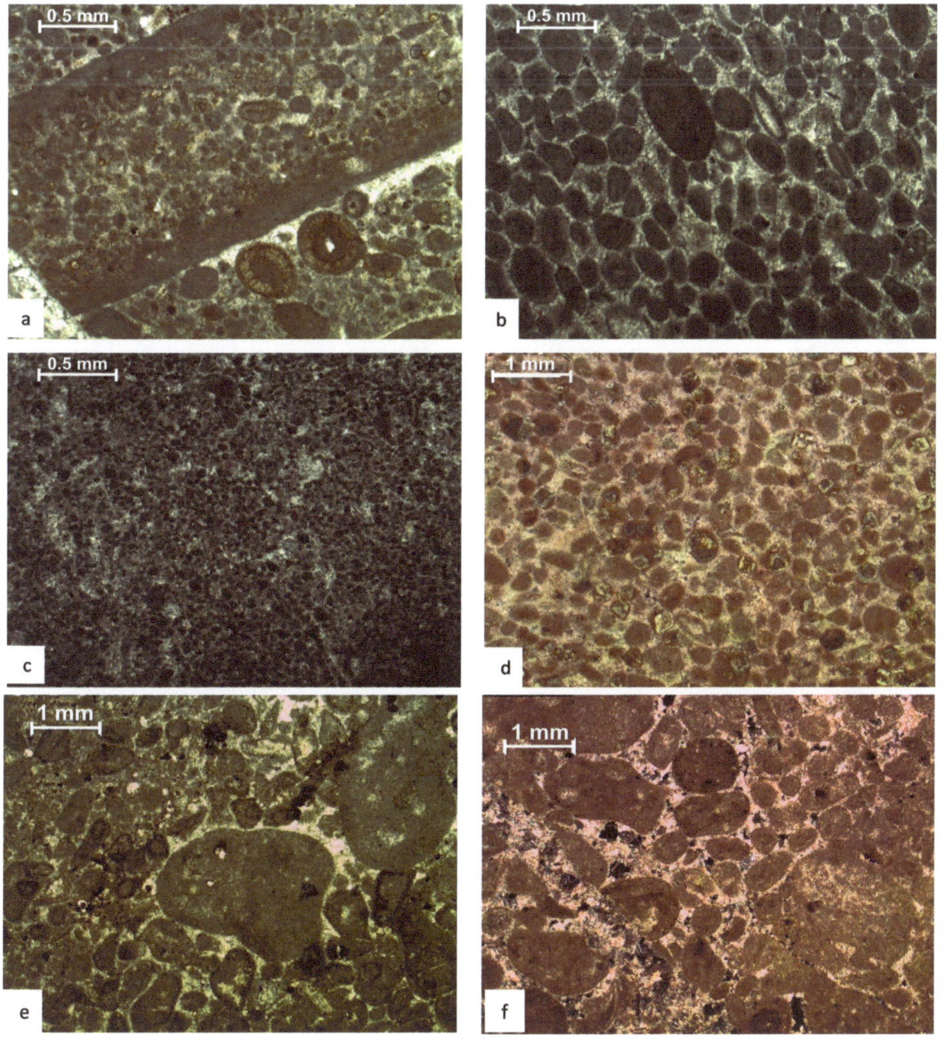

Figure 4. Photomicrographs of microfacies of Samana Suk Formation: (**a**,**b**) ooidal grainstone, (**c**,**d**) peloidal grainstone, and (**e**,**f**) intraclastic peloidal grainstone.

4.2.5. Peloidal Grainstone (MF-5)

This microfacies accounts for 20% of the overall thickness of the Mera section and 10% of Por Miana. On the outcrop, this facies consists of stratiform yellowish dolomite with alternating layers of light and dark limestone. It is prevalent in the third and fourth depositional cycles of the Mera section, whereas it is found randomly in the Por Miana section. In thin sections, this facies is composed of 90% peloids, 6% cement, and 4% quartz. Peloids are micritic, cements are blocky and syntaxial (Figure 4c,d), and in various thin sections, tectonic/post-diagenetic stylolites were also identified. Along with stylolite, dolomite rhombs ferroan in their centers were also observed. In the first stage of diagenesis, both peloids and bioclasts are micritized. In the second stage, twin, equant, and blocky types of cement are found in various thin sections. The third stage consists of stylolites and pervasive dolomite next to the stylolite, followed by burrowed dolomite.

4.2.6. Intraclastic Peloidal Grainstone (MF-6)

This facies accounts for 1% and 4.5% in the Mera Rehmat and Por Miana sections, respectively. On the outcrop, this facies is light gray limestone with minor patches of dolomite and bedding parallel stylolites. It is found at the bottom of the first depositional cycle, and in other dispersed areas of the subsequent cycles, but in Por Miana, it is found at the bottom of the section rather than at the top. This facies comprises 30% bioclasts, 40% intraclasts, 10% micritic peloids, and 20% cement among grains, according to thin section analysis (Figure 4e,f). Diagenetic stages show micritic ooids, cement, stylolite in unstained areas, and pervasive dolomite rhombs.

4.2.7. Peloidal Ooidal Grainstone (MF-7)

This facies comprises 1% of the Por Miana sections and 9% of the Mera Rehmat section. On the outcrop, this microfacies consists of medium to thickly bedded hard and compact limestone with occasional crossbedding and brecciation in parts with patchy dolomite and bedded parallel stylolites. It is abundant in all depositional cycles and at various horizons in the Mera Rehmat section, while it is found in the lower to the middle part of the Por Miana section. In the studied thin sections, it is composed of 80% peloids and 10% ooids (Figure 5a,b). It also contains biota such as miliolides, gastropods, pelecypods, and brachiopods. In the first stage of diagenesis, micrite, micritic ooids, and peloids are prevalent; in the second stage, equant cement is found among grains, and twin calcite veins run parallel to the bedding plane; in the third stage, tectonic stylolites are evident, along which pervasive dolomite rhombs are found; in the fourth stage, selective dolomitization takes place which could be related to Himalayan tectonics; and in the final stage, dedolomitization of dolomite occurs.

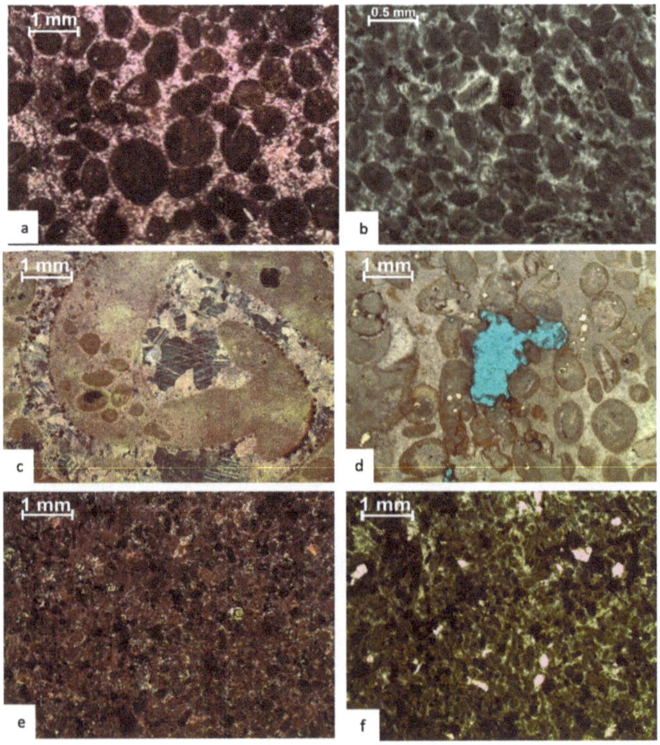

Figure 5. Photomicrographs of microfacies of Samana Suk Formation: (**a**,**b**) peloidal ooidal grainstone, (**c**,**d**) ooidal grainstone, and (**e**,**f**) peloidal packstone.

4.2.8. Peloidal Wackestone (MF-8)

This microfacies is present in the Mera Rehmat section (1%) and in the Por Miana section (7%). On the outcrop, this microfacies displays characteristics such as thin to thickly bedded gray peloidal and oolitic limestone with yellow dolomite patches. Petrographic studies reveal that this microfacies is composed of 40% peloids, 30% matrix, 10% cement, and 10% dolomite. There are 5% quartz crystals scattered in the matrix (Figure 5c). In the first stage of diagenesis, this facies manifests as micritic peloids and matrix. Blocky cement in veins is the second diagenetic stage, tectonic stylolite with dolomite inclusions, whereas the third-stage dolomite rhombs have cloudy cores (Figure 5d). This facies is composed of bands of alternating peloids, ooids, and micrite.

4.2.9. Peloidal Packstone (MF-9)

This facies comprises 1% of the overall thickness of the Mera Rehmat layer and 10% of the Por Miana section. On the outcrop, this microfacies exhibits light gray, sandy limestone with parallel dolomite bedding. This microfacies is predominant in the first and second cycles of the Mera Rehmat section and dispersed throughout the Por Miana region. Peloids account for 85% of the microfacies, while dolomite rhombs make up the remaining 15%. Peloids and bioclasts are micritic and cemented by equant cement, while bioclasts are calcitized by twin calcite and dedolomitized scattered dolomite rhombs (Figure 5a–f). Extensive bioturbation, abundant peloids, rare smaller benthic forams, gastropods, and pelecypod fragments were also observed.

4.3. Diagenetic Processes

Carbonate rocks are very susceptible to post-depositional diageneses, such as cementation, recrystallization, compaction, and dolomitization [57]. In the examined sections, Jurassic carbonate deposits underwent a variety of diagenetic processes, including micritization, cementation, compaction (physical and chemical), dissolution, and dolomitization.

Following is a description of the various diagenetic processes impacting the carbonates under study.

4.3.1. Micritization

Micritization was readily observed in ooids and bioclasts during thin-section analysis at various Samana Suk Fm levels. Micritization has cut through the entire cortex in the ooids, following the radial pattern. Early stages of micritization in the examined sections indicate the formation of micrite envelopes surrounding grains. Certain microfacies are partly or completely micritized (Figure 6a,b), and the ooids within them are only distinguishable due to their spherical shapes and weak relics. The majority of the abundant peloids in the Samana Suk Formation seem to be micritized. In the thin section, ooids and bioclasts in various states of micritization are identifiable.

4.3.2. Cementation

According to sedimentological characteristics, numerous generations of calcite cement with little dolomite input were identified. In the Samana Suk Formation, the six cement types listed below have been identified:

(a) The most common type of cement was the equant calcite spar, which filled pores and skeletal chambers, and veins (Figure 6c–e);
(b) Granular blocky or equant calcite cement: The crystal boundaries in this cement are frequently irregular to curved, but occasionally straight. This cement resulted from aragonite leaching since it is present in both intergranular pores and molds. The crystal's size increased from the mold's perimeter to its center. Occasionally, crystals exhibit intense twinning;
(c) Coarse, blocky calcite cement is the most prevalent type of cement in the Samana Suk Fm. Clear, coarse to very coarse calcite cement (Figure 6e). In thin sections, it appeared to have straight boundaries on one side and curved boundaries on the other. It

completely fills all accessible intergranular pores. Sometimes, thin isopachous cement is used to line molds, and the remaining space is filled with a single calcite crystal;

(d) This cement was evident in compacted grainstones with syntaxial overgrowths. Syntaxial overgrowth (Figure 6f) is difficult to distinguish from poikilotopic cement and frequently replaces bioclasts;

(e) Poikilotopic cement: This form of cement in the Samana Suk Formation is a highly coarse-grained, poikilotopic (Figure 7a), clear, blocky, calcite cement that fills pore spaces surrounding multiple grains. It has a diameter of several millimeters;

(f) Ferroan dolomite cement; molds and intergranular pores include dark brown to reddish brown, anhedral to subhedral, coarsely crystalline ferroan and non-ferroan dolomite and twin calcite (Figure 7b–d).

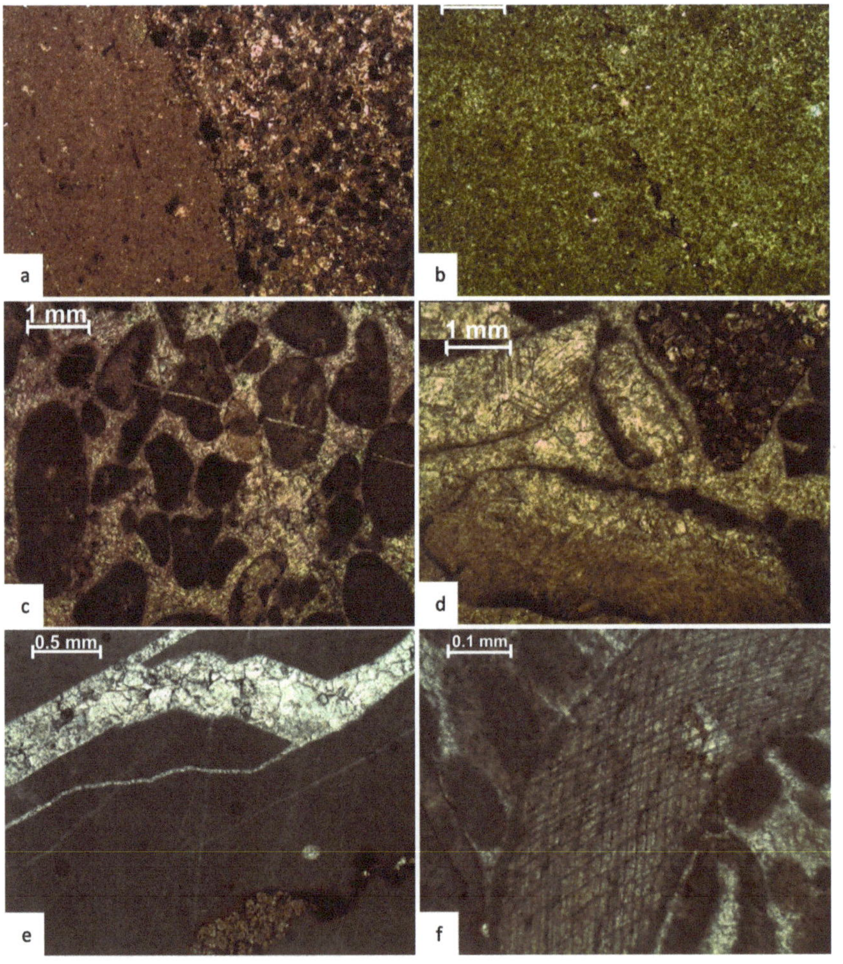

Figure 6. Photomicrographs of diagenetic fabric (**a**,**b**) micritization and cementation, (**c**,**d**) equant cement, (**e**) blocky cement, and (**f**) syntaxial cement.

Figure 7. Photomicrographs of diagenetic fabric (**a**) poikilotopic cement, (**b**) ferroan dolomite cement, (**c**,**d**) twin calcite, and (**e**,**f**) dolomitization.

4.3.3. Dissolution

Dissolution is common in the examined intervals, indicating late-stage porosity development (Figure 8). Moldic (Figure 8c), vuggy, and inter- and intra-particle (Figure 8a,b) porosity are all types of dissolution-associated porosity.

4.3.4. Mechanical and Chemical Compaction

The depth-dependent porosity distribution and compaction trends are influenced by various textures, diagenetic conditions, and mineralogies [58,59]. Various horizons record the detailed distribution of these features in individual microfacies. In the investigated section, broken and cracked ooids along the lamellae and broken cortical layers are common. Mudstone (micrite) is often converted into microspar owing to compaction; however, compaction is not uniform across all microfacies (Figure 9b).

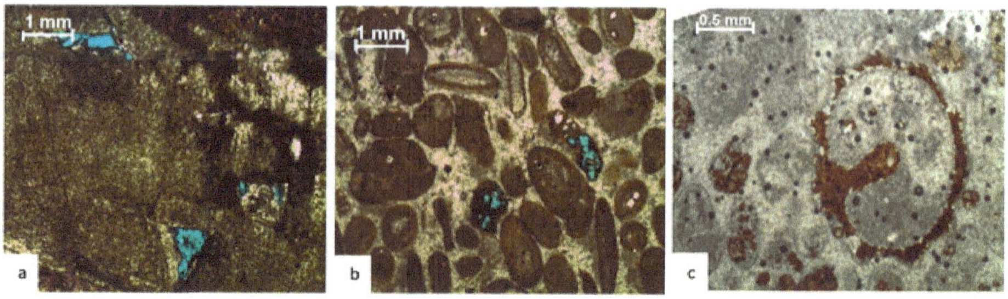

Figure 8. Photomicrographs of diagenetic fabric showing dissolution (**a**) inter-particle porosity, (**b**) intra-particle porosity, and (**c**) moldic porosity.

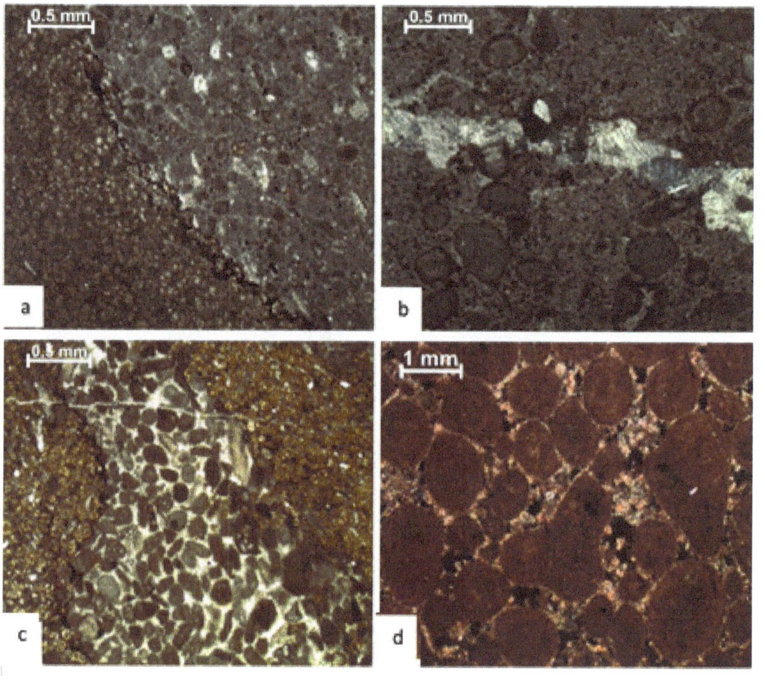

Figure 9. Photomicrographs showing mechanical and chemical compaction (**a**) sutured seams, (**b**) broken micritic envelope, (**c**) grain interpenetration, and (**d**) sutured contacts.

Sutured seams (Figure 9a), microstylolites, and stylolites identified in sections constitute the pressure solution (Figure 9a,c). Sutured seams and microstylolites are mostly horizontal and parallel to bedding planes, indicating that overburden stresses most likely caused the pressure solution phenomenon. The initial pressure solution is concentrated at grain contacts in grainstone (Figure 9d), resulting in sutured contacts and grain interpenetration (Figure 9c).

4.3.5. Neomorphism

Calcite largely replaced aragonitic allochems in the section studied (i.e., shell fragments). Petrographic investigations demonstrated that microcrystalline calcite recrystallized into equant calcite spars of varying crystal sizes. Aggregational neomorphism results from partial to total recrystallization of the skeletal components. The different skeletal grains, echinoderms, foraminifera, algal grains, and other bioclasts coated with micritic envelopes and exhibiting aggradation neomorphism, suggestive of subaerial diagenesis, support this process.

4.3.6. Dolomitization

Pore-filling dolomite plays a major role in the diagenetic process in the section (Figure 10). The Samana Suk Formation contains dolomite in several forms: (a) layered thin beds; (b) burrowed dolomite (Figure 10e); (c) patches, clusters, matrix replacive and dispersed dolomite crystals (Figure 10c,d); (d) grain-selective (Figure 10f) and matrix-selective dolomite (Figure 10a); (e) molds and fracture-filling dolomite; and (f) pervasive dolomite. The parent rock is occasionally visible in the dolomitized layer (Figure 10b). Fine to medium-crystalline, anhedral to subhedral mosaics of dolomite crystals pervade several thin sections (Figure 7e,f). This widespread dolomite formed early in the diagenetic period.

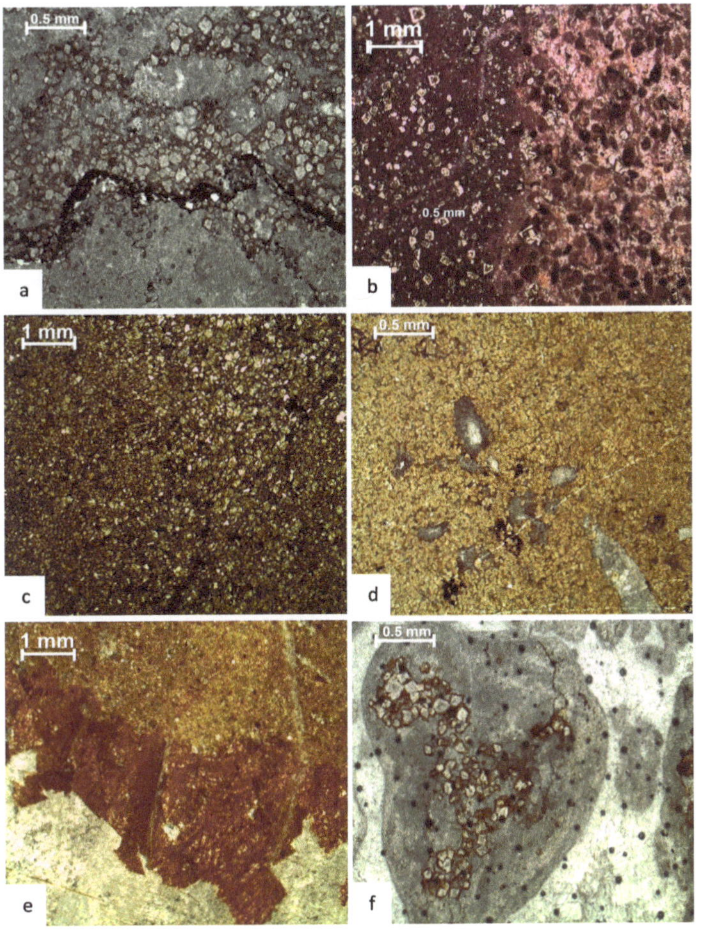

Figure 10. Photomicrographs reflecting different types of dolomitization (**a**) pervasive dolomitization, (**b**) layered, (**c**,**d**) matrix repulsive, (**e**) burrowed dolomite, and (**f**) grain selective.

The dolomite crystals that have replaced the allochems are rich in inclusions, but the crystals formed in the intergranular spaces are finer and coarser. Crystal size grows toward the pore center in molds. Very fine crystalline subhedral dolomite was used to fill the intergranular pore spaces. Dolomite is also observed as cement in the slightly dolomitized mudstones.

4.4. Sequence Stratigraphy

The Samana Suk Formation represents deposition in three different environments: open marine, lagoon, and shoreline/coastline. According to the microfacies and sequence stratigraphic data, the Samana Suk Formation is composed of a 2nd-order local cycle, which is equivalent to the global sequences reported by Haq et al. [60]. Similarly, the Samana Suk Formation's 3rd-order sequences comprise five High-stand Systems Tracts (HSTs), five Transgressive Systems Tracts (TST) in the Mera Rehmat section (Figure 11), and three TST and two Highstand Systems in the Por Miana section (Figure 12). The Samana Suk Formation has recorded third-order system tracts (Figures 11 and 12).

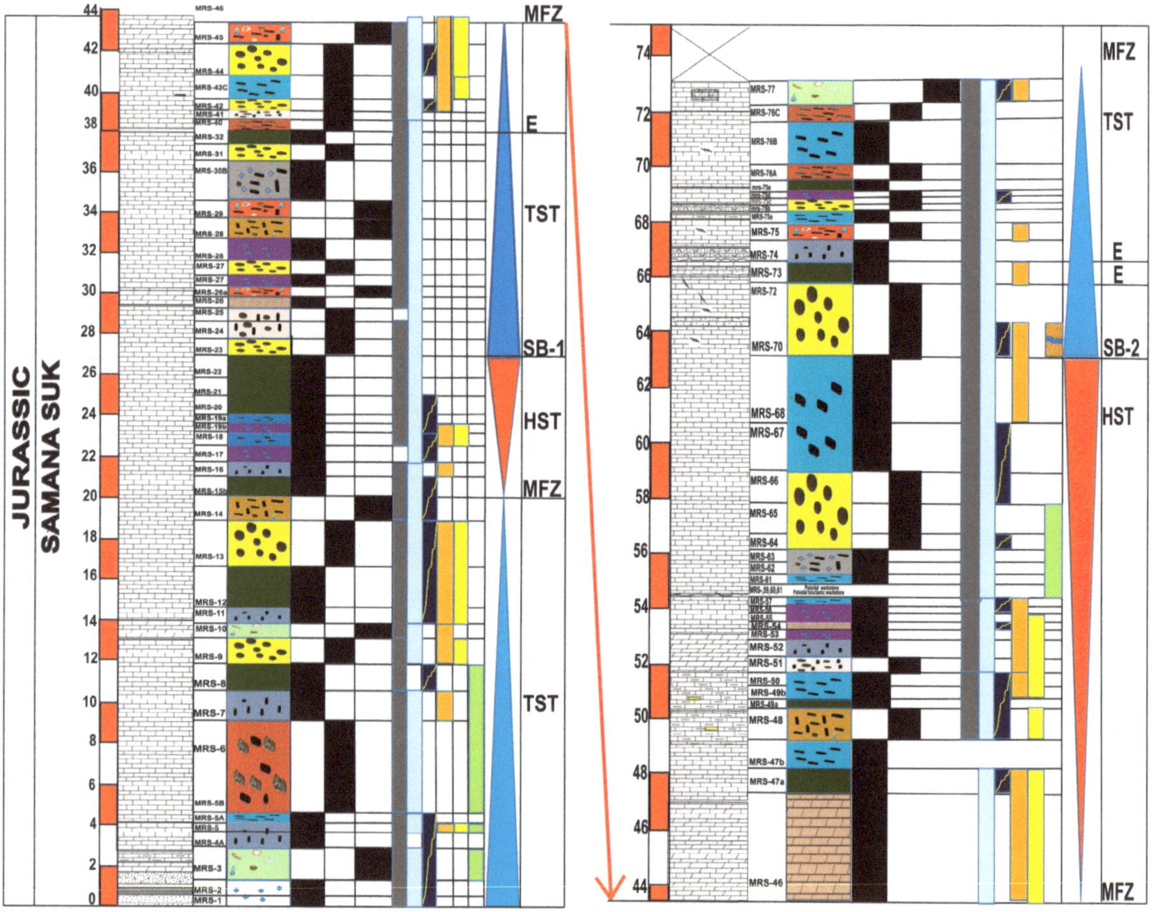

Figure 11. *Cont.*

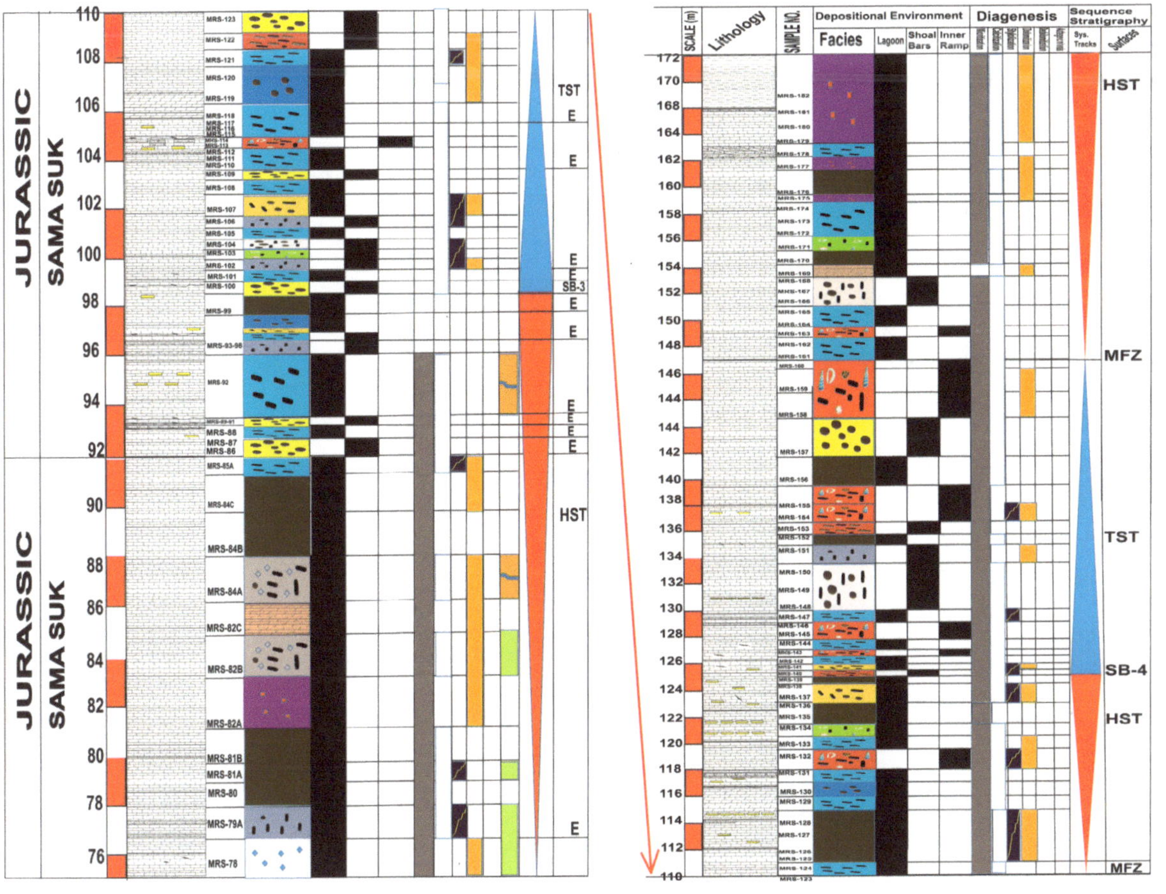

Figure 11. Litholog showing microfacies distribution, diagenetic and sequence stratigraphic attributes of Mera Rehmat section.

4.4.1. Depositional Sequence 1

A TST near the base of the Mera Rehmat section makes sharp contact with the underlying Datta FM. This section begins with siliciclastic wackestone facies displaying underlying formation inclusions, followed by a lagoonal to inner ramp setting (Figure 13).

Micritization and cementation are prevalent diagenetic features with stylolites and authigenic content in alternate bands of grainstone, wackestone, and mudstone with minor peloidal intraclastic facies. While the uppermost ooidal grainstone is the Maximum Flooding Zone (MFZ), dolomitization and dedolomitization occur at the end of this transgression cycle. The MFZ is represented by the topmost ooidal grainstone in the Mera Rehmat section. The TST in Por Miana is characterized by the repetitive deposition of peloidal grainstone and intraclastic peloidal grainstone, as well as lagoonal facies such as peloidal packstone and wackestone. Micritization and calcitization are early diagenetic features, whereas stylization and dolomitization are late diagenetic features, with minimal authigenic content in the intraclastic peloidal grainstone as the last diagenetic phase. Por Miana is distinguished by the presence of dolomite at the base, which is followed by intraclastic peloidal grainstone. The packstone and wackestone facies exhibit micritization, cementation, dolomitization, and stylolization. Dedolomitization and authigenic minerals were also observed in intra-clastic peloidal grainstone microfacies (Figure 14).

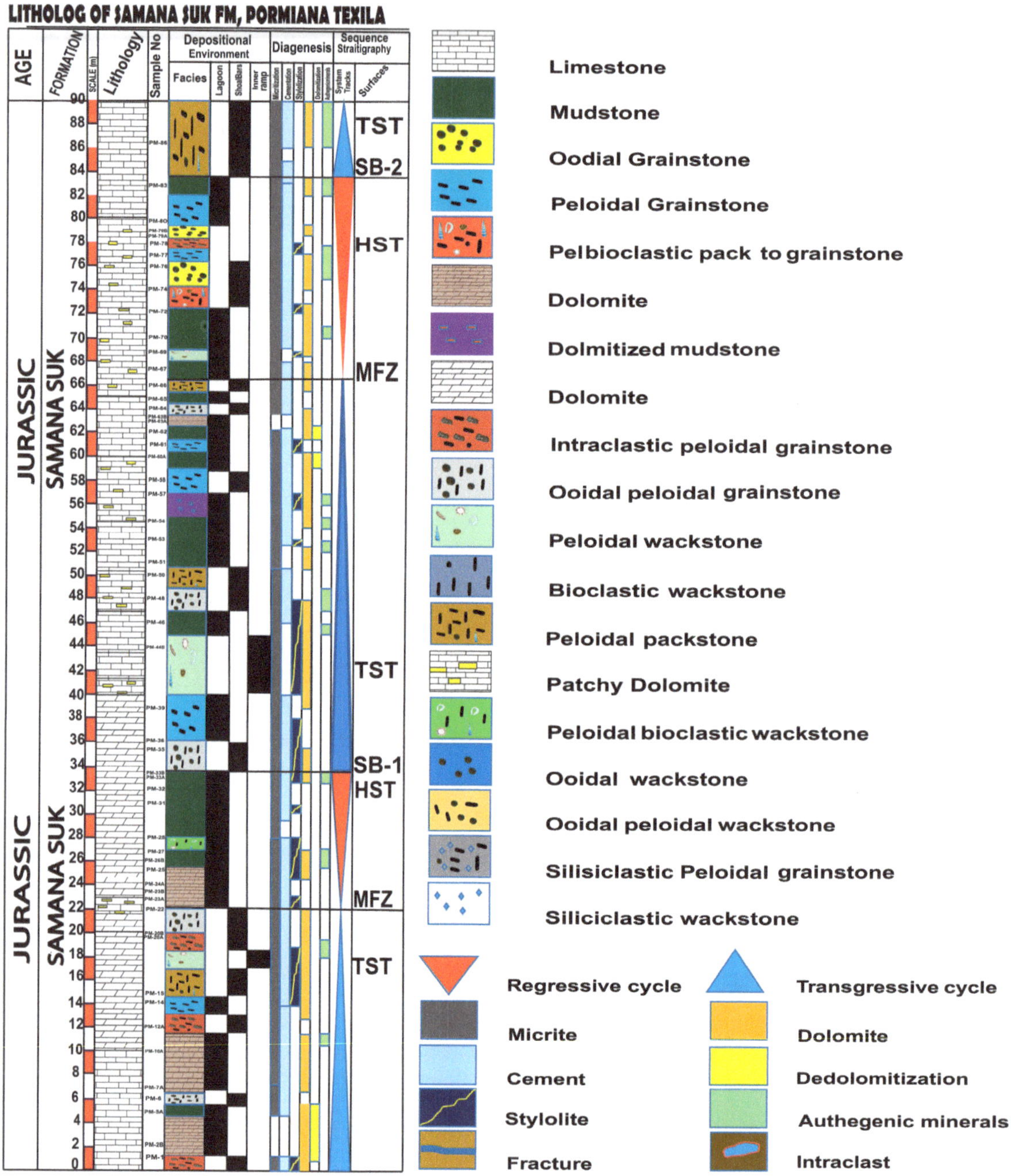

Figure 12. Litholog showing microfacies distribution, diagenetic, and stratigraphic sequence attributes in the Por Miana section.

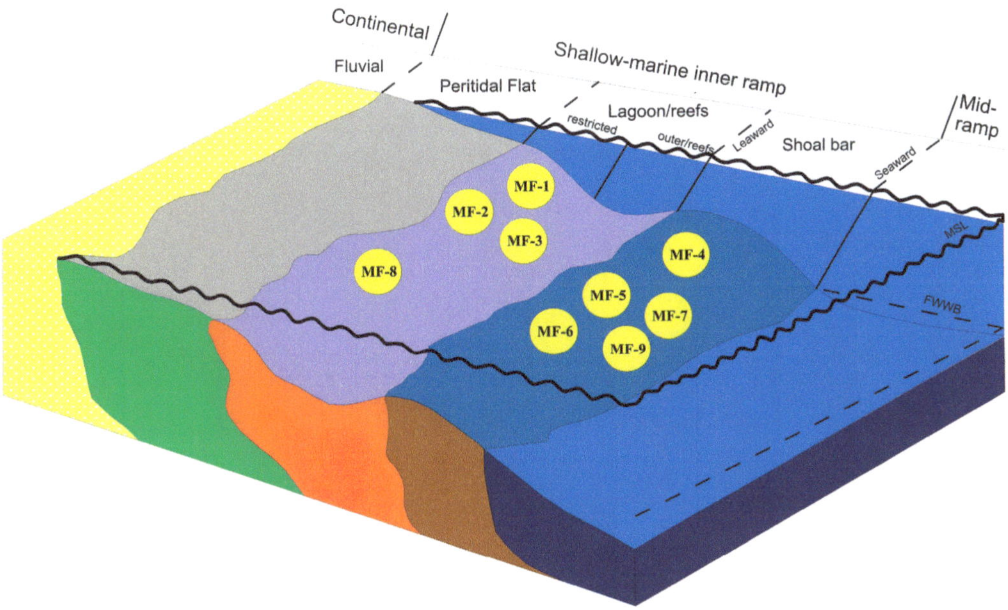

Figure 13. Hypothetical depositional model for Samana Suk Formation.

4.4.2. Depositional Sequence 2

The Mera Rehmat's HST of DS2 is characterized by the deposition of mudstones and repeated dolomitized mudstone and peloidal grainstone facies of the lagoon. These facies have been micritized, calcitized, and stylolized. In the last diagenetic stage, dolomitized mudstone is dolomitized and dedolomitized. Dolomite, mudstone, and bioclastic peloidal and bioclastic wackestones characterize the HST in the Por Miana region. The lagoonal environment defines this area (Figure 12)—micritized, calcitized, and dolomitized mudstone and dolomite facies containing authigenic inclusions in mudstone. Micritization, calcitization, and stylolization are early stages of diagenesis in peloidal bioclastic wackestone. The Mera Rehmat section's TST of the depositional sequence is described by the deposition of ooidal grainstone, ooidal peloidal grainstone, dolomitized mudstone, peloidal packstone, and pel-bioclastic packstone to grainstone facies, which represent the lagoon to inner ramp environment and sea level transgression. These facies are micritized and calcitized, whereas the highest flooding zone is dolomitized ooidal grainstone at the top of the cycle. The TST of DS2 in the Por Miana section is characterized by the presence of peloidal grainstone, ooidal peloidal grainstone, mudstone, dolomitized mudstone, and peloidal packstone facies, all of which are micritized except for dolomite. The peloidal wackestone, dolomite, and mudstone facies lack cementation. Stylolite is observed at the base of TST in peloidal grainstone, peloidal ooidal grainstone, peloidal wackestone, and mudstone facies, and in lesser amounts at the top. Dolomitization often occurs in mudstone, peloidal wackestone, and dolomite facies. Dolomitized mudstone facies towards the top have some authigenic material (Figure 14).

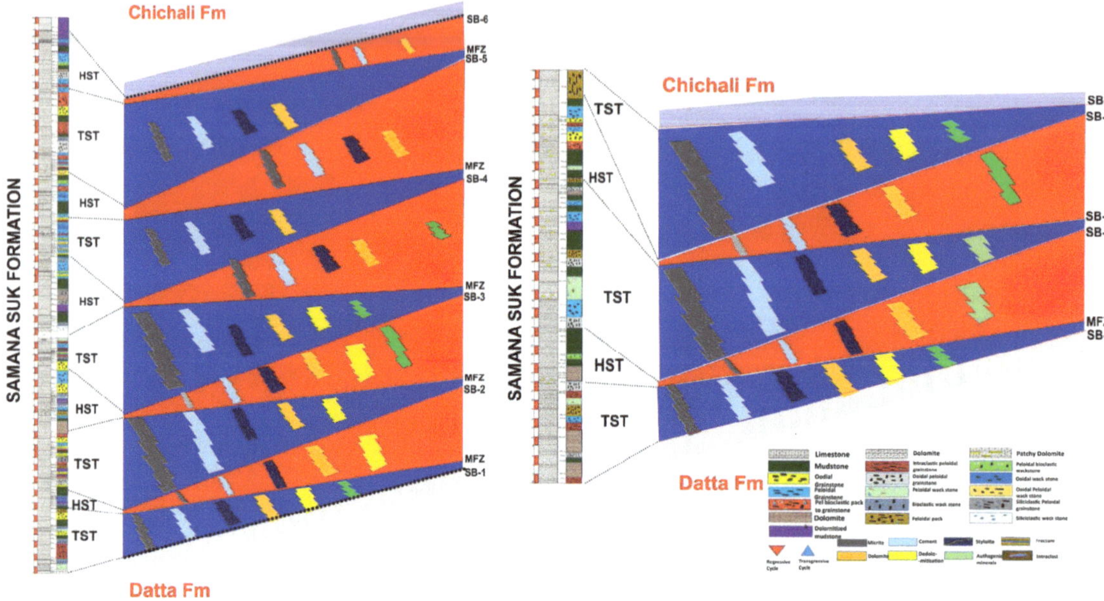

Figure 14. The model represents the overall interplay of sequence stratigraphic architecture of the Samana Suk Formation.

4.4.3. Depositional Sequence 3

Dolomite at the bottom of the HST of DS3 in the Mera Rehmat section and peloidal and ooidal grainstones that look similar to lagoonal and shoal bars set it apart. All facies are micritized and calcitized. Peloidal and ooidal grainstones are stylolized, dolomitized, and dedolomitized at the base of the cycle. Por Miana's High Systems Trajectory represents the peloidal grainstone, ooidal grainstone, mudstone, pel-bioclastic pack-to-grainstone, and intraclastic peloidal grainstone, which characterize the depositional environment ranging from lagoon to shoal bar. All facies are micritized and calcitized except for peloidal wackestone, which lacks cement. The mudstone and ooidal grainstone facies are dolomitized and include authigenic minerals. A sequence boundary is marked at the top of HST due to the presence of an erosive surface. The transgressive systems tract (TST) of DS3 in the Mera Rehmat section is marked by repeated ooidal grainstone, peloidal grainstone, mudstone, intraclastic peloidal grainstone, and the topmost peloidal wackestone facies. From lagoonal mud to the inner ramp's peloidal wackestone facies, these facies ultimately mark the depositional environment. The top is marked by a maximum flooding surface. The TST of DS3 in the section is marked by a single peloidal packstone facie that is micritized, calcitized, dolomitized and includes authigenic content (Figure 14).

4.4.4. Depositional Sequence 4

In the Mera Rehmat section, the fourth cycle of the Middle Jurassic Samana Suk Formation comprises a TST with peloidal grainstone, ooidal grainstone, peloidal bioclastic grainstone, and intraclastic peloidal grainstone facies. Micritization and cementation episodes affect these grainstones, while dolomites occur at three horizons. Stylolites are observed in peloidal and ooidal grainstones at specific horizons within the TST. The HST in the Mera Rehmat section consists of basal thick mudstone facies, followed by peloidal grainstone, ooidal wackestone, pelbioclastic pack to grainstone, and peloidal ooidal grainstone facies. The topmost layer represents a thin mudstone facies marking the sequence boundary. Dolomites and stylolites are confined to the peloidal ooidal grainstone and pelbioclastic pack to grainstone facies within the HST. It is important to note that this sequence is not observed in the Por Miana section, indicating lateral facies variations within the Samana Suk Formation. These precise observations provide valuable insights into the sedimentological and diagenetic characteristics of the Middle Jurassic carbonates in the Mera Rehmat section.

4.4.5. Depositional Sequence 5

In the Mera Rehmat section, the TST of the fourth cycle in the Middle Jurassic Samana Suk Formation comprises the following facies: six units of pelbioclastic pack to grainstone, three units of peloidal grainstone, ooidal grainstone, mudstone, intraclastic peloidal grainstone, bioclastic grainstone, and peloidal ooidal grainstone. Stylolites and dolomites are observed at three horizons within the pelbioclastic pack to grainstone facies towards the top, at the base of the intraclastic peloidal grainstone facies, and in the middle part of the peloidal grainstone facies. Above the transgressive surface, the HST in the Mera Rehmat section includes the following facies: peloidal grainstone, dolomitized mudstone, dolomite, mudstone, peloidal grainstone, and pelbioclastic pack to grainstone. Dolomites are also present in the pelbioclastic pack to grainstone facies, which marks the MFZ. It is important to note that this specific sequence is not observed in the Por Miana section, indicating lateral facies variations within the Samana Suk Formation. These precise observations provide valuable insights into the facies distribution, diagenetic features such as stylolites and dolomites, and the stratigraphic architecture of the Mera Rehmat section in the Middle Jurassic Samana Suk Formation.

5. Discussion

5.1. Depositional Environment

The variety of microfacies within the Samana Suk Formation suggests a complex depositional history with different energy conditions, ranging from hypersaline supratidal to high-energy marine environments, highlighting the dynamic nature of the depositional setting.

MF-1: Mudstone facies-1 often follows transgressions in sequence stratigraphy (Figures 11 and 12). The mudstone texture, dolomitized lime mud matrix, and absence of marine biota all point to low-energy depositional conditions. The presence of lime mud in microfacies suggests a calm water condition. This non-laminated, unfossiliferous, homogenous micritic mudstone microfacies is characteristic of a low-energy supratidal environment, most likely mudflats or tidal ponds. This facies coincides with Flügel and Munnecke's [55] SMF-23 and Wilson's FZ-8 [53].

MF-2: These mudstones are often deposited along the edge of shoals, where energy shifts are frequent, and the resultant sedimentary facies assemblage consists of alternating grainstone and mudstone layers [61]. This facies is similar to SMF-23 of Flügel and Munnecke [55] and FZ-8 of Wilson [53].

MF-3: Diagenetic dolomite rhombs replaced the matrix. Low faunal contents suggest a tidal flat-deposited lime-mudstone parent rock. This facies matches Flügel and Munnecke [55] SMF-23 and Wilson FZ-8 Wilson [53].

MF-4: Ooids mostly form on high-energy shoals of the inner shelf within the platform interior and are limited to near-coast, marginal marine environments. Ooids are first deposited along with gastropods in the low-energy lagoon setting. Encrusted and micritized ooids form grapestone fabric. Later, ooids and intraclasts are reworked to the high-energy inner shelf and deposited above Fair Weather Wave Base (FWWB) in typical marine conditions. This facies matches Flügel and Munnecke [55] SMF-23 and Wilson FZ-8 Wilson [53].

MF-5: This facies is deposited at lower water depths than the packstone facies in a high-energy shallow subtidal shoal environment. It is interpreted as a carbonate shoal deposit above the normal wave base. Grainstone texture and broken fragments of bioclasts imply high-energy environments, whereas peloids indicate a low-energy environment. Thus, lagoonal tidal inlets with high-energy settings can deposit microfacies. The bioclasts imply a shallow environment with normal salinity and may have been transferred to lagoonal tidal inlets by storm surges or waves. This facies corresponds to SMF 16 of Wilson [53] and Flügel and Munnecke [55].

MF-6: Many peloids indicate low energy. Bioclasts, pelecypods, brachiopods, and other skeletal fragments suggest a shallow, normal-salinity marine environment. Intraclasts, bioclasts, and ooids indicate high energy above the fair-weather wave base. These allochems and fossils indicate that the microfacies is deposited in the near-shore zone of the inner shelf during storms. The peloids of low-energy conditions are reworked into such high-energy settings. The facies correspond to SMF 16 of Wilson [53] and Flügel and Munnecke [55].

MF-7: Micritic envelopes and selective micritization have been observed. Due to biogenic encrustations, micritized ooids occur in protected lagoonal environments. Peloids, milliolids, and gastropods are also indicative of lagoonal environments. The presence of bioclasts, pelecypods, brachiopods, and other skeletal fragments suggests a shallow marine environment with normal salinity. The ooids develop in the high-energy shoal environments of the inner shelf's interior, which are restricted to near-coast marginal marine environments. The dominance of sparite further supports the high-energy condition. Initially, ooids are reworked and deposited in the low-energy lagoon with miliolids and gastropods. The ooids are encrusted and micritized to form a fabric of grapestone. Later, the ooids and intraclasts are redistributed to the inner shelf with high energy and deposited in typical marine conditions above the FWWB [62]. The facies match [53] and [55] SMF 16.

MF-8: The rich lime mud matrix with limited fauna suggests a lagoonal environment with low circulation [62]. The facies match [53] and [55] in SMF 16.

MF-9: Extensive bioturbation, abundant peloids, rare smaller benthic forams, gastropods, and pelecypod fragments in the micritic matrix suggest a low-energy, shallow subtidal lagoonal environment. Pseudo-ooids imply sporadic, moderate energy. Iron oxides surrounding grains suggest slow sedimentation. This facies implies warm, moderate to highly agitated, shallow water up to 5 m in the shoal (Figures 11 and 12). The facies match [53] and [55] in SMF 16.

The Samana Suk Formation exhibits a diverse range of microfacies, each indicative of specific depositional environments. The presence of MF-1, characterized by non-laminated, unfossiliferous mudstone with dolomitized lime mud matrix, calcite-filled evaporative molds, and absence of marine biota, suggests a low-energy hypersaline deposition in a supratidal environment, possibly mudflats or tidal ponds [22]. MF-2, which comprises alternating grainstone and mudstone layers, is typically found along the edges of shoals where frequent energy shifts occur [63]. MF-3, with diagenetic dolomite rhombs and low faunal content, suggests deposition on tidal flats. MF-4 corresponds to high-energy shoals within the platform interior, featuring ooides and grapestone fabric, indicating deposition in near-coast, marginal marine environments [64]. MF-5 is interpreted as a carbonate shoal deposit in a high-energy shallow subtidal shoal environment, with grainstone texture, broken fragments of bioclasts, and peloids indicating both high- and low-energy environments. MF-6 represents the near-shore zone of the inner shelf during storms, with peloids, bioclasts, pelecypods, and ooids suggesting high-energy deposition. MF-7 shows micritic envelopes, selective micritization, and encrusted ooids, indicating protected lagoonal environments [55]. MF-8 suggests a lagoonal environment with low circulation, while MF-9 implies a low-energy, shallow subtidal lagoonal setting with extensive bioturbation, peloids, benthic foraminifera, and gastropods [65].

5.2. *Diagenetic History and Sequence Stratigraphy*

The diagenetic history of the Samana Suk Formation is characterized by a series of processes and products that have influenced the sedimentary rocks. After sediment deposition and early grain settlement, a microbial activity played a significant role in filling grain pores with micrite and forming micrite envelopes around ooids and other allochems. Endolithic fungi, algae, and bacteria are believed to be responsible for this micritization process [59,66,67]. The extensive micritization observed in the formation indicates a slow sedimentation rate, allowing ample time for blue-green algae to transform grains before burial [68]. Micritization varies across different horizons, with some mudstones exhibiting more extensive micritization compared to grainstones and dolostones. Thin-section studies have revealed ooids and bioclasts in various stages of micritization, often displaying a radial pattern that extends throughout the entire cortex. Bacterial boring can lead to the complete loss of primary particle texture and subsequent micritization [69,70].

Additionally, diagenetic processes such as calcitization of dolomite, dedolomitization, and early aragonite dissolution have played a role in the formation. Calcitization occurs due to changes in pore water chemistry from marine to meteoric composition, resulting in the transformation of dolomite to calcite [71]. Dissolution of aragonite skeletal detritus, mainly gastropods and bivalves, has been observed, with isopachous bladed and blocky types of cement filling moldic pores, indicating early dissolution. Pressure solution diagenesis, characterized by chemical compaction, occurs as a result of gravity loading or unilateral tectonic stress, leading to the development of microstylolites, sutured seams, and stylolites. Pressure solution dissolution at grain contacts and solution seams or stylolites causes calcite to re-precipitate as cement in surrounding limestone, reducing porosity [72–75].

Dolomitization is another important diagenetic process observed in the Samana Suk Formation. It is associated with the evaporation of marine pore water, particularly in near-shore environments and the mixed meteoric/marine pore water zone. Dolomitization occurs during a relative sea-level fall when the mixed marine/meteoric pore-water zone shifts landwards. These highly dolomitized zones can impede vertical hydrocarbon flow due to their tight nature and the presence of organic matter from decayed cyanobacteria [13].

Early diagenetic processes and products in the formation vary depending on the system tract. The shelf regions experience the formation of carbonate sediments, often cemented by marine aragonite and/or high-Mg calcite rims and pore-filling cement, during the HST. In the more exposed sections of the ramp and platform sediments, meteoric water flow influences the diagenetic processes, while deeper and more buried regions may undergo marine pore-water diagenesis. The variation in pore-water composition is attributed to the "floating" of meteoric waters over denser marine pore waters [76]. As the HST deposits become subaerial, meteoric water percolation leads to the dissolution of marine grains and types of cement, forming karstic structures as they reach the sequence boundary [77,78]. Carbonate deposits along the TST and approaching the maximum flooding surface (MFS) exhibit increased marine carbonate types of cement, such as aragonite/high-Mg calcite rims and syntaxial overgrowths, as well as dolomitization. This is attributed to the migration of the marine pore-water zone into the basin and the mixing of meteoric and seawater in the sediments [15].

The diagenetic sequence of the Samana Suk Formation Involves four settings: marine, mixed marine-meteoric, burial, and uplifting [64]. Micritization occurred during early marine diagenesis, influenced by endolithic algal borings [62,79]. Micritization varies across horizons, with extensive micritization in mudstones. Ooids and bioclasts show radial micritization patterns. Dolomitization likely happened in mixed marine-meteoric water. Calcitization of dolomite to calcite is common below the sequence boundary. Early dissolution of aragonite skeletal debris is observed, and moldic pores are filled with isopachous bladed and blocky cements. Pressure-solution diagenesis, driven by gravity and tectonic stress, leads to chemical compaction, microstylolite development, and concentration of dolomite and quartz along seams. Burial diagenesis is marked by fractures, grain distortion, stylolites, and cementation. Dissolution and cementation occur during both initial deposition and subsequent uplift [80].

Sequence stratigraphy is crucial for understanding the deposition and pore development in carbonates and is essential for reservoir evaluation. Carbonates are more susceptible to sequential changes compared to siliciclastics due to their unstable mineralogy, making even small sea-level fluctuations capable of altering pore-water chemistry and pore types [20]. The Samana Suk Formation consists of three depositional settings: open marine, lagoon, and beach. Shah [34] suggested that the formation was deposited around 10 million years ago (170–160 Ma) based on biostratigraphic evidence. The presence of different facies at varying depths within the vertical stratigraphic column indicates sea-level fluctuations and the occurrence of both deeper and shallower environments. Additionally, the formation exhibits HST and TST, which were marked based on microfacies and field evidence. The formation comprises calm lagoonal deposits, including Mudstone, Dolo-Mudstone, and Bioclastic Mud-Wackestone microfacies, indicating a sea-level lowstand during HST. A transition to retrogradational Mudstone deposition signifies the onset of TST.

6. Conclusions

This study conducted a detailed analysis of the sedimentology, diagenesis, and sequence stratigraphy of the Middle Jurassic Samana Suk Formation in the Hazara Basin of northern Pakistan. Two sections, Mera Rehmat and Por Miana, were specifically studied to unravel the complex geological processes within the formation. The examination of microfacies in the studied sections identified nine distinct depositional textures, ranging from mudstone to wackestone, packstone, and grainstone. These variations indicate the presence of different inner ramp environments, including open marine, lagoon, and coastal settings. Petrographic investigations provided insights into the diagenetic processes that have affected the formation. The identified diagenetic processes include micritization, cementation, dissolution, compaction, neomorphism, and dolomitization. Six different cementation types were identified, and the patterns of dolomitization varied, particularly highlighting lagoonal environments and mudstone replacement. This study also conducted a stratigraphic sequence analysis, which revealed interesting patterns within the Samana Suk Formation. The high-stand system tract was characterized by mudstones, pelloidal grainstones, and dolomitized mudstones, indicating periods of high sea level. On the other hand, the transgressive system tract displayed ooidal grainstones, pelloidal packstones, and pel-bioclastic grainstones, representing transgression and inundation of previously exposed areas. One significant finding of this study was the impact of diagenesis on reservoir quality parameters, specifically porosity, and permeability. The various diagenetic processes, cementation types, and dolomitization patterns significantly altered the pore network within the formation. This highlights the importance of considering diagenesis when assessing the Samana Suk Formation as a potential hydrocarbon reservoir. Overall, this research provides a comprehensive understanding of the sedimentology, diagenesis, and sequence stratigraphy of the Middle Jurassic Samana Suk Formation. The findings contribute to our knowledge of similar carbonate reservoirs globally and can enhance the exploration and development of hydrocarbon resources in comparable depositional environments. The results presented in this study are precise, coherent, consistent, and clear, providing valuable insights into the geological processes within the formation.

Author Contributions: Conceptualization, M.M.S. and S.Q.; methodology, S.Q., M.M.S. and H.T.J.; software, S.Q., M.M.S., H.T.J. and A.S.; validation, M.M.S. and H.T.J.; formal analysis, S.Q., M.M.S and A.S.; investigation, S.Q., M.M.S., H.T.J., G.K., A.S. and E.B.; resources, M.M.S.; data curation, S.Q. and M.M.S.; writing—original draft preparation, S.Q. and M.M.S.; writing—review and editing, H.T.J., M.M.S., G.K., A.S. and E.B.; visualization, M.M.S.; supervision, M.M.S.; project administration, M.M.S.; funding acquisition, G.K., and H.T.J. All authors have read and agreed to the published version of the manuscript.

Funding: This research received no external funding.

Institutional Review Board Statement: Not applicable.

Informed Consent Statement: Not applicable.

Data Availability Statement: The data used in this work is available on request to the corresponding author(s).

Conflicts of Interest: The authors declare no conflict of interest.

References

1. Lucia, F.J. Petrophysical rock properties. In *Carbonate Reservoir Characterization: An Integrated Approach*; Springer: Berlin/Heidelberg, Germany, 2007; pp. 1–27.
2. Ahr, W.M. *Geology of Carbonate Reservoirs: The Identification, Description and Characterization of Hydrocarbon Reservoirs in Carbonate Rocks*; John Wiley & Sons: Hoboken, NJ, USA, 2011.
3. Beigi, M.; Jafarian, A.; Javanbakht, M.; Wanas, H.A.; Mattern, F.; Tabatabaei, A. Facies analysis, diagenesis and sequence stratigraphy of the carbonate-evaporite succession of the Upper Jurassic Surmeh Formation: Impacts on reservoir quality (Salman Oil Field, Persian Gulf, Iran). *J. Afr. Earth Sci.* **2017**, *129*, 179–194. [CrossRef]
4. Gaupp, R.; Matter, A.; Platt, J.; Ramseyer, K.; Walzebuck, J. Diagenesis and fluid evolution of deeply buried Permian (Rotliegende) gas reservoirs, northwest Germany. *AAPG Bull.* **1993**, *77*, 1111–1128.
5. Mazzullo, S.J. Diagenesis in a sequence-stratigraphic setting: Porosity evolution in periplatform carbonate reservoirs, Permian Basin, Texas and New Mexico. *J. Pet. Sci. Eng.* **1994**, *11*, 311–322. [CrossRef]
6. Hiatt, E.E.; Kyser, T.K. Links between depositional and diagenetic processes in basin analysis: Porosity and permeability evolution in sedimentary rocks. In *Fluids Basin Evolution*; Short Course; Mineralogical Association of Canada: Québec City, QC, Canada, 2000; Volume 28, pp. 63–92.
7. Janjuhah, H.T.; Salim, A.M.A.; Ghosh, D.P.; Wahid, A. Diagenetic process and their effect on reservoir quality in Miocene carbonate reservoir, Offshore, Sarawak, Malaysia. In *ICIPEG 2016, Proceedings of the International Conference on Integrated Petroleum Engineering and Geosciences, Kuala Lumpur, Malaysia, 15–17 August 2016*; Springer: Berlin/Heidelberg, Germany, 2017; pp. 545–558.
8. Posamentier, H.W.; Vail, P.R. Eustatic controls on clastic deposition II—Sequence and systems tract models. In *Sea-Level Changes: An Integrated Approach*; Wilgus, C.K., Hastings, B.S., Posamentier, H., Van Wagoner, J., Ross, C.A., Kendall, C.G., Eds.; Society for Sedimentary Geology: Tulsa, OK, USA, 1988.
9. Van Wagoner, J.C.; Mitchum, R.M.; Campion, K.M.; Rahmanian, V.D. *Siliciclastic Sequence Stratigraphy in Well Logs, Cores, and Outcrops: Concepts for High-Resolution Correlation of Time and Facies*; American Association of Petroleum Geologists: Tulsa, OK, USA, 1990.
10. Emery, D.; Myers, K.J. *Sequence Stratigraphy*; Black Ltd.: Oxford, UK, 1996.
11. Posamentier, H.W.; Allen, G.P. *Siliciclastic Sequence Stratigraphy—Concepts and Applications*; SEPM Concepts in Sedimentology and Paleontology; SEPM Society for Sedimentary Geology: Tulsa, OK, USA, 1999; Volume 7, 210p.
12. Rahimpour-Bonab, H.; Mehrabi, H.; Enayati-Bidgoli, A.H.; Omidvar, M. Coupled imprints of tropical climate and recurring emergence on reservoir evolution of a mid Cretaceous carbonate ramp, Zagros Basin, southwest Iran. *Cretac. Res.* **2012**, *37*, 15–34. [CrossRef]
13. Taghavi, A.A.; Mørk, A.; Emadi, M.A. Sequence stratigraphically controlled diagenesis governs reservoir quality in the carbonate Dehluran Field, southwest Iran. *Pet. Geosci.* **2006**, *12*, 115–126. [CrossRef]
14. Morad, S.; Ketzer, J.M.; De Ros, L.F. Spatial and temporal distribution of diagenetic alterations in siliciclastic rocks: Implications for mass transfer in sedimentary basins. *Sedimentology* **2000**, *47*, 95–120. [CrossRef]
15. Tucker, M.E. Carbonate diagenesis and sequence stratigraphy. In *Sedimentology Review 1*; Wright, V.P., Ed.; Blackwell Scientific Publications: Oxford, UK, 1993; pp. 51–72.
16. South, D.L.; Talbot, M.R. The sequence stratigraphic framework of carbonate diagenesis within transgressive fan-delta deposits: Sant Llorenç del Munt fan-delta complex, SE Ebro Basin, NE Spain. *Sediment. Geol.* **2000**, *138*, 179–198. [CrossRef]
17. Ketzer, J.M.; Holz, M.; Morad, S.; Al-Aasm, I.S. Sequence stratigraphic distribution of diagenetic alterations in coal-bearing, paralic sandstones: Evidence from the Rio Bonito Formation (early Permian), southern Brazil. *Sedimentology* **2003**, *50*, 855–877. [CrossRef]
18. Moss, S.; Tucker, M.E. Diagenesis of Barremian-Aptian platform carbonates (the Urgonian Limestone Formation of SE France): Near-surface and shallow-burial diagenesis. *Sedimentology* **1995**, *42*, 853–874. [CrossRef]
19. Booler, J.; Tucker, M.E. Distribution and geometry of facies and early diagenesis: The key to accommodation space variation and sequence stratigraphy: Upper Cretaceous Congost Carbonate platform, Spanish Pyrenees. *Sediment. Geol.* **2002**, *146*, 225–247. [CrossRef]
20. Morad, S.; Al-Aasm, I.S.; Nader, F.H.; Ceriani, A.; Gasparrini, M.; Mansurbeg, H. Impact of diagenesis on the spatial and temporal distribution of reservoir quality in the Jurassic Arab D and C members, offshore Abu Dhabi oilfield, United Arab Emirates. *GeoArabia* **2012**, *17*, 17–56. [CrossRef]
21. Janjuhah, H.T.; Alansari, A.; Ghosh, D.P.; Bashir, Y. New approach towards the classification of microporosity in Miocene carbonate rocks, Central Luconia, offshore Sarawak, Malaysia. *J. Nat. Gas Geosci.* **2018**, *3*, 119–133. [CrossRef]
22. Shah, M.M.; Rahim, H.U.; Hassan, A.; Mustafa, M.R.; Ahmad, I. Facies control on selective dolomitization and its impact on reservoir heterogeneities in the Samana Suk Formation (middle Jurassic), Southern Hazara Basin (NW Himalaya, Pakistan): An outcrop analogue. *Geosci. J.* **2020**, *24*, 295–314. [CrossRef]
23. Rahim, H.-u.; Qamar, S.; Shah, M.M.; Corbella, M.; Martín-Martín, J.D.; Janjuhah, H.T.; Navarro-Ciurana, D.; Lianou, V.; Kontakiotis, G. Processes Associated with Multiphase Dolomitization and Other Related Diagenetic Events in the Jurassic Samana Suk Formation, Himalayan Foreland Basin, NW Pakistan. *Minerals* **2022**, *12*, 1320. [CrossRef]

24. Rahim, H.U.; Shah, M.M.; Corbella, M.; Navarro-Ciurana, D. Diagenetic evolution and associated dolomitization events in the middle Jurassic Samana Suk Formation, Lesser Himalayan Hill Ranges, NW Pakistan. *Carbonates Evaporites* **2020**, *35*, 101. [CrossRef]
25. Yeats, R.S.; Hussain, A. Timing of structural events in the Himalayan foothills of northwestern Pakistan. *Geol. Soc. Am. Bull.* **1987**, *99*, 161–176. [CrossRef]
26. DiPietro, J.A.; Pogue, K.R. Tectonostratigraphic subdivisions of the Himalaya: A view from the west. *Tectonics* **2004**, *23*, TC5001. [CrossRef]
27. Shah, M.T.; Moon, C.J. Mineralogy, geochemistry and genesis of the ferromanganese ores from Hazara area, NW Himalayas, northern Pakistan. *J. Asian Earth Sci.* **2004**, *23*, 1–15. [CrossRef]
28. Rehman, H.U.; Seno, T.; Yamamoto, H.; Khan, T. Timing of collision of the Kohistan–Ladakh Arc with India and Asia: Debate. *Isl. Arc* **2011**, *20*, 308–328. [CrossRef]
29. Qasim, M.; Ding, L.; Khan, M.A.; Jadoon, I.A.K.; Haneef, M.; Baral, U.; Cai, F.; Wang, H.; Yue, Y. Tectonic implications of detrital zircon ages from lesser Himalayan Mesozoic-Cenozoic strata, Pakistan. *Geochem. Geophys. Geosystems* **2018**, *19*, 1636–1659. [CrossRef]
30. Qureshi, K.A.; Arif, M.; Basit, A.; Ahmad, S.; Janjuhah, H.T.; Kontakiotis, G. Sedimentological Controls on the Reservoir Characteristics of the Mid-Triassic Tredian Formation in the Salt and Trans-Indus Surghar Ranges, Pakistan: Integration of Outcrop, Petrographic, and SEM Analyses. *J. Mar. Sci. Eng.* **2023**, *11*, 1019. [CrossRef]
31. Yin, A. Cenozoic tectonic evolution of the Himalayan orogen as constrained by along-strike variation of structural geometry, exhumation history, and foreland sedimentation. *Earth-Sci. Rev.* **2006**, *76*, 1–131. [CrossRef]
32. Chatterjee, S.; Goswami, A.; Scotese, C.R. The longest voyage: Tectonic, magmatic, and paleoclimatic evolution of the Indian plate during its northward flight from Gondwana to Asia. *Gondwana Res.* **2013**, *23*, 238–267. [CrossRef]
33. Han, Z.; Hu, X.; Li, J.; Garzanti, E. Jurassic carbonate microfacies and relative sea-level changes in the Tethys Himalaya (southern Tibet). *Palaeogeogr. Palaeoclimatol. Palaeoecol.* **2016**, *456*, 1–20. [CrossRef]
34. Shah, S.M.I. Stratigraphy of Pakistan (memoirs of the geological survey of Pakistan). *Geol. Surv. Pak.* **2009**, *22*, 381.
35. Mehmood, M.; Naseem, A.A.; Saleem, M.; Rehman, J.u.; Kontakiotis, G.; Janjuhah, H.T.; Khan, E.U.; Antonarakou, A.; Khan, I.; Rehman, A.u. Sedimentary Facies, Architectural Elements, and Depositional Environments of the Maastrichtian Pab Formation in the Rakhi Gorge, Eastern Sulaiman Ranges, Pakistan. *J. Mar. Sci. Eng.* **2023**, *11*, 726. [CrossRef]
36. Ahsan, N.; Chaudhry, M.N. Geology of Hettangian to middle Eocene rocks of Hazara and Kashmir basins, Northwest lesser Himalayas, Pakistan. *Geol. Bull. Panjab Univ.* **2008**, *43*, 131–152.
37. Hussain, H.S.; Fayaz, M.; Haneef, M.; Hanif, M.; Jan, I.U.; Gul, B. Microfacies and diagenetic-fabric of the Samana Suk Formation at Harnoi Section, Abbottabad, Khyber Pakhtunkhwa, Pakistan. *J. Himal. Earth Sci.* **2013**, *46*, 79.
38. Ali, S.K.; Lashari, R.A.; Sahito, A.G.; Kontakiotis, G.; Janjuhah, H.T.; Mughal, M.S.; Bilal, A.; Mehmood, T.; Majeed, K.U. Sedimentological and Petrographical Characterization of the Cambrian Abbottabad Formation in Kamsar Section, Muzaffarabad Area: Implications for Proto-Tethys Ocean Evolution. *J. Mar. Sci. Eng.* **2023**, *11*, 526. [CrossRef]
39. Bilal, A.; Yang, R.; Janjuhah, H.T.; Mughal, M.S.; Li, Y.; Kontakiotis, G.; Lenhardt, N. Microfacies analysis of the Palaeocene Lockhart limestone on the eastern margin of the Upper Indus Basin (Pakistan): Implications for the depositional environment and reservoir characteristics. *Depos. Rec.* **2023**, *9*, 152–173. [CrossRef]
40. Basharat, M.; Qasim, M.; Shafique, M.; Hameed, N.; Riaz, M.T.; Khan, M.R. Regolith thickness modeling using a GIS approach for landslide distribution analysis, NW Himalayas. *J. Mt. Sci.* **2018**, *15*, 2466–2479. [CrossRef]
41. Garzanti, E. Himalayan ironstones, "superplumes", and the breakup of Gondwana. *Geology* **1993**, *21*, 105–108. [CrossRef]
42. Bilal, A.; Yang, R.; Mughal, M.S.; Janjuhah, H.T.; Zaheer, M.; Kontakiotis, G. Sedimentology and Diagenesis of the Early–Middle Eocene Carbonate Deposits of the Ceno-Tethys Ocean. *J. Mar. Sci. Eng.* **2022**, *10*, 1794. [CrossRef]
43. Jan, M.Q.; Windley, B.F.; Khan, A. The Waziristan ophiolite, Pakistan; general geology and chemistry of chromite and associated phases. *Econ. Geol.* **1985**, *80*, 294–306. [CrossRef]
44. Hanif, M.; Imraz, M.; Ali, F.; Haneef, M.; Saboor, A.; Iqbal, S.; Ahmad, S. The inner ramp facies of the Thanetian Lockhart Formation, western Salt Range, Indus Basin, Pakistan. *Arab. J. Geosci.* **2014**, *7*, 4911–4926. [CrossRef]
45. Ahmad, I.; Shah, M.M.; Janjuhah, H.T.; Trave, A.; Antonarakou, A.; Kontakiotis, G. Multiphase Diagenetic Processes and Their Impact on Reservoir Character of the Late Triassic (Rhaetian) Kingriali Formation, Upper Indus Basin, Pakistan. *Minerals* **2022**, *12*, 1049. [CrossRef]
46. Shahzad, A.; Munir, M.u.H.; Adatte, T.; Riaz, M.T.; Basharat, M.; Ahmed, K.S.; Sarfraz, Y.; Khan, J.; Mughal, M.S. Multi-proxy approach of the stratigraphy, geochemistry, and sedimentology of Eocene Margalla Hill Limestone: Case study from Muzaffarabad area, Sub-Himalayas, Pakistan. *Geol. J.* **2022**, *57*, 186–207. [CrossRef]
47. Shahzad, A.; Khan, J.; Hanif, M.; Baumgartner-Mora, C.; Sarfraz, Y.; Ahmed, K.S.; Riaz, M.T.; Baumgartner, P.O.; ul Hassan Munir, M.; Wazir, A. Eocene nannofossils and paleoenvironmental reconstruction of the Kuldana Formation in Yadgar area, Muzaffarabad, northern Pakistan. *Palaeoworld* **2023**. [CrossRef]
48. Critelli, S.; Garzanti, E. Provenance of the lower Tertiary Murree redbeds (Hazara-Kashmir Syntaxis, Pakistan) and initial rising of the Himalayas. *Sediment. Geol.* **1994**, *89*, 265–284. [CrossRef]

49. Ali, S.K.; Janjuhah, H.T.; Shahzad, S.M.; Kontakiotis, G.; Saleem, M.H.; Khan, U.; Zarkogiannis, S.D.; Makri, P.; Antonarakou, A. Depositional Sedimentary Facies, Stratigraphic Control, Paleoecological Constraints, and Paleogeographic Reconstruction of Late Permian Chhidru Formation (Western Salt Range, Pakistan). *J. Mar. Sci. Eng.* **2021**, *9*, 1372. [CrossRef]
50. Dickson, J.A.D. A modified staining technique for carbonates in thin section. *Nature* **1965**, *205*, 587. [CrossRef]
51. Dunham, R.J. Classification of carbonate rocks according to depositional textures. In *Classification of Carbonate Rocks—A Symposium*; AAPG Datapages Inc.: Tulsa, OK, USA, 1962.
52. Embry, A.F.; Klovan, J.E. A late Devonian reef tract on northeastern Banks Island, NWT. *Bull. Can. Pet. Geol.* **1971**, *19*, 730–781.
53. Wilson, J.L. The lower carboniferous Waulsortian facies. In *Carbonate Facies in Geologic History*; Springer: Berlin/Heidelberg, Germany, 1975; pp. 148–168.
54. Buxton, M.W.N.; Pedley, H.M. Short Paper: A standardized model for Tethyan Tertiary carbonate ramps. *J. Geol. Soc.* **1989**, *146*, 746–748. [CrossRef]
55. Flügel, E.; Munnecke, A. *Microfacies of Carbonate Rocks: Analysis, Interpretation and Application*; Springer: Berlin/Heidelberg, Germany, 2010; Volume 976.
56. Plumley, W.J.; Risley, G.A.; Graves Jr, R.W.; Kaley, M.E. Energy index for limestone interpretation and classification. In *Classification of Carbonate Rocks—A Symposium*; AAPG Datapages Inc.: Tulsa, OK, USA, 1962.
57. Janjuhah, H.T.; Alansari, A. Offshore carbonate facies characterization and reservoir quality of Miocene rocks in the southern margin of South China Sea. *Acta Geol. Sin. Engl. Ed.* **2020**, *94*, 1547–1561. [CrossRef]
58. Lee, E.Y.; Kominz, M.; Reuning, L.; Gallagher, S.J.; Takayanagi, H.; Ishiwa, T.; Knierzinger, W.; Wagreich, M. Quantitative compaction trends of Miocene to Holocene carbonates off the west coast of Australia. *Aust. J. Earth Sci.* **2021**, *68*, 1149–1161. [CrossRef]
59. Janjuhah, H.T.; Alansari, A.; Santha, P.R. Interrelationship between facies association, diagenetic alteration and reservoir properties evolution in the Middle Miocene carbonate build up, Central Luconia, Offshore Sarawak, Malaysia. *Arab. J. Sci. Eng.* **2019**, *44*, 341–356. [CrossRef]
60. Haq, B.U.; Hardenbol, J.; Vail, P.R. Mesozoic and Cenozoic chronostratigraphy and cycles of sea-level change. In *Sea-Level Changes: An Integrated Approach*; Wilgus, C.K., Hastings, B.S., Posamentier, H., Van Wagoner, J., Ross, C.A., Kendall, C.G., Eds.; Society for Sedimentary Geology: Tulsa, OK, USA, 1988.
61. Shah, M.M.; Ahmed, I. Diagenetic Studies and Its Implications on the Reservoir Character of the Anisian-Norian (Triassic) Kingriali Formation, Salt Range (Pakistan). In Proceedings of the AAPG/SEG International Conference & Exhibition, Cancun, Mexico, 6–9 September 2016.
62. Wadood, S.A.; Boli, G.; Xiaowen, Z.; Hussain, I.; Yimin, W. Recent development in the application of analytical techniques for the traceability and authenticity of food of plant origin. *Microchem. J.* **2020**, *152*, 104295. [CrossRef]
63. Ngia, N.R.; Hu, M.; Gao, D. Hydrocarbon reservoir development in reef and shoal complexes of the Lower Ordovician carbonate successions in the Tazhong Uplift in central Tarim basin, NW China: Constraints from microfacies characteristics and sequence stratigraphy. *J. Pet. Explor. Prod. Technol.* **2020**, *10*, 2693–2720. [CrossRef]
64. Wadood, B.; Khan, S.; Li, H.; Liu, Y.; Ahmad, S.; Jiao, X. Sequence Stratigraphic Framework of the Jurassic Samana Suk Carbonate Formation, North Pakistan: Implications for Reservoir Potential. *Arab. J. Sci. Eng.* **2021**, *46*, 525–542. [CrossRef]
65. Karami-Movahed, F.; Aleali, M.; Ghazanfari, P. Facies Analysis, Depositional Environment and Diagenetic Features of the Qom Formation in the Saran Semnan, Central Iran. *Open J. Geol.* **2016**, *6*, 349–362. [CrossRef]
66. Bathurst, R.G.C. Boring algae, micrite envelopes and lithification of molluscan biosparites. *Geol. J.* **1966**, *5*, 15–32. [CrossRef]
67. Kobluk, D.R.; Risk, M.J. Micritization and carbonate-grain binding by endolithic algae. *AAPG Bull.* **1977**, *61*, 1069–1082.
68. Smosna, R.; Koehler, B. Tidal Origin of a Mississippian Oolite on the West Virginia Dome: Chapter 11. In *Mississippian Oolites and Modern Analogs*; American Association of Petroleum Geologists: Tulsa, OK, USA, 1993.
69. Perry, C.T. Grain susceptibility to the effects of microboring: Implications for the preservation of skeletal carbonates. *Sedimentology* **1998**, *45*, 39–51. [CrossRef]
70. Janjuhah, H.T.; Kontakiotis, G.; Wahid, A.; Khan, D.M.; Zarkogiannis, S.D.; Antonarakou, A. Integrated porosity classification and quantification scheme for enhanced carbonate reservoir quality: Implications from the miocene malaysian carbonates. *J. Mar. Sci. Eng.* **2021**, *9*, 1410. [CrossRef]
71. Janjuhah, H.T.; Salim, A.M.A.; Alansari, A.; Ghosh, D.P. Presence of microporosity in Miocene carbonate platform, Central Luconia, offshore Sarawak, Malaysia. *Arab. J. Geosci.* **2018**, *11*, 204. [CrossRef]
72. Bathurst, R.G.C. *Carbonate Sediments and Their Diagenesis*; Elsevier: Amsterdam, The Netherlands, 1972.
73. Halley, R.B.; Scholle, P.A. Radiaxial fibrous calcite as early-burial, open-system cement: Isotopic evidence from Permian of China. *AAPG Bull.* **1985**, *69*, 261–262.
74. Moore, C.H. Diagenetic environments of porosity modification and tools for their recognition in the geologic record. *Carbonate Reserv. Porosity Evol. Diagenesis A Seq. Stratigr. Framew. Dev. Sedimentol.* **2001**, *55*, 61–88.
75. Salah, M.K.; Janjuhah, H.; Sanjuan, J.; Maalouf, E. Impact of diagenesis and pore aspects on the petrophysical and elastic properties of carbonate rocks from southern Lebanon. *Bull. Eng. Geol. Environ.* **2023**, *82*, 67. [CrossRef]
76. Hitchon, B.; Friedman, I. Geochemistry and origin of formation waters in the western Canada sedimentary basin—I. Stable isotopes of hydrogen and oxygen. *Geochim. Cosmochim. Acta* **1969**, *33*, 1321–1349. [CrossRef]

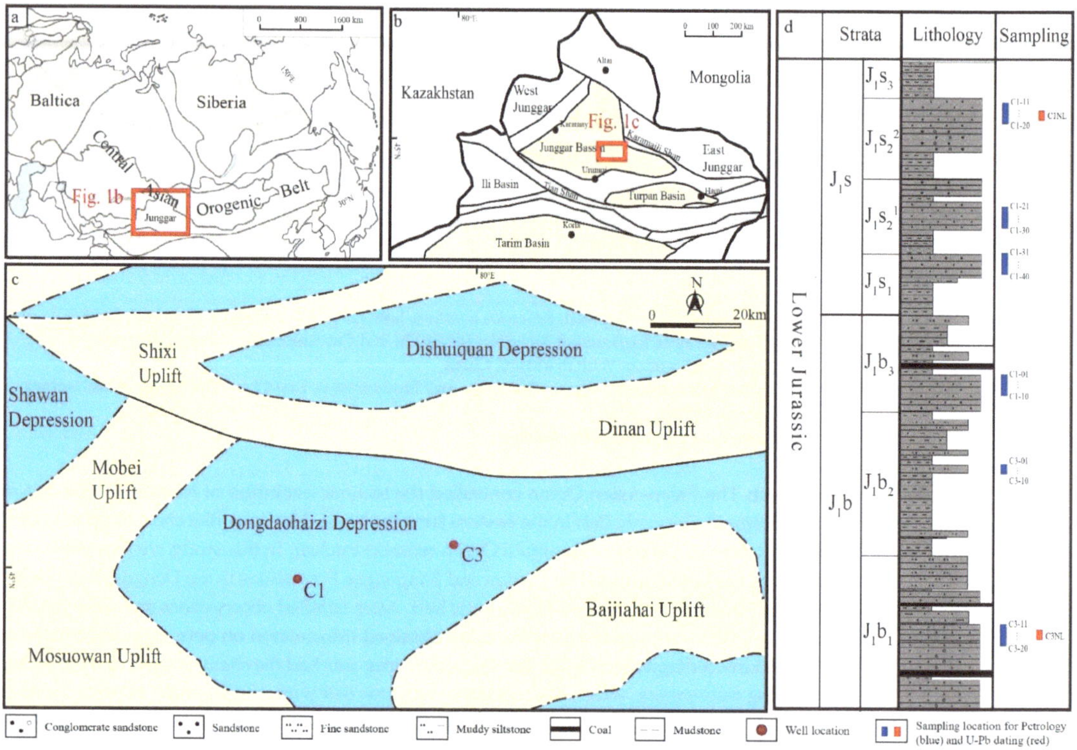

Figure 1. (**a–c**) Geological map of Eurosia Plate, Junggar Basin, and Dongdaohaizi area, respectively. (**d**) Chronostratigraphy and lithology section of Lower Jurassic in Dongdaohaizi Depression. Abbreviations: J_1b = Lower Jurassic Badaowan Formation, J_1s = Lower Jurassic Sangonghe Formation.

The Karamaili Ocean is located between the Kazakhstan-Junggar and Siberian plates and is an important branch of the Paleo-Asian Ocean. Its subduction represents the subduction, collision, and accretion of the Junggar-Tairm block onto the Siberian plate, forming the Karamaili tectonic belt in the eastern Junggar area (Figure 1b) [6,7,11–14]. Investigating the evolutionary history of the Karamaili Ocean provides valuable insights into the broader evolution of the Paleo-Asian Ocean.

Previous studies on the Karamaili Ocean are mainly based on the outcrops of the eastern Junggar orogenic belt. The outcrops are limited, dispersed, heavily fragmented, and subjected to significant metamorphism and deformation by the intense tectonic activity of the orogenic belt, resulting in highly disputed chronological constraints for the evolution of the Karamaili Ocean. For example, the duration and the timing of the final closure of the Karamaili Ocean have been a topic of debate in the geological community due to the varying ages reported by different methods and data sources, ranging from the Late Devonian to the Late Permian. Some studies have employed geochronological data to suggest a Late Carboniferous or Early Permian closure of the Karamaili Ocean, based on the ages of ophiolites, arc magmatism, and collisional granites along the suture zone [15–17]. In other studies, paleomagnetic data have been utilized in some studies to support a Late Devonian or Early Carboniferous closure of the Karamaili Ocean, citing the similarity of paleolatitudes between Siberia and Tarim [18–20]. Therefore, more reliable data reflecting the overall evolutionary development of the ocean basin are needed.

The subduction of the ocean basin transported a large amount of ocean basin volcanic and clastic material from the ocean-land boundary to the basin through the source-sink system [21]. By comparing the properties of sediments in the stable continental basin nearest to the margin where the ocean basin closed, we can obtain a complete understanding of the transformation history of the ocean.

The Junggar Basin was formed by the assemblage of volcanic island arcs and accretionary wedges and was a stable basin with a Paleozoic basement in the early Jurassic. The Dongdaohaizi Depression is located in the eastern part of the Junggar Basin, adjacent to the orogenic belt of eastern Junggar, and has incorporated a large amount of sediment from the ancient ocean during the basin-margin orogeny (Figure 1b,c). According to a recent study [22], a large number of shallow braided rivers developed in the basin during the Early to Middle Jurassic, with sediment from the basin margin being transported more than 200 km, effectively ensuring the breadth and volume of the sediment supply.

Currently, quantitative analysis has become the main direction of provenance analysis. Through U-Pb isotopic dating analysis of characteristic minerals such as zircons, geological activities of the source region for the clastic sediments present in the Dongdaohaizi depression can be examined [23]. Therefore, this study takes the deep-drilling cores of the Lower Jurassic Badaowan Formation and Sangonghe Formation sandstones in the Dongdaohaizi Depression as the research object. Based on the study of petrological characteristics in combination with the LA-ICP-MS detrital zircon U-Pb geochronological study, the age characteristics of the Badaowan and Sangonghe formations were determined, and the provenance characteristics and the evolution of the Karamaili Ocean were further revealed accordingly.

2. Geological Setting and Sampling

The Junggar Basin is located in the northern part of Xinjiang, China, at the intersection of the Kazakhstan Plate, Siberian Plate, and Tarim Plate, and is the eastward extension of the Kazakhstan Plate (Figure 1a,b) [24,25]. The basin is generally wider in the south and narrower in the north and has a triangular distribution, consisting of the primary tectonic units of the Ulungur Depression, Luliang Uplift, Central Depression, eastern Uplift, western Uplift, and Southern Margin Thrust Belt. The Dongdaohaizi Depression belongs to the secondary tectonic unit of the central depression of the Junggar Basin. The study area is located in the middle of the Dongdaohaizi Depression, with the Mosuowan Uplift to the west, the Baijiahai Uplift to the east, and the Fukang Depression to the south (Figure 1c). The Dongdaohaizi Depression was formed in the early stage of the Hercynian period and developed a large set of marine shales and marls with widespread distribution of calc-alkaline volcanic rocks. Strong earthquake and volcanic activity, accompanied by faulting, occurred in the Early Carboniferous to Early Permian, providing the basis for the internal structure of the central depression [26]. The structures in the study area mainly developed in the Devonian basement. The basement is then covered with a huge thickness of Mesozoic (>5000 m) and Cenozoic (>3000 m) rocks. the tectonic movements during the Late Hercynian and Yanshanian played a decisive role in the structural deformation of the study area [27]. Interlayer faults trending nearly northeast, dominated by normal faults with a short horizontal extension developed during the Jurassic.

The Jurassic strata in the Dongdaohaizi Depression include the Badaowan Formation, Sangonghe Formation, Xishanyao Formation, and Toutunhe Formation, while the Qigu Formation is not developed in this area. Among them, the Badaowan Formation, Sangonghe Formation, and Xishanyao Formation include coal-bearing strata. The Xishanyao Formation is mainly composed of fluvial deposits, with sandstone thickness accounting for 50% of the stratum thickness. The maximum thickness of a single layer is 14 m, and the general thickness ranges from 2 m to 8 m, with significant lateral thickness variations. The Toutunhe Formation is dominated by continental (lake-swamp) deposits and is composed of sandstone, red and gray sandy mudstone in ascending order.

The Lower Jurassic Badaowan Formation and Sangonghe Formation are the target intervals of this study in the Dongdaohaizi Sag area (Figure 1d). The Badaowan Formation is primarily composed of fluvial and deltaic deposits. Sandstone beds account for 45% of the formation thickness, with a maximum single-layer thickness of 30 m and a general thickness range of 4 m to 22 m. The Badaowan Formation exhibits distinct cyclicity, with a coarse-fine-coarse sequence from bottom to top. The lithological assemblage consists mainly of conglomerate, sandstone, mudstone, and coal seams. In the Early Jurassic, the Indosinian movement, caused by the collision of eastern Asia blocks, led to the formation of the Sikezhuang Sag, resulting in a regional angular unconformity between the Jurassic and Triassic strata. The sedimentary center during the deposition of the Lower Jurassic Badaowan Formation was located in the Changji Sag and the Penyijingxi Sag, with a maximum thickness of 1200 m. The lithology of the formation can be divided into two parts: the upper part is composed of interbedded mud and sand, and the middle and lower parts are composed of gray-green or gray-black mudstone interbedded with conglomerate or pebbly sandstone. The Badaowan Formation is the first major coal-bearing formation developed in the Junggar Basin and also constitutes the first major hydrocarbon source rock and reservoir in the basin. The first and third members of the Badaowan Formation are mainly sandstone, while the second member is mainly mudstone.

The Sangonghe Formation is primarily composed of gray sandstone, fine-grained sandstone, and mudstone containing siltstone, with tuff and quartz as the dominant sandstone, followed by feldspar, with small amounts of illite and smectite minerals. The sandstones exhibit moderate-to-well sorting and sub-angular to sub-rounded grains, and the mudstone-calcium mudstone facies shows a matrix-supported texture. The sandstone layers account for 37% of the formation thickness, with a maximum single-layer thickness of 14 m and a general thickness range of 2 m to 10 m. The sedimentary environment is primarily deltaic deposits, and it is similar to the sedimentary area of the Badaowan Formation. The sedimentary center was located in the central Junggar Basin, and the sedimentary thickness was also approximately 1200 m. The lithology of the formation can be divided into two parts: the middle and upper members are interbedded sand-mudstone, and the lower member is composed of medium-grained conglomerate and sandstone rocks, forming the main hydrocarbon source rock and reservoir Jurassic in the Junggar Basin.

In this study, 60 samples were collected from lower Jurassic Badaowan and Sangonghe Formations for petrological and whole-rock analysis (Figure 1d, blue bars; Table S1). Samples C1-1 to C1-10 were collected from Badaowan Formation upper member sandstones at well C1; samples C3-1 to C10 were collected from Badaowan Formation middle member fine sandstones at well C3; samples C3-11 to C3-20 were collected from Badaowan Formation lower member sandstones at well C3; samples C1-11 to C1-20 were collected from Sangonghe Formation middle member upper sub-member sandstones at well C1; samples C1-21 to C30 were collected from Sangonghe Formation middle member lower sub-member sandstones at well C1; samples C1-31 to C40 were collected from Sangonghe Formation lower member sandstones at well C1. In addition, samples C1NL and C3NL were collected from Sangonghe Formation middle member upper sub-member sandstones at well C1 and Badaowan Formation lower member sandstones at well C3 for detrital zircon U-Pb dating (Figure 1d, red bars; Table S2).

3. Petrology and Methods

3.1. Petrologic Characteristics

Samples from Badaowan and Sangonghe formations are mainly medium-fine sandstones (Figures 1d and 2a). The detrital components of the formations are mainly composed of quartz, feldspar, and lithic fragments. The sandstone structures are well-developed, with sub-angular to sub-rounded grains and moderate sorting. The types of cement are diverse, with mudstone being the dominant matrix, while calcite is the main cement mineral, with occasional occurrences of dolomite. The primary mode of cementation is compactional cementation.

Figure 2. (**a**) Representative photograph of sandstone samples from Badaowan Formation (**a1–a3**) and Sangonghe Formation (**a4–a6**), respectively. Cross-polarized light photomicrograph of sandstone samples from Badaowan Formation (**b1–b9**) and Sangonghe Formation (**c1–c6**), respectively. Q = quartz, Qp = polycrystalline quartz; Pl = plagioclase, Mc = microcline, Lv = volcanic rock fragments, g = granite clasts, Lmp = quartz-mica schist clasts, Lmm = slate clasts, Pth = perthitic feldspar.

Based on mineralogical analyses of slices of the Badaowan and Sangonghe Formations, the components and their proportions are summarized in Table S1. The compositional features of the detrital rocks in the study area were analyzed in conjunction with observations of the samples and slices. The samples are mainly grey, feldspathic litharenite with a fine to medium grain size and blocky structure (Figure 2(a1–a3)). Quartz is one of the main detrital components in the sandstones of the study area, with a content range of 40% to 60%. It is well-rounded and mature, with angular to sub-angular shapes being common (Figure 2(b1)). Single crystal and polycrystalline quartz grains are also present (Figure 2(b2)), with some exhibiting undulose extinction. Feldspar content ranges from 12% to 20%, with sub-angular shapes being common. Plagioclase and potassium feldspar are the main types of feldspar present, with widespread alteration and a dirty appearance. Sericite and kaolinite are also abundant, with polysynthetic twins visible in some plagioclase grains (Figure 2(b3)) and grid twinning visible in some microcline grains (Figure 2(b4)). The lithic fragment content is typically 30% to 40%, with volcanic lithic fragments being the most common, especially tuffaceous lithic fragments. Sedimentary and metamorphic lithic fragments are less common, with occasional occurrences of mudstone, shale, quartzite, and phyllite lithic fragments (Figure 2(b5–b9)). Tuffaceous lithic fragments are composed of angular to sub-angular feldspar and quartz crystals with varying degrees of devitrification. Chert lithic fragments are mainly composed of microcrystalline or cryptocrystalline quartz. Phyllite lithic fragments are mainly composed of sericite, with a schistose texture and phyllitic structure.

The lithological and mineralogical characteristics of the Sangonghe Formation in the study area were summarized and analyzed based on the thin section data (Table S1). Microscopic observations reveal that the Sangonghe Formation is predominantly composed of plagioclase-rich lithic sandstone, lithic sandstone, and minor plagioclase-lithic sandstone. The lithic fragments are dominated by igneous and metamorphic rocks, and exhibit medium to fine grain size and blocky texture. Quartz, which accounts for 30% to 50% of the Sangonghe Formation, shows angular to sub-angular shapes with straight to concave-convex contacts, and some quartz grains exhibit secondary enlargement (Figure 2(c1)), undulose extinction, and embayment structures, indicating their derivation from igneous rocks (Figure 2(c2,c3)). Feldspar fragments, including plagioclase, striped feldspar, and potassium feldspar, exhibit dirty surfaces and widespread alteration, with intense sericitization and clay mineralization (Figure 2(c3)).

The lithic components of the Sangonghe Formation sandstones are predominantly composed of igneous and metamorphic rocks, indicating a low compositional maturity and a short transport distance, which suggests a control from the eastern Karamaili Mountain provenance. Some samples mainly consist of lithic sandstone, with igneous rock lithic fragments as the main component and low compositional maturity, indicating a short transport distance and control from the eastern Karamaili Mountain provenance. Microscopic observations also reveal abundant tuffaceous lithic fragments and a small amount of basaltic lithic fragments in the Sangonghe Formation (Figure 2(c5,c6)). Overall, the Sangonghe Formation samples have a lower lithic content and tuffaceous lithic content than the Badaowan Formation and exhibit higher porosity.

3.2. Methods

3.2.1. Zircon U-Pb Dating

Samples C1NL (~2 kg) and C3NL (~2.5 kg) were collected, and samples with relatively uniform color and lithology were selected. These samples were sent to Chengxin Geological Services Company in Langfang, Hebei for zircon separation to obtain sufficient zircons for testing and analysis. The zircons were made into targets at Beijing Zirconia Navigation Technology Co., Ltd., Beijing, China. The zircon samples were then subjected to transmitted and reflected light and Cathodoluminescence (CL) photography to observe their structure. The choice of CL images for the analysis is the Chromal CL 2 cathodoluminescence probe, which is configured on the JEOL JSM-6510A scanning electron microscope. The analysis conditions are as follows: acceleration voltage of 10 kV, beam current of SS65, and a working distance of 14 mm. These experiments were worked out at the Experimental Center of the School of Earth Sciences, Yangtze University.

The zircon U-Pb isotopic dating analysis was conducted at the Experimental Center of the School of Earth Sciences, Yangtze University. The NWR193HE laser ablation system (Elemental Scientific Lasers LLC, Bozeman, MT, USA) produced by ESI in the United States was used. The isotope ratios and element contents of the samples were calculated using the GLITTER program (ver4.0, Macquarie University), and the common Pb correction was performed using the method developed by Andersen [28].

Due to the presence of a large amount of radiogenic Pb in some zircons, the $^{207}Pb/^{206}Pb$ surface age was used for detrital zircons with U-Pb ages greater than 1000 Ma, while the $^{206}Pb/^{238}U$ surface age was used for zircons with U-Pb ages less than 1000 Ma [29,30]. The age-weighted average and age concordia diagram were calculated using the Isoplot program [31]. The standard deviation of single measurements of zircon U-Th-Pb isotope ratios and age data was 2σ, and the weighted average age was calculated using 2σ.

3.2.2. Whole-Rock Analysis

The composition of rock fragments is a direct expression of the combination of rock types and properties of source rocks. Microscopic analysis of the rock fragment composition in a study area can, to a certain extent, match the composition of the detrital material from the source rocks. The Dickinson triangle diagram is a widely used method for source rock discrimination studies, which can determine the properties of the source rocks and their tectonic background by analyzing the sandstone fragment composition, including the content of quartz, feldspar, and lithic fragments. In this study, we analyzed the component characteristics of the detrital rock samples from the study area by using the Dickinson triangle diagram based on the statistical analysis of the detrital zircon particles in the samples.

Fine-grained, medium-coarse-grained, and coarse-grained sandstone samples were selected from the specimens, and the weathered surface layer was removed first. The fresh rock portions were then made into thin rock sections. The square mesh intersection method was used to statistically count the detrital particles [32], and the statistical method of Gazzi-Dickinson was referenced [33,34]. The ternary provenance diagram in the Gazzi-Dickinson method aims to determine the provenance of the clastic material in analyzed sandstones. Provenance diagrams show percentage ratios of monomineral quartz, feldspar, and lithic fragments in each analyzed sample. Based on their composition, samples are plotted in different provenance areas. To ensure the accuracy of the results, samples with a matrix volume fraction greater than 25% were excluded from the statistical and plotting analyses, and the average particle size of the sandstone samples included in the statistics and plot analysis was limited to between medium-coarse-grained (including conglomerate) and arithmetic particle size 0.2–2 mm. The rock samples were used only for reference, and statistical errors due to the influence of detrital particle size on mineral composition properties were minimized as much as possible. A total of 300–400 statistical points were used for each thin section, and the caliper was used to control the spacing during the counting process. The grid spacing was twice the average particle size. In other words, a particle point was counted every time the caliper was moved along the same measurement line by a distance of twice the average particle size. After counting one line, the caliper was moved vertically by a distance of twice the average particle size, and the second line was counted, and so on. The counted detrital particle types included single-crystal quartz, polycrystalline quartz, plagioclase, potassium feldspar, sedimentary rock debris, metamorphic rock debris, volcanic rock debris, and secondary mineral particles such as mica, hornblende, and epidote. The specific types of detrital particles and their modal contents are shown in Table 1.

Table 1. Detrital particle compositions and calculation of detrital particles.

Detrital Particle Composition	Calculation of Detrital Particles
Q = Qm + Qp = total quartz Qm = monocrystalline quartz Qp = poly-crystalline quartz	QFL%Q = 100 × Q/(Q + F + L) QFL%F = 100 × F/Q + F + L) QFL%L = 100 × L/(Q + F + L) QmFLt%Qm = 100 × Qm/(Qm + F + Lt)
F = total feldspar	
Lt = L + Qp = total lithic content L = Ls + Lm + Lv = total unstable lithic fragments Ls = sedimentary lithic fragments Lm = metamorphic lithic fragments Lv = volcanic lithic fragments	

Note: Q = total monocrystalline (Qm) and polycrystalline (Qp) quartz grains; F = total potassium feldspar (K) and plagioclase (P) grains; L = aphanitic unstable lithic grains, including sedimentary (Ls), metamorphic (Lm) and volcanic (Lv) fragments.

4. Results

4.1. Badaowan Formation

The zircon morphology of the sandstone from the Badaowan Formation is mainly short columnar and irregular grains, with a grain size ranging mostly from 70 to 180 μm and an average size of 126 μm. Most of the zircons are angular, but some are well-rounded, suggesting that they may have been transported over a long distance. Cathodoluminescence (CL) images of the zircons from the Badaowan Formation reveal that most zircons have weak luminescence and typical oscillatory zoning features, indicating that they are mainly magmatic zircons (Figure 3a). The Th/U values of the zircons are generally greater than 0.4, with a range of 0.4–1.6. Zircons with a Th/U ratio greater than 0.4 accounts for approximately 90% of the zircons (Table S2), reflecting their magmatic origin. However, a few zircons exhibit strong cathodoluminescence intensity without zoning or with weak or patchy zoning (Figure 3a), indicating that they are metamorphic zircons. A small number of zircons also have Th/U values less than 0.1, reflecting their metamorphic origin, while others have Th/U values between 0.1 and 0.4 (Table S2).

The valid zircon U-Pb ages from Badaowan Formation are mainly at a range of 200–600 Ma, with a concordance of more than 94% (Figure 4a). Detrital zircon ages display three prominent peaks on the U-Pb age spectrum: 280–330 Ma, 400–440 Ma, and 490–520 Ma (Figure 4b), with a minimum age of 241 ± 2 Ma and a maximum age of 1992 ± 17 Ma (Figure 4a and Table S2). The detrital zircons are mainly derived from the Permian and Devonian, followed by the Cambrian, Carboniferous, and Silurian, with a few Triassic and Proterozoic zircons.

The sandstones from the Badaowan Formation are concentrated in a relatively narrow area. The QmFLt ternary diagram shows that the samples are mainly distributed in the transitional and back-arc source regions, with a few falling on the edges of the accretionary orogenic and transitional arcs, indicating that the source region contains a large number of volcanic rocks and intermediate to acidic intrusive rocks (Figure 5a). On the QtFL ternary diagram, the samples are mainly distributed in the recycled orogenic source region, with a larger proportion of oceanic material and a center of gravity that leans towards oceanic material, indicating that the source region is a collisional suture or fold-thrust belt (Figure 5b). The QmFLt auxiliary diagram indicates that the provenance area is basically a recycled orogenic belt, and the high chert-to-quartz ratio indicates that the provenance area belongs to the collision suture zone or fold-thrust zone (Figure 5c). The QpLvLs ternary diagram suggests that the sandstone samples are almost all from the collisional suture zone and fold-thrust belt source region, based on the high ratio of quartz to feldspar (Figure 5d).

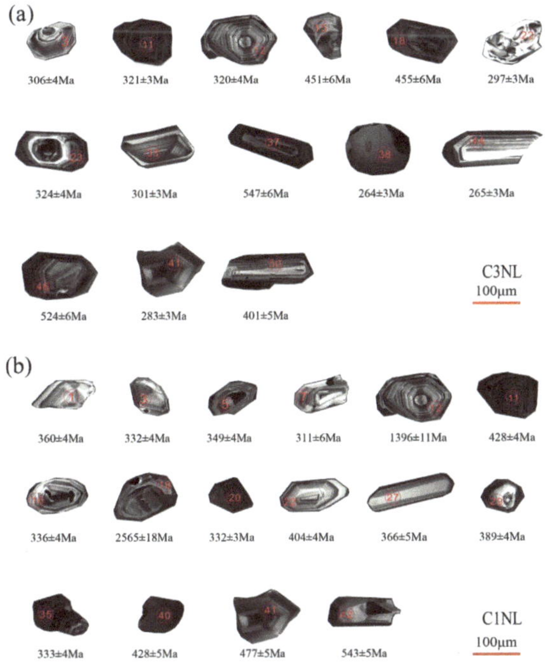

Figure 3. (**a**,**b**) Representative detrital zircons of Badaowan and Sangonghe Formation samples, respectively.

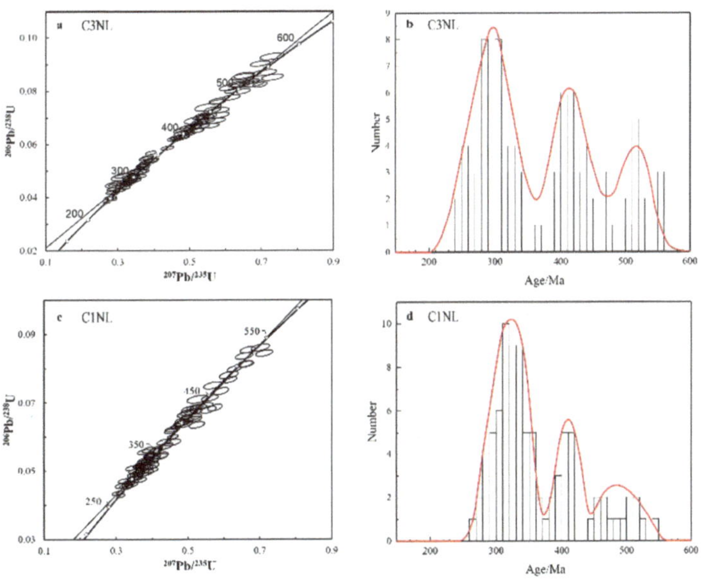

Figure 4. (**a**,**b**) Concordia diagram showing zircon LA-ICPMS U-Pb dating results and histogram of the detrital zircon ages for Badaowan Formation, respectively. (**c**,**d**) Concordia diagram showing zircon LA-ICPMS U-Pb dating results and histogram of the detrital zircon ages for Sangonghe Formation, respectively. Data are shown in Table S2.

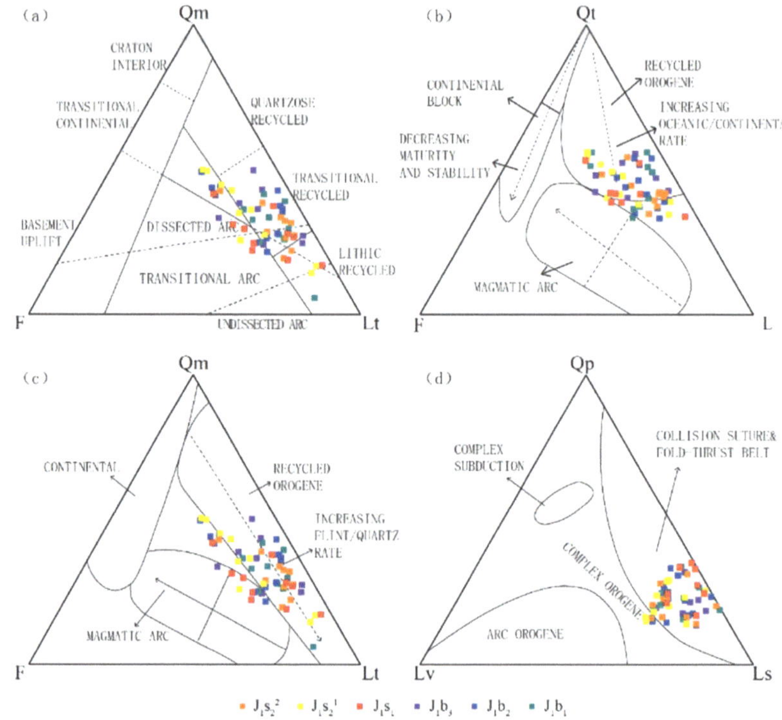

Figure 5. The (**a**) Qm-F-Lt, (**b**) Qt-F-L, (**c**) Qm-F-Lt and (**d**) Qp-Lv-Ls ternary plots of relative proportions of framework-grain compositions for sandstones from Badaowan and Sangonghe Formations in Dongdaohaizi Depression. The data and abbreviations are shown in Table S1.

4.2. Sangonghe Formation

The zircon morphology of the sandstones from the Sangonghe Formation is mainly short-columnar and irregular-granular, with some zircons showing good rounding and indicating long-distance transportation (Figure 3b). The average grain size is 120 μm, and the majority of zircons exhibit typical oscillatory zoning, suggesting magmatic origin with a Th/U ratio greater than 0.4. The Th/U ratio of zircons from the Sangonghe Formation ranges from 0.4 to 1.8, with about 90% of zircons having a ratio greater than 0.4 (Table S2). A small number of metamorphic zircons is also present with a Th/U ratio generally less than 0.1. From the CL images of zircons, it is evident that magmatic zircons exhibit weak cathodoluminescence intensity and typical oscillatory zoning, while metamorphic zircons generally show strong cathodoluminescence intensity with no or weak zoning (Figure 3b).

The valid zircon U-Pb ages range mainly from 200 to 600 Ma, with the youngest age being 270 $\pm$ 3 Ma and the oldest age being 2565 $\pm$ 18 Ma (Figure 4c and Table S2). The concordance of the zircon U-Pb ages is over 93% (Figure 4c). The detrital zircons from the second segment of the Sangonghe Formation are mainly distributed in three intervals, 280–350 Ma (Artinskian–Tournaisian), 380–430 Ma (Givetian–Homerian), and 460–520 Ma (Darrtwilian–Fortunian) (Figure 4d), with the Carboniferous being the main source and followed by the Permian and Devonian. A small number of zircons from the Cambrian, Ordovician, Silurian, and Precambrian are also present.

The QmFLt composition from the Sangonghe Formation sandstones is mainly distributed in the transitional and back-arc provenance areas, as well as the adjacent cutting island arc and transitional island arc margins, indicating a complex provenance with multiple episodes of tectonic activity (Figure 5a). Based on the QmFLt ternary diagram, the provenance of the sandstones is mainly from the retro-arc orogen, indicating that the source area is a volcanic and intermediate intrusive rock-dominated island arc (Figure 5b). According to the QtFL ternary diagram, the samples are mainly distributed in the recycled orogen provenance area, with a large proportion of oceanic source material, suggesting an oceanic subduction background with some mixed provenance characteristics (Figure 5c). From the QpLvLs ternary diagram, the sandstone samples are almost exclusively located in the collisional suture and fold-thrust belt provenance areas, with only a few showing mixed orogenic characteristics (Figure 5d).

5. Discussion

5.1. Provenances of the Dongdaohaizi Depression in Lower Jurassic

Zircon distribution in basin sediments is widespread and has strong stability, which can preserve a large amount of source information. CL images of detrital zircon can clearly reveal information such as internal belts and metamorphic growth, which is beneficial for analyzing the genesis of the parent rock. Combined with zircon U-Pb ages, it can accurately indicate the geological age of the parent rock, and improve the precision of comparison [35–38]. The age combination of zircon in sand bodies supplied by the same provenance must be the same, while there are obvious differences in the age combination of zircon in sand bodies supplied by different provenances. By analyzing the U-Pb ages of detrital zircon in the Badaowan Formation of the study area and determining the genesis of zircon using CL images, the age of the parent rock can be determined, and by comparing with the provenance system, the material source of the sediment can be clarified.

The potential source area of the study area mainly includes the orogenic belts surrounding the basin, such as the eastern Karamaili Mountain, the western Zayir Mountain, and the southern Bogda Mountain. Overall, the zircon ages in the potential source area are mainly concentrated in the range of 200–600 Ma. Among them, the age ranges of the intermediate acidic-acidic eruptive rock, volcanoclastic rock, amphibolite, and other rocks in the Karamaili area include 250–300, 350–380, 400–413, and 475–520, respectively. Additionally, ages 300–350, 388–420, and 485–510 were obtained from rock samples of the Ophiolite belt: basalt, diabase, and gabbro [39–41]. In the Lower Jurassic clastic zircon U-Pb age spectrum of the Dongdaohaizi Depression in this study, three peaks were observed: 280–340 Ma, 400–430 Ma, and 480–520 Ma (Figure 4), which have similar peak intervals with the clastic zircon U-Pb ages in the Karamaili area. It especially contains a peak with high similarity to Karamaili at ~500 Ma, while the Zayir Mountain and the western section of Bogda Mountain do not have this peak age, indicating significant differences. In addition, the Bogda Mountain had not completely uplifted in the Early Jurassic, so the possibility of the Bogda Mountain supplying sources to the study area can be excluded [37,39–44]. Therefore, it is likely that the sediments in the study area represent material brought in by the closure of the Karamaili Ocean.

Additionally, the ternary diagram of the Lower Jurassic sandstones in the Dongdaohaizi Depression shows that the primary source background is composed of transitional reworked orogenic belts, which are consistent with the tectonic background of the Karamaili area (Figure 5). This reflects the processes of island arc collision accompanied by folding and thrusting at the plate margin. Previous studies have found that the Carboniferous was affected by the evolution of the Paleotethys tectonic domain, and the basement was formed by the accretion of volcanic island arcs [45–47]. The most important island arc belt is the Karakorum–Altai–Darbute belt [48,49]. However, the Luliang uplift was a continuous burial background during the Jurassic period and was not a source. The most recent source is the Karakorum-Altai region, so it is the Karakorum-Altai provenance. In addition, the rock debris is mainly composed of volcanic rock debris, and there is a large number of tuffaceous debris, while the proportion of sedimentary and metamorphic rock debris is relatively small, indicating that the main source of the basin-margin Karakorum orogeny is volcanic rocks. Previous studies of sedimentary features and ancient rivers in the northwest indicate that they have shallow-water traction flow characteristics and the ability to transport over long distances [22]. Therefore, we believe that the Lower Jurassic in the Dongdaohaizi Depression received a large amount of Karamaili provenance sediment from the eastern margin of the Junggar basin, which can be used to study the evolutionary process of the Karamaili Ocean.

Based on the ages of sandstone detrital zircons from the Badaowan Formation to the Sangonghe Formation, the three main peaks of zircon age spectra are consistent between them and exhibit a characteristic age range of 490–520 Ma, which is similar to the mother rocks from Karamaili region. Therefore, the study area in the Lower Jurassic is mainly supplied with sediment from a stable source region in the Karamaili Orogenic Belt. It is worth noting that among the zircon ages of the Badaowan Formation, 36 zircons are from the Permian period, accounting for 36% of the total, while 24 zircons are from the Carboniferous period, accounting for 24% of the total (Table S2). The zircon results of the Sangonghe Formation show that 18 zircons are from the Permian period, accounting for 18% of the total, while 50 zircons are from the Carboniferous period, accounting for 50% of the total (Table S2). From the early to late Jurassic, the Permian zircons decreased, while the Carboniferous zircons increased significantly, indicating a trend of younger to older ages in the stratigraphic sequence and suggesting that older rocks from the deeper source region have continued to erode and expose over time. In addition, the sedimentary sequence fines upwards during the lower Jurassic, which may indicate an increasing distance of sources from sediment to the basin caused by basin extension.

Based on the Dickinson provenance analysis, it is suggested that the source area of the Kelamayi Mountains during the Lower Jurassic was subjected to a background of collisional sutures and folding-thrusting, indicating a continuous compression and uplift environment. In previous traditional geological studies in the eastern Junggar Basin, based on methods such as the tectonic stratigraphic layering [50], the changing of sedimentary cycles and migration of sedimentary centers [51,52], and the planar changes of sedimentary facies [53], it was determined that the Early Jurassic was a stage of intra-cratonic sagging. By analyzing the original basin properties through the geometric characteristics of the tectonic structure profile, it was considered that the basin was a sag-type basin caused by peripheral squeezing [48,54–56]. From the burial history of the basin, geochemical analysis, and field geological profile observations, it was suggested that the basin was an intracratonic sag basin [57,58]. Although there are still discrepancies in the evidence obtained in some areas, overall, the Early Jurassic Junggar Basin was a stage of compression and sagging under the control of the Yanshanian movement caused by the closure of the Okhotsk Ocean during the Jurassic–Cretaceous.

In the Zhantenggou area of the eastern Junggar Basin, the unconformable contact between the Jurassic Badaowan Formation and the Triassic Xiaoquangou Group ($J_1b/T_{2\text{-}3}xq$) was discovered, indicating a significant tectonic uplift event during the Late Triassic to Early Jurassic with clear regional tectonic background and geological responses such as folding and faulting, representing the continuous tectonic uplift and erosion of the eastern Junggar orogenic belt [59]. The tectonic uplift event during the Late Triassic to Early Jurassic (210 to 180 Ma) was identified through in-situ apatite fission track analysis [60]. Similarly, the tectonic uplift event during the Middle Triassic to Early Jurassic (240 to 180 Ma) was recognized through seismic profiles combined with apatite fission track data analysis [61]. The evidence above indicates the continuous tectonic uplift and erosion of the eastern Junggar orogenic belt. The continuous tectonic uplift resulted in the continuous exposure of deeper strata in the source area, and older rock debris was transported to the study area.

During the late Triassic to early Jurassic, the grid-like tectonic pattern of the sedimentary layers was filled in and the basin began to contract under the influence of the Indonesian movement, leading to a shrinking lacustrine basin and distribution of swamps [48,62]. In the late Early Jurassic, the basin base subsided again, the lacus expanded, and coal seams developed in the shallow lacus and swamp sedimentary environment of the basin. In the early Middle to Late Jurassic, the mountain-building activity began to resume, and the peripheral thrust belts oscillated back and forth and thrust into the basin, causing the basin to shrink inward and resulting in large-scale peat swamp sedimentation throughout the basin. Based on field outcrop and core observations, various facies markers were identified, and it was found that braid-deltaic plains and frontiers with a large number of branching river channels developed from the Karamaili area of East Junggar to the study area [63–65]. Under this environment, the river channels continuously eroded and destroyed the mountains, causing the drainage area to increase and older layers to be exposed and transported into the study area.

5.2. Tectonic Implications

The analysis results show that all the sample points in the QFL diagram fall into the area of the retro-orogenic belt (Figure 5), and the samples are biased toward the continental source area. Meanwhile, the low content of plagioclase and the high content of quartz (including chert) in the clastic rocks indicate that their source area mainly comes from the retro-orogenic belt. In the QmFLt diagram, the samples are mainly distributed in the source areas of quartz retro-orogenic and transitional retro-orogenic (6-b), which are mostly close to the magma island arc source area, indicating that the source area contains a large number of volcanic rocks and intermediate-acid intrusive rocks. Both the QFL and QmFLt diagrams indicate that the source area is basically a retro-orogenic belt, indicating that the source area belongs to a collisional suture zone or a fold-thrust belt. Based on the regional geological background and the location of the sample points, the source area in the study area is an evolution process of the cut island arc and transitional arc towards the transitional retro-orogenic belt and the detrital retro-orogenic belt (collisional orogenic belt). Combining the characteristics of the peak age of zircon, it reflects the process of multiple oceanic basin extensions, subductions, collisions, and compressional over thrusting in the Eastern Junggar region.

In this study, zircon grains from 460–540 Ma were used to indicate the timing of subduction and accretion of the Karamaili Ocean in the late Cambrian to early Ordovician, with peak magmatism occurring around 520 Ma. In the late Cambrian, the northern branch of the East Junggar Ocean, the paleo-Karamaili Ocean, subducted and was consumed during the Devonian, forming the preliminary Karamaili orogenic belt. Some previous studies also support this conclusion. The age range of 490–520 Ma is unique to the Karamaili region and is likely derived from the reworked part of the Paleozoic strata [53]. Ophiolites in the East Junggar Altay area formed in the Cambrian-early Ordovician and include diorite (489 $\pm$ 4 Ma [66])) and gabbro (503 $\pm$ 7 Ma and 496 $\pm$ 6 Ma [67]). Jian [66] obtained two SHRIMP U-Pb zircon ages of diorite from the Zhaheba ophiolite in the East Junggar region, which was 489 $\pm$ 4 Ma, and they considered this to be the formation age of the ophiolite. Fang [68] obtained a peak age of 514 $\pm$ 5 Ma for the diorite from the Karamaili ophiolite in the East Junggar region using LA-ICP-MS zircon U-Pb dating. Combining the ages of the Zhaheba and Altay ophiolites obtained in previous studies (489 $\pm$ 4 Ma and 503 $\pm$ 7 Ma [66,67]), the formation age of the Beitaishan ophiolite was determined to be 494 $\pm$ 3 Ma. QFL diagram analysis results indicate that the samples all fall within the orogenic belt of reworked strata (Figure 5), with samples tending towards an oceanic source, while feldspar content is low and quartz (including chert) content is high, indicating that the Karamaili region is mainly derived from the reworked orogenic belt. It is a mantle material from oceanic subduction to continental arc with volcanic island arc characteristics, forming recycled oceanic debris.

The latest zircon dating results from this study constrain the extension and convergence processes of the limited ocean basin of Karamaili Ocean from the Late Silurian to Early Carboniferous periods to be within 360–440 Ma. Jian [66] obtained two groups of SHRIMP U-Pb ages of 406 $\pm$ 4 Ma from the zircons of gabbros in the Zhahaba ophiolite complex in eastern Junggar and suggested that the occurrence of this age group represented a magmatic thermal event. Zhang [69] obtained a zircon age of 409 $\pm$ 9 Ma from the gabbros in Zhahaba, and based on previous research results, they concluded that the Zhahaba-Almantai region in eastern Junggar contained Middle Ordovician to Early Devonian ophiolites, reflecting the background of continuous oceanic subduction. Fang [68] studied the formation age of the Karamaili ophiolite in eastern Junggar and obtained a peak zircon age of 406 ± 2 Ma by LA-ICP-MS U-Pb dating of gabbros, which they believed represented the age of the Karamaili ophiolite formation. Liu [70] also found a younger group of zircons (412 ± 13 Ma) in the gabbros, which have ages similar to the granite porphyry intrusions (407 ± 2 Ma) in the nearby basalts. They speculated that these ages may be related to magmatic thermal events in the island arc area under compression during the convergence and closure of the Zhahaba-Almantai-North Tashan ocean basin. The geochemical characteristics of the granite porphyries discovered in the North Tashan ophiolitic mélange suggest that they may have formed in an island arc area with thickened crust during the subduction closure of the ocean basin, and therefore, the formation age of the granite porphyries at 407 ± 2 Ma marks the convergence and closure of the Almantai ocean basin, entering the collisional tectonic evolution stage from the Late Silurian to Early Devonian periods. The triangular diagrams reflecting the re-collision orogen indicate that the source area belongs to a collisional suture zone. The obtained age peak at ~400 Ma is similar to the results of these studies, reflecting the tectonic events of convergence and closure from the Late Silurian to Early Devonian periods, followed by the collisional tectonic evolution stage. During this time, volcanic and magmatic activities were relatively active, and the volcanic rocks of the island arc were weathered and eroded to form clastic rocks through nearby transport and accumulation.

The zircon age range of 240–340 Ma indicates the complete disappearance of the Eastern Junggar Ocean during the late Carboniferous to the Triassic period [71]. The LA-ICP-MS zircon U-Pb age of the Laoyemiao rhyolite shows that the crystallization age of the rhyolite is (311.6 ± 3.1) Ma, indicating its formation in the late Carboniferous. The rhyolite has A-type granite characteristics, with high Ga/Al, Zr, Nb, Ga, Y, and Ce, low Sr, and a significant negative Eu anomaly. It is classified as an A2-type granite and formed in a post-collisional environment, marking the complete closure of the basin in the eastern Junggar area [72]. Detailed petrological, geochemical, zircon U-Pb chronology and Sr-Nd-Pb isotope composition studies were conducted on the late Carboniferous bimodal volcanic in the Karamaili orogenic belt in eastern Junggar. The volcanic rocks were considered to have formed in the early late Carboniferous at 320.2 ± 4.2 Ma. The bimodal volcanic rocks were produced in a post-collisional tectonic environment, where the sinking of the oceanic crust caused an upwelling of the asthenosphere, resulting in partial melting of the overlying mantle, and the intrusion of the lower crust caused partial melting, leading to the eruption of the bimodal volcanic rocks. The occurrence of the bimodal volcanic rocks marks the end of the orogenic process in the Karamaili area of eastern Junggar [73]. The age and petrogeochemistry of the alkaline granite in the Ashudas area of the northeast margin of eastern Junggar were investigated by SHRIMP zircon U-Pb dating, indicating a crystallization age of (319.3 ± 2.5) Ma, with typical A-type granite characteristics. Its formation marked the end of the orogenic phase in the northern margin of eastern Junggar and the beginning of an intra-plate extensional environment. The Tawibask granite porphyry has high SiO2, quasi-aluminous-weakly peraluminous, and calc-alkaline I-type granite characteristics, with weighted average ages of 301.3 ± 2.5 Ma (MSWD = 0.33) and 310.7 ± 3.6 Ma (MSWD = 0.75), indicating that the late Carboniferous (310~301 Ma) in the eastern Junggar region was in a post-collisional tectonic setting with mixed crust-mantle and incremental growth characteristics [74]. The Dickinson triangle reflects a retro-arc orogenic belt, indicating typical collisional suture and fold-thrust belt source characteristics, indicating that the Karamaili area of eastern Junggar has entered a comprehensive and sustained compressive uplift background. This study demonstrates a peak in the Late Carboniferous to Early Permian, indicating the end of the eastern Junggar orogeny. During the Jurassic period, the pre-existing multiple sets of volcanic rock weathering debris and peripheral volcanic debris rocks were uplifted to the surface, weathered, and eroded again, becoming the source rocks of Jurassic detrital rocks [49,75,76]. In the Jurassic, the closure of the Paleo-Tethys Ocean and the opening of the Neo-Tethys Ocean affected western China and central Asia, which were in the stage of basin extension and faulting. Under this background, the Karamaili area continued to uplift in the Early Jurassic, developing mountain basins, and providing a large amount of detrital material for the area [77,78].

In summary, the detrital zircon age spectrum of the Early Jurassic in the Dongdaohaizi Depression fully reflects the tectonic characteristics of the provenance area in the Karamaili Mountains region. The ages record the East Junggar Ocean's extension-collapse cycles during the Cambrian to Early Silurian; the limited Karamaili Ocean's subduction and convergence during the Late Silurian to Early Carboniferous; and the complete disappearance of the Paleo-Asian Ocean during the Late Carboniferous to Triassic. Then the North Tianshan Ocean basin underwent a transformation from an oceanic basin to a foreland basin during the convergent and compressive stages of the extension-collapse cycles. This comprehensive record witnessed the complete evolution process of the Paleo-Asian Ocean and provides valuable insights into the tectonic evolution of the study area.

6. Conclusions

(1) The sedimentary characteristics of the Lower Jurassic in the Dongdaohaizi Depression of the Junggar Basin indicate a transitional and episodic provenance from a collisional orogen with an oceanic-continental subduction background.

(2) The age ranges of the Badaowan Formation are 240–340 Ma, 360–440 Ma, and 480–540 Ma, while those of the Sangonghe Formation are 260–340 Ma, 370–440 Ma, and 460–520 Ma. The youngest detrital zircon age is 241 ± 2 Ma, indicating the earliest limit of sedimentation in the Early Permian.

(3) During the Early Jurassic, the consistent subduction of the residual East Junggar Ocean induced continuous uplift in the Karamaili region, resulting in an increasing exposure of deep-seated rocks to provide sedimentary material.

(4) The Karamaili Ocean, as part of the Paleo-Asian Ocean, experienced three evolutionary stages: Cambrian-Early Silurian (460–540 Ma), Late Silurian-Early Carboniferous (360–440 Ma), and Late Carboniferous-Triassic (240–340 Ma).

Supplementary Materials: The following supporting information can be downloaded at: https://www.mdpi.com/article/10.3390/jmse11071375/s1, Table S1: Components and contents of Badaowan and Sangonghe Formation samples in the Dongdaohaizi Depression; Table S2: LA-ICP-MS zircon U-Pb dating results of sandstones from the Badaowan and Sangonghe formations in Dongdaohaizi Depression, Junggar basin.

Author Contributions: Conceptualization, Y.G. and F.L.; methodology, G.Z. and R.G.; writing—original draft preparation, Y.G. and F.L.; writing—review and editing, Y.G., S.L. (Songtao Li) and S.L. (Shengqian Liu); investigation, Z.Z., S.C. and Z.X.; visualization, L.L. and H.Z. All authors have read and agreed to the published version of the manuscript.

Funding: This research was funded by Sinopec Science and Technology Projects (P21077-2; YKK2212; YJQ2301).

Institutional Review Board Statement: Not applicable.

Informed Consent Statement: Not applicable.

Acknowledgments: We acknowledge the Sinopec Science and Technology Projects (P21077-2; YKK2212; YJQ2301) for funding this study. We thank Xiaoliang We for the fantastic devices. Thanks to the 4 anonymous reviewers for their valuable suggestions.

Conflicts of Interest: The authors declare no conflict of interest.

References

1. Jahn, B.-M.; Wu, F.; Chen, B. Massive granitoid generation in Central Asia: Nd isotope evidence and implication for continental growth in the Phanerozoic. *Epis. J. Int. Geosci.* **2000**, *23*, 82–92. [CrossRef] [PubMed]
2. Kroener, A.; Rojas-Agramonte, Y. The Altaids as Seen by Eduard Suess, and Present Thinking on the Late Mesoproterozoic to Palaeozoic Evolution of Central Asia. *Austrian J. Earth Sci.* **2014**, *107*, 156–168.

3. Kovalenko, V.I.; Yarmolyuk, V.V.; Kovach, V.P.; Kotov, A.B.; Kozakov, I.K.; Salnikova, E.B.; Larin, A.M. Isotope provinces, mechanisms of generation and sources of the continental crust in the Central Asian mobile belt: Geological and isotopic evidence. *J. Asian Earth Sci.* **2004**, *23*, 605–627. [CrossRef]
4. Safonova, I.Y. Geochemical evolution of intraplate magmatism in the Paleo-Asian Ocean from the Late Neoproterozoic to the Early Cambrian. *Petrology* **2008**, *16*, 492. [CrossRef]
5. Xiao, W.; Santosh, M. The western Central Asian Orogenic Belt: A window to accretionary orogenesis and continental growth. *Gondwana Res.* **2014**, *25*, 1429–1444. [CrossRef]
6. Xiao, W.; Han, C.; Yuan, C.; Sun, M.; Lin, S.; Chen, H.; Li, Z.; Li, J.; Sun, S. Middle Cambrian to Permian subduction-related accretionary orogenesis of Northern Xinjiang, NW China: Implications for the tectonic evolution of central Asia. *J. Asian Earth Sci.* **2008**, *32*, 102–117. [CrossRef]
7. Xiao, W.; Windley, B.F.; Badarch, G.; Sun, S.; Li, J.; Qin, K.; Wang, Z. Palaeozoic accretionary and convergent tectonics of the southern Altaids: Implications for the growth of Central Asia. *J. Geol. Soc.* **2004**, *161*, 339–342. [CrossRef]
8. Sengör, A.M.C.; Natal'In, B.A. Turkic-type orogeny and its role in the making of the continental crust. *Annu. Rev. Earth Planet. Sci.* **1996**, *24*, 263–337. [CrossRef]
9. Şengör, A.M.C.; Natal'in, B.A. Phanerozoic analogues of Archaean oceanic basement fragments: Altaid ophiolites and ophirags. *Dev. Precambrian Geol.* **2004**, *13*, 675–726.
10. Yakubchuk, A.; Seltmann, R.; Shatov, V.; Cole, A. The Altaids: Tectonic evolution and metallogeny. *SEG Discov.* **2001**, *1*, 1–14. [CrossRef]
11. Allen, M.B.; Windley, B.F.; Zhang, C. Palaeozoic collisional tectonics and magmatism of the Chinese Tien Shan, central Asia. *Tectonophysics* **1993**, *220*, 89–115. [CrossRef]
12. Windley, B.F.; Allen, M.B.; Zhang, C.; Zhao, Z.Y.; Wang, G.R. Paleozoic accretion and Cenozoic redeformation of the Chinese Tien Shan Range, central Asia. *Geology* **1990**, *18*, 128–131. [CrossRef]
13. Şengör, A.M.C.; Natal'in, B.A.; Burtman, V.S. Evolution of the Altaid tectonic collage and Palaeozoic crustal growth in Eurasia. *Nature* **1993**, *364*, 299–307. [CrossRef]
14. Coleman, R.G. Continental growth of northwest China. *Tectonics* **1989**, *8*, 621–635. [CrossRef]
15. Liu, X.; Xiao, W.; Xu, J.; Castillo, P.R.; Shi, Y. Geochemical signature and rock associations of ocean ridge-subduction: Evidence from the Karamaili Paleo-Asian ophiolite in east Junggar, NW China. *Gondwana Res.* **2017**, *48*, 34–49. [CrossRef]
16. Li, D.; He, D.; Sun, M.; Zhang, L. The Role of Arc-Arc Collision in Accretionary Orogenesis: Insights From ~320 Ma Tectono-sedimentary Transition in the Karamaili Area, NW China. *Tectonics* **2020**, *39*, e2019TC005623. [CrossRef]
17. Zhang, B.; Chen, C.; Gong, X.; Yalikun, Y.; Kaheman, K. The Kamusite A2-type granites in the Karamaili tectonic belt, Xinjiang (NW China): Tracing staged postcollisional delamination in the eastern Junggar. *Geol. Mag.* **2020**, *158*, 723–748. [CrossRef]
18. Gilder, S.; Zhao, X.; Coe, R.; Meng, Z.; Courtillot, V.; Besse, J. Paleomagnetism and tectonics of the southern Tarim basin, northwestern China. *J. Geophys. Res. Solid Earth* **1996**, *101*, 22015–22031. [CrossRef]
19. Gao, Y.; Zhang, S.; Zhao, H.; Ren, Q.; Yang, T.; Wu, H.; Li, H. North China block underwent simultaneous true polar wander and tectonic convergence in late Jurassic: New paleomagnetic constraints. *Earth Planet. Sci. Lett.* **2021**, *567*, 117012. [CrossRef]
20. Sharps, R.W. *Paleomagnetism and Tectonics of Northwest China*; Stanford University: Stanford, CA, USA, 1990.
21. Zhang, L.; Dengfa, H.; Zejun, Y.; Di, L.I. Tectonic relationship between the Kelameili range and the Dajing depression: Insights into the Carboniferous tectonic-sedimentary framework. *Pet. Explor. Dev.* **2020**, *47*, 30–45. [CrossRef]
22. Zheng, S. Sedimentary Pattern of the Shallow-Water Delta in the Sangonghe Formation of Central Junggar Basin and Its Significance for Hydrocarbon Exploration. *Spec. Oil Gas Reserv.* **2019**, *26*, 87–93.
23. Chew, D.M.; Spikings, R.A. Geochronology and Thermochronology Using Apatite: Time and Temperature, Lower Crust to Surface. *Elements* **2015**, *11*, 189–194. [CrossRef]
24. Feng, Y.; Coleman, R.G.; Tilton, G.; Xiao, X. Tectonic evolution of the west Junggar region, Xinjiang, China. *Tectonics* **1989**, *8*, 729–752. [CrossRef]
25. Buckman, S.; Aitchison, J.C. Tectonic evolution of Palaeozoic terranes in west Junggar, Xinjiang, NW China. *Geol. Soc. Lond. Spec. Publ.* **2004**, *226*, 101–129. [CrossRef]
26. Gao, R.; Xiao, L.; Pirajno, F.; Wang, G.-c.; He, X.-x.; Yang, G.; Yan, S.-w. Carboniferous–Permian extensive magmatism in the West Junggar, Xinjiang, northwestern China: Its geochemistry, geochronology, and petrogenesis. *Lithos* **2014**, *204*, 125–143. [CrossRef]
27. Chen, Y.; Xiaogan, C.; Han, Z.; Li, C.; Ma, Y.; Wang, G. Fault characteristics and control on hydrocarbon accumulation of middle-shallow layers in the slope zone of Mahu sag, Junggar Basin, NW China. *Pet. Explor. Dev.* **2018**, *45*, 1050–1060. [CrossRef]
28. Andersen, T. Correction of common lead in U-Pb analyses that do not report 204Pb. *Chem. Geol.* **2002**, *192*, 59–79. [CrossRef]
29. Sircombe, K.N. Tracing provenance through the isotope ages of lit-toral and sedimentary detrital zircon, eastern Australia. *Sediment. Geol.* **1999**, *124*, 47–67. [CrossRef]
30. Sudar, V.; Aljinović, D.; Smirčić, D.; Barudžija, U. Composition and provenance of Gröden sandstone from the Velebit Mts. *Rudarsko-Geološko-Naftni Zbornik* **2016**, *31*, 53–67. [CrossRef]
31. Ludwig, K. RISOPLOT 2.49: A Geochronological Toolkit for Microsoft Excel. *Berkeley Geochronol. Cent. Spec. Publ.* **2001**, *1a*, 1–58.
32. Li, Z.; Li, R.; Sun, S.; Jiang, M.; Zhang, W. Detrital composition and provenance tectonic attributes of Jurassic sandstones, south Hefei basin. *Acta Petrol. Sin.* **1999**, *15*, 438–445.

33. Ingersoll, R.V.; Bullard, T.F.; Ford, R.L. The effect of grain size on detrital modes: A testof the Gazzi-Dickinson point-counting method. *J. Sediment. Petrol.* **1984**, *54*, 103–116.
34. Dickinson, W.R. Interpreting detrital models of graywacke and arkose. *J. Sediment Petrol.* **1970**, *40*, 695–707.
35. Guangwen, H.; Jiayong, P.; Guangnan, H.; Fei, X.; Dehai, W.; Fujun, Z. Zircon Chronological Characteristics and Geological Significance of Sedimentary Rocks of Badaowan Formation in Mengqiguer Area, Xinjiang. *Xinjiang Geol.* **2021**, *39*, 210–219.
36. Ni, M. Source Analysis and Tectonic Significance of Middle-lowermesozoic in Northwest Margin of Junggar Basin: Evidence Fromdetrital Zircon U-pb Age. Master's Thesis, China University of Petroleum (Beijing), Beijing, China, 2019.
37. Wang, X.; Cai, K.; Sun, M.; Xiao, W.; Xia, X.; Wan, B.; Bao, Z.; Wang, Y. Two contrasting late Paleozoicmagmatic episodes in the northwestern Chinese Tianshan Belt, NWChina: Implication for tectonic transition from plate convergence tointra-plate adjustment during accretionary orogenesis. *J. Ofasian Earth Sci.* **2017**, *153*, 118–138. [CrossRef]
38. Hartanto, P.; Alam, B.Y.C.S.; Lubis, R.F.; Hendarmawan, H. The Origin and Quality of the Groundwater of the Rawadanau Basin in Serang Banten. *Indonesia. Rud. Geološko Naft. Zb.* **2021**, *36*, 11–24. [CrossRef]
39. Liu, X. The Evolution of Dynamic Structural Belt in Kalamaili Mountains in the Junggar Basin., 2016. Ph.D. Thesis, Southwest Petroleum University, Chengdu, China, 2016.
40. Zhang, Y.; Pe-Piper, G.; Piper, D.J.W.; Guo, Z. Early Carboniferous collision of the Kalamaili orogenic belt, North Xinjiang, and its implications: Evidence from molasse deposits. *Bulletin* **2013**, *125*, 932–944. [CrossRef]
41. Zhang, Y.; Guo, Z.; Pe-Piper, G.; Piper, D.J.W. Geochemistry and petrogenesis of Early Carboniferous volcanic rocks in East Junggar, North Xinjiang: Implications for post-collisional magmatism and geodynamic process. *Gondwana Res.* **2015**, *28*, 1466–1481. [CrossRef]
42. Ji, H.; Tao, H.; Wang, Q. Early to middle Jurassic tectonicevolution of the Bogda Mountains, Northwest China: Evidence fromsedimentology and detrital zircon geochronology. *J. Asian Earth Sci.* **2018**, *153*, 57–74. [CrossRef]
43. Han, B.F.; Guo, Z.J.; Zhang, Z.C. Age, geochemistry, andtectonic implications of a late Paleozoic stitching pluton in the northTian Shan suture zone, western China. *Geol. Soc. Ofamerica Bull.* **2010**, *122*, 627–640. [CrossRef]
44. Wali, G.; Wang, B.; Cluzel, D. Carboniferous-Early Permianmagmatic evolution of the Bogda Range (Xinjiang, NW China): Implications for the Late Paleozoic accretionary tectonics of the SWCentral Asian orogenic belt. *J. Asian Earth Sci.* **2018**, *153*, 238–251. [CrossRef]
45. Filippova, I.B.; Bush, V.A.; Didenko, A.N. Middle Paleozoic subduction belts: Theleading factor in the formation of the Central Asian fold-and-thrust belt. *Russ. J. Earth Sci.* **2001**, *3*, 405–426. [CrossRef]
46. Mossakovsky, A.A.; Ruzhentsev, S.V.; Samygin, S.G.; Kheraskova, T.N. The CentralAsian fold belt: Geodynamic evolution and formation history. *Geotectonics* **1993**, *26*, 455–473.
47. Xiao, W.J.; Windley, B.F.; Allen, M.F.; Han, C.M. Paleozoic multiple accretionary andcollisional tectonics of the Chinese Tianshan orogenic collage. *Gondwana Res.* **2013**, *23*, 1316–1341. [CrossRef]
48. He, D.; Zhang, L.; Wu, S.; Li, D.; Zhen, Y. Tectonic evolution stages and features of the Junggar Basin. *Oil Gas Geol.* **2018**, *39*, 845–861.
49. Li, D.; He, D.; Fan, C.; Xiang, K.; Jin, L. Early Permian Post-collisional Magmatic Events, East Junggar: Constraints From Zircon Shrimp U-pb Age, Geochemistry And Hf Isotope Of Rhyolite In The Yundukala Area. *Acta Petrol. Sin.* **2013**, *1*, 317–337.
50. Chen, X.; Lu, H.; Shu, L.; Wang, H.; Zhang, G. Study on Tectonic Evolution of Junggar Basin. *Geol. J. China Univ.* **2002**, *3*, 257–267.
51. Chen, F.; Zhang, G.; Chen, Z. Unconformity Analysis and Its Significance in The Study of Continental Basin Tectonics. *Geoscience* **2004**, *3*, 269–275.
52. Qu, G.-s.; MA, Z.-j.; Chen, X.-f.; Li, T.; Zhang, N. On structures and evolutions in Junggar Basin. *Xinjiang Pet. Geol.* **2009**, *30*, 1–5.
53. Zhang, C.; He, D.; Wu, X.; Shi, X. Formation and evolution of multicycle superimposed basins in Junggar Basin. *China Pet. Explor.* **2006**, *11*, 47.
54. He, D.; Zhou, L.; Tang, Y.; Wu, X.; Du, S. Characteristics of unconformity between the Xishanyao Formation and Toutunhe Formation of Middle Jurassic in Junggar Basin and its significance in petroleum exploration. *J. Palaeogeogr.* **2007**, *4*, 387–396.
55. Wu, K.; Paton, D.; Zha, M. Unconformity structures controlling stratigraphic reservoirs in the north-west margin of Junggar basin, North-west China. *Front. Earth Sci.* **2013**, *7*, 55–64. [CrossRef]
56. Han, X.; Ayiguli, W.U. Tectonic Framework and Stucutral Patterns in Peripheral Eastern Junggar Basin. *Xinjiang Pet. Geol.* **2001**, *22*, 202.
57. Wang, S.; Zou, C.; Hou, L.; Wei, Y.; Luo, X.; Guo, Z. Petroleum geological characteristics of gas accumulation in carboniferous volcanics and prospect for gas exploration, Eastern Junggar Basin. *Earth Sci. Front.* **2013**, *2*, 226–236.
58. Cai, Z.; Chen, F.; Jia, Z. Types and Tectonic Evolution of Junger Basin. *Earth Sci. Front.* **2003**, *7*, 431–440.
59. Nie, L.; Ma, J.; Tang, X. Meso-Cenozoic tectonic events and their petroleum geological significance in Zhangpenggou area, eastern Junggar Basin. *Lithol. Reserv.* **2023**, *35*, 81–91.
60. Song, J.; Qin, M.; Cai, Y.; Guo, Q.; He, Z.; Liu, Z.; Cao, X.; Chen, Z. Uplift-Denudation of Orogenic Belts Control on the Formation of Sandstone Type Uranium(U) Deposits in Eastern Junggar, Northwest China: Implications from Apatite Fission Track (AFT). *Earth Sci.* **2019**, *44*, 3910–3925.
61. Wu, Z.; Han, X.; JI, H. Mesozoic-Ceno-zoic tectonic events of eastem Junggar Basin, NW China andtheir significance for uranium mineralization; Insights fromseismic profiling and AFT dating analysis. *Ore Geol. Rev.* **2021**, *139*, 104488. [CrossRef]

62. Xinzhi, Z.; Linhao, F.; Tao, W.; Yunfa, M.; Mingzhen, Z.; Xuelian, W.; Peizong, L.; Xiujuan, H.; Ruoyuan, Q.; Han, Y. Palynological assemblages and palaeoclimate across the Triassic-Jurassic boundary in the Haojiagou section, southern Junggar Basin. *Chin. J. Geol.* **2022**, *57*, 1161–1176.
63. Song, C.; Cheng, C.; Qiao, Y. Sedimentary Characteristics of Jurassic Toutunhe Formation in Block 4 Central Junggar Basin. *Adv. Geosci.* **2019**, *9*, 1229–1236. [CrossRef]
64. Xia, T. Reservoir Characteristics of the Badaowan Formation in Central Junggar Basin. Master's Thesis, China University of Petroleum (Qingdao), Qingdao, China, 2018.
65. Zhu, X.; Li, S.; Wu, D.; Zhu, S.; Dong, Y.; Zhao, D.; Wang, X.; Zhang, Q. Sedimentary characteristics of shallow-water braided delta of the Jurassic, Junggar basin, Western China. *J. Pet. Sci. Eng.* **2017**, *149*, 591–602. [CrossRef]
66. Jian, P.; Liu, D.-y.; Zhang, Q.; Zhang, F.-q.; Shi, Y.-r.; Shi, G.-h.; Zhang, L.-q.; Tao, H. Shrimp Dating of Ophiolite And Leucocratic Rocks Within Ophiolite. *Earth Sci. Front.* **2003**, *4*, 439–456.
67. Xiao, W.; Windley, B.F.; Yan, Q.; Qin, K.; Chen, H.; Yuan, C.; Sun, M.; Li, J.; Sun, S. SHRIMP zircon age of the Aermantai ophiolite in the North Xinjiang area, China and its tectonic implications. *Acta Geol. Sin.* **2006**, *80*, 32–37.
68. Fang, A.; Wang, S.; Zhang, J.; Zang, M.; Fang, J.; Hu, J. The U-Pb ages of zircons from the gabbro in the Kalamaili ophiolite, North Xinjiang and its tectonic significances. *Chin. J. Geol.* **2015**, *50*, 140–154.
69. Zhang, Y. The Study About the Composition, the Structure, Theemplacement Mechanism and Significance of Tectonic of the Armantai Ophiolite Melange, East Junggar, Xinjiang, China. Ph.D. Thesis, Chang'an University, Xi'an, China, 2013.
70. Liu, Y. Early Paleozoic ophiolite SHRIMP U-Pb dating and its Genetic Types & Tectonic Setting in eastern Junggar, Xinjiang. Ph.D. Thesis, China University of Geosciences, Beijing, China, 2016.
71. Xiong, S.; Zhang, Z.; Li, G.; Liu, R.; Hua, X.; Zhao, F.; Zhou, P.; Li, G. Zircon U-Pb Dating, Geochemical Characteristics of Alkali-granites in Laoyemiao Area, Eastern Junggar, and Geological Significance. *Geol. Rev.* **2019**, *65*, 221–231.
72. Luo, T.; Chen, S.; Liao, Q.A.; Chen, J.; Hu, C.; Wang, F.; Tian, J.; Wu, W. Geochronology, Geochemistry and Geological Significance of the Late Carboniferous Bimodal Volcanic Rocks in the Eastern Junggar. *Earth Sci.* **2016**, *41*, 1845–1862.
73. Alimujiang, A.; Gong, X.; Li, X.; Zheng, F.; Shang, M.; Muzhapaer, M. LA-ICP-MS Zircon U-Pb Dating, Geochemical Characteristics and Tectonic Significance of Ashudas Late Carboniferous Alkaline Granite in Northeast Margin of Eastern Junggar, Xinjiang. *Xinjiang Geol.* **2020**, *38*, 184–191.
74. Wei, T.; Rukui, L.; Ling, G.; Yang, L.; Gang, H.; Feng, Y. Zircon U-Pb dating, geochemistry and tectonic significance of the volcanic rocks in Liushugou Formation, East Junggar. *Chin. J. Geol.* **2022**, *57*, 782–808.
75. Jin, J.; Wang, J.; Liu, J.; Tuxun'ayi, T.; Ji, H.; Jia, H.; Fang, C.; Li, C. Characteristic and distribution of lithology-lithofacies of deep Carboniferous reservoirs in the Ke-Bai fractured zone of the northwestern margin in Junggar Basin. *J. Northwest Univ.* **2018**, *48*, 238–245.
76. Liu, J.; Yang, Y. Comparison of Sedimentary Characteristics on Both Sides of Karamaili Ophiolite Belt in Eastern Junggar, Xinjiang. *Adv. Geosci.* **2020**, *10*, 758–770. [CrossRef]
77. Zheng, M.; Tian, A.; Yang, T.; He, W.; Chen, L.; Wu, H.; Ding, J. Structural evolution and hydrocarbon accumulation in the eastern Junggar Basin. *Oil Gas Geol.* **2018**, *39*, 907–917.
78. Li, D.; He, D.; Tang, Y.; Fan, C.; Kong, Y. Genesis of Early Carboniferous volcanic rocks of the Di'nan uplift in Junggar Basin: Constraints to the closure time of Kalamaili ocean. *Acta Petrol. Sin.* **2015**, *28*, 2340–2354.

Journal of
Marine Science and Engineering

MDPI

Article

Ordovician Tsunamis: Summary of Hypotheses and Implications for Geoheritage Resources

Dmitry A. Ruban [1,*] and Natalia N. Yashalova [2]

1 Department of Organization and Technologies of Service Activities, Institute of Tourism, Service and Creative Industries, Southern Federal University, 23-ja Linija Street 43, Rostov-on-Don 344019, Rostov Oblast, Russia
2 Department of Economics and Management, Business School, Cherepovets State University, Sovetskiy Avenue 10, Cherepovets 162600, Vologda Region, Russia; natalij2005@mail.ru
* Correspondence: ruban-d@mail.ru

Abstract: Ordovician tsunamis have been hypothesized for more than 25 years, but the related knowledge is yet to be systematized. The published sources bearing the pieces of this knowledge were collected in the course of the specialized bibliographical survey, and these pieces of evidence were summarized and interpreted with special attention to the spatiotemporal distribution of Ordovician tsunamis. It is found that the latter were reported from many places of the world (24 localities are established), which represent the Gondwana periphery, some isolated continental blocks, and terranes. Tsunamis were hypothesized for all epochs of the considered period, but the evidence is especially numerous for the Middle Ordovician. The degree of certainty of the interpretations of these tsunamis is chiefly moderate. It appears that only a tiny portion of Ordovician tsunamis is known, but the amount of available information is expected, taking into account the possibility of finding very ancient tsunamis. The outcomes of this study contribute to a better awareness of the world's geoheritage resources. The established localities representing Ordovician tsunamis can be considered potential geosites, two of which are promising start points for further, field-based research.

Keywords: early Paleozoic; geosites; natural hazards; resources management; systematic review

Citation: Ruban, D.A.; Yashalova, N.N. Ordovician Tsunamis: Summary of Hypotheses and Implications for Geoheritage Resources. *J. Mar. Sci. Eng.* **2023**, *11*, 1764. https://doi.org/10.3390/jmse11091764

Academic Editor: Markes E. Johnson

Received: 27 August 2023
Revised: 7 September 2023
Accepted: 8 September 2023
Published: 9 September 2023

1. Introduction

The Ordovician, which was the second Paleozoic period with three epochs and seven stages, started at 485.4 Ma and ended at 443.8 Ma [1–3]. It was marked by outstanding biotic changes such as the long-term biodiversification event [4–6] and the strong mass extinction [7–9] and the notable abiotic phenomena such as the very high global sea level [10–12], the coexistence of the single supercontinent (Gondwana) with several smaller continents and terranes [13–15], the outstanding palaeotemperature shifts [16–20], and the extraterrestrial impacts [21,22]. However, many aspects of the Ordovician world are yet to be fully realized, and the related information needs regular systematization and reinterpretation.

In the modern world, tsunamis are rather common events with different triggers [23]. Apparently, they were similarly common and diverse in geological history, including the Ordovician. They can be interpreted on the basis of the careful analysis of sedimentary structures or the modeling of tsunami-triggering processes. Nonetheless, the evidence of pre-Quaternary tsunamis is very limited due to their poor preservation, the methodological difficulties, and, probably, the unawareness of geologists about their indicators in the sedimentary successions [24–28]. In particular, Pratt [29] stated that the tsunami horizons could occur more widely in the stratigraphical successions than expected. If so, the question is how to "catch" them. Ordovician tsunamis have been hypothesized time and time again (see the literature review below), but the related information remains scattered in the huge volume of literature sources. This is why it appears very important to gather it together and systematize it. Moreover, Ruban [27] explained that it is really important to pay attention to the quality (better put, certainty) of the evidence of ancient tsunamis. In some cases, they were proposed too briefly and without detailed argumentation, or the researchers were unable to choose between tsunamis and severe storms. In some other cases, clear arguments for the existence of tsunamis were presented. Indeed, all judgments of pre-Quaternary tsunamis are hypothetic, but some are more hypothetic than others due to the lack of strong evidence.

The scarcity of the knowledge on very ancient tsunamis makes the localities, critical to their studies, unique and important to research, education, and tourism, i.e., geoheritage sites (geosites sensu [30,31]). Tsunamis impress the public, and this kind of natural hazard is well popularized and related to tourist activities [32–34]. As a result, the idea of tsunamis from the Deep Past should be both interesting and clear to even non-specialists. In other words, their localities are potential geosites and geotourist attractions. A better understanding of pre-Quaternary tsunamis contributes to the exploration of geoheritage resources. Taking into account the demand for the latter in contemporary society, the need for these resources makes the ancient tsunami studies even more urgent.

The objective of this present paper is to characterize Ordovician tsunamis in light of the evidence collected from the literature. Their spatiotemporal distribution is emphasized. The outcomes of such a study seem to be important not only theoretically but also in regard to finding new, world-class geosites. This study focuses on a single period of the Paleozoic Era for reasons such as the representativeness and the heterogeneity of the literature and the potential for event correlation, which differ between the periods. Moreover, the Ordovician was the really specific (see above) time interval of the Earth's evolution, and thus, it deserves special attention. It also appears reasonable to synthesize the knowledge step by step (period by period) to avoid delays and the related outdating of the information.

2. Materials and Methods

This present work is based on the existing literature and its systematic review. The main procedures were as follows. First, the "Scopus" bibliographical database, which boasts significant completeness and historical "depth" [35–37], was employed. The published sources mentioning Ordovician (and its subdivisions) and tsunamis (also tsunamites) in their titles, abstracts, and keywords were collected. Second, the opportunities of full-text search of such articles with the major publishers' databases (e.g., "ScienceDirect" by "Elsevier" and "Springer Link" by "Springer") were used. A direct search on Internet was also undertaken. Third, the lists of references in the collected sources were analyzed carefully to find some additional sources. In all cases, the content of the found sources was checked briefly to exclude the irrelevant ones. As the outcome of these efforts, a representative (if not comprehensive) sample was formed. It includes the sources mentioning Ordovician tsunamis and/or their deposits (tsunamites), even if very briefly. Indeed, the topic is highly specific, and thus, it is not surprising that the total number of sources is small.

The information from the collected sources was analyzed in two steps. First, the lines of evidence were summarized. For this purpose, the localities of the Ordovician tsunamis (one locality represents one relatively small area and one relatively "thin" time slice) were established. For each of them, the location, the age, the character of evidence, and the sources were indicated. The location and the age were considered generally as they are mentioned in the analyzed sources (if even the chronostratigraphic units are outdated). The character of evidence reflects the consideration of tsunamis in the sources. It was established as follows. Ancient tsunamis cannot be observed or evidenced with some historical documents, and thus, they can only be interpreted, i.e., hypothesized. If so, the character of evidence depends on the "quality" of the mentioned hypotheses. The latter can be well developed (formulated extensively and chiefly without alternative explanations), preferred (chosen among two or more possible explanations), alternative (considered as a possible alternative to any other explanations), simply stated, and debatable (with some criticism). They can also be presented in previous publications extensively or briefly. The related peculiarities of the information in the analyzed sources were established for the purposes of this present study. The triggers of the hypothesized tsunamis, as proposed in the original sources, were also indicated.

Second, the summarized lines of evidence collected from the existing literature were interpreted. The major palaeotectonic domain (affinity to supercontinents, large continental blocks, and terranes) was established for all localities. For this purpose, the information from the analyzed sources was used together with the consideration of the "fresh" plate tectonic reconstructions [13–15]. Indeed, some generalizations and assumptions were unavoidable due to the differences in these reconstructions. The age of the localities was justified against the present version of the international geological time scale of the Ordovician [1,2]. In several cases, the original information did not permit to attribute the localities to the particular stages of this period, and, therefore, the interpreted age was limited to the level of epochs (this should not be too large a generalization taking into account how old are Ordovician deposits). The degree of certainty of the interpretations of tsunamis offered in the original sources was established. The summarized information of the character of evidence allows scoring the localities, where one score signifies the lowest degree of certainty (when the hypothesis is debatable), and five scores signify the highest degree of certainty (when the hypothesis is well-developed and argued extensively). Importantly, the authors of this present work do not express their own opinions on the quality of the previous research and hypotheses and do not evaluate them critically (this task is left for future studies, which will require additional field-based information and methods that are yet to be developed). All these interpretations allow judgments of the spatiotemporal distribution of Ordovician tsunamis and the quality of the related information.

3. Results

3.1. Summary of the Evidence Collected from the Literature

First of all, it is reasonable to summarize the information collected from the existing literature. The published evidence of Ordovician tsunamis is rather heterogeneous (Table 1). Nonetheless, it allows for establishing two dozen localities where these natural hazards took place (several times in some localities). It appears that significant research attention was paid previously to eastern Laurentia and northwestern Baltica, which are the focus of several publications.

Table 1. Summarized evidence of Ordovician tsunamis.

ID *	Location	Age	Character of Evidence	Source(s)
	(How Indicated in the Original Sources)			
1	Cincinnati Arch/Basin	Katian, Mohawkian–Cincinnatian	Alternative hypothesis, proposed briefly; seismicity-triggered tsunami	[24,38–41]
2	Armorican Massif	Arenig–Llanvirn	Preferred hypothesis, argued briefly; volcanism-triggered tsunami	[42]
3	Hesperian Massif			
4	North Wales	Caradoc	Well-developed hypothesis, moderate argumentation; combined volcanism and landslide-triggered tsunami	[43]
5	Illinois Basin	Sandbian	Alternative hypothesis, proposed briefly; impact-triggered tsunami	[44]
6 (11)	Baltoscandian Basin	Darriwilian	Debatable hypothesis, rather extensive evidence; seismicity-triggered tsunami	[45,46]
7	Baltoscandian Basin	Dapingian	Well-developed hypothesis; extensive argumentation; seismicity- or impact-triggered tsunami	[47]

Table 1. *Cont.*

ID *	Location (How Indicated in the Original Sources)	Age	Character of Evidence	Source(s)
8	Baltoscandian Basin	455 Ma	Simply stated hypothesis, argued briefly; impact-triggered tsunami	[48]
9	Montana	Earliest Ordovician **	Preferred hypothesis, argued extensively; seismicity-triggered tsunami	[29]
10	Western Precordillera	Early Ordovician	Preferred hypothesis, argued extensively; seismicity-triggered tsunami	[49]
11 (6)	Estonia	Darriwilian	Well developed but critically reconsidered, moderate argumentation; complex mechanism that remains debatable	[50,51]
12	South Wales			
13	North England			
14	Yukon			
15	Western Newfoundland			
16	Western Ireland			
17	Central Newfoundland			
18	Western Norway			
19	Korea			
20	Western Precordillera			
21	Central Precordillera			
22	New South Wales			
23	Northwestern Argentina			
24	Western Yunnan			

Notes: * It cannot be excluded that some localities with the same age coincide: both numbers are indicated in such cases; ** the sedimentary succession, for which tsunamis were interpreted, embraces the middle Cambrian–earliest Ordovician, and the attention was paid to its Cambrian part in the original source.

There is no ready "receipt" to hypothesize ancient tsunamis, and the related research employs state-of-the-art techniques. Tsunamis were hypothesized by the previous researchers with two approaches, which can be defined provisionally as direct identification and indirect assumption. The direct identification means that some peculiarities of sedimentary successions (rock sequences, specific lithologies, and structures) allowed judgments about the possibility of a tsunami. For instance, Pratt et al. [49] interpreted tsunamis for the La Silla Formation of the Western Precordillera (Argentina) on the basis of the very accurate analysis of the sedimentary signatures of unusual erosive events and also argued why the alternative explanations (tidal channels and severe storms) should be excluded. The indirect assumption means that tsunamis were noted as the expected consequence of some other, also interpreted geological processes. This means that tsunamis are assumed possible because their triggers are interpreted. For instance, Põldsaar et al. [47] proposed a tsunami after the analysis of the Volkhov Oil Collector in the Baltoscandian Basin when they discussed the possible trigger of the registered turbidite formation; importantly, these authors based their proposal on the comparison of what is known about tsunamis (also present) in similar settings. Additionally, some sources considered possible tsunamis but only referred to the previous studies. Principally, this may mean that the authors of the following source found the precedent hypothesis reasonable to a certain degree. For instance, Lam et al. [40] noted tsunamis interpreted previously in the Late Ordovician sedimentary successions of the Cincinnati Basin. It is also reasonable to note the work by Meinhold et al. [50], who discussed the earlier proposal by Parnell [51] of the possible impact-triggered, global-scale episode of mass wasting in the Middle Ordovician. This is an example of when the initial proposal was challenged later.

Although the published information on Ordovician tsunamis seems to be limited, one should note that it is rather diverse (Table 1). Previous researchers focused on different places of the world and different time slices. The ages of the deposits with tsunami-related signatures were often established accurately. The character of the evidence differs, but even more important is that the really different triggers of tsunamis were discussed (Table 1).

3.2. Interpretation of the Collected Lines of Evidence

The treatment of the collected lines of the evidence (Table 2) allows several inferences. These are original interpretations offered in this present work, with some examples. First, it appears that Ordovician tsunamis struck the periphery of Gondwana, isolated landmasses, and islands (or microcontinents) related to terranes and their groups or chains, such as Avalonia or Precordillera. Second, possible tsunamis took place in all epochs of the Ordovician, especially the Middle Ordovician. Third, the degree of certainty of the tsunami proposals is chiefly moderate, and the very certain and very uncertain lines of evidence are equally rare (Table 2).

Table 2. Interpreted evidence of Ordovician tsunamis (based on Table 1).

ID *	Domain **	Epoch	Degree of Certainty ***	Trigger
1	Laurentia	Late Ordovician	2	Seismicity
2	Gondwana	Middle Ordovician	3	Volcanism
3	Gondwana			
4	terranes 1	Late Ordovician	4	Complex
5	Laurentia	Late Ordovician	2	Impact
6 (11)	Baltica	Middle Ordovician	1	Seismicity
7	Baltica	Middle Ordovician	5	Questionable
8	Baltica	Late Ordovician	3	Impact
9	Laurentia	Early Ordovician	4	Seismicity
10	terranes 2	Early Ordovician	4	Seismicity
11 (6)	Baltica	Middle Ordovician	3	Questionable
12	terranes 1			
13	terranes 1			
14	Laurentia			
15	Laurentia			
16	terranes 1			
17	terranes 1			
18	Baltica			
19	North China			
20	terranes 2			
21	terranes 2			
22	Gondwana			
23	terranes 2			
24	South China			

Notes: * It cannot be excluded that some localities with the same age coincide: both numbers are indicated in such cases; ** some localities represent terranes and their chains: as their composition and configuration remain debatable, two groups of terranes are distinguished provisionally, namely terranes 1 (first of all, Avalonia) between Baltica, Gondwana, and Laurentia and terranes 2 (first of all Precordillera) along the American periphery of the Panthalassa; *** 1 marks the lowest degree of certainty, and 5 marks the highest degree of certainty (see methodological explanations above).

The distribution of the hypothesized Ordovician tsunamis in time is uneven (Figure 1). A few events with the seismic trigger were reported from the Early Ordovician but with significant certainty. With regard to the Middle Ordovician tsunamis, they seem to be numerous (among those reported), had different triggers, and were established with different but chiefly moderate certainty. Importantly, the majority of the localities of this age are taken from the works by Meinhold et al. [50] and Parnell [51]. Their exclusion would make the situation in the Middle Ordovician similar to that of the Late Ordovician. As for the latter, four localities represent the related tsunamis with different triggers (half of them could be caused by extraterrestrial impacts); the degree of certainty also differs, and it is relatively low in half of the cases. Generally, one can note that the higher number of localities and, thus, the concentration of the research do not enlarge the degree of certainty of the available information. On average, the Early Ordovician tsunamis seem better argued than the Middle–Late Ordovician tsunamis.

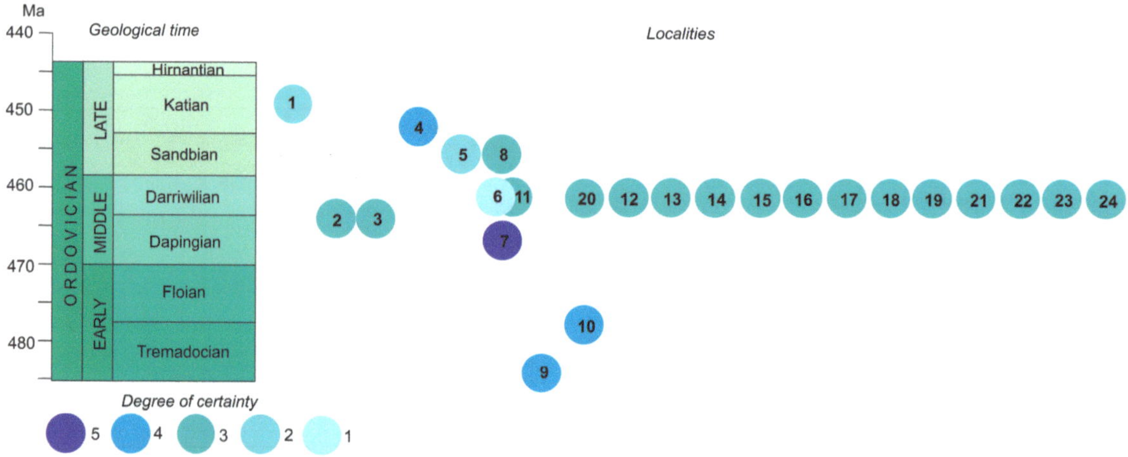

Figure 1. The distribution of the hypothesized Ordovician tsunamis in geological time. See Table 1 for an explanation of the locality numbers. The geological time scale follows [2].

The distribution of the hypothesized Ordovician tsunamis in the paleospace is notable. Although the localities are found in many parts of the world (both present and past—see Tables 1 and 2), such a situation is typical to only Middle Ordovician (Figure 2). The Early Ordovician localities are found in the American domains, and the Late Ordovician localities concentrate in Laurentia, Baltica, and the "nearby" terranes (Figure 2). Even if the Middle Ordovician tsunamis were hypothesized with a moderate degree of certainty (Figure 1), they appear to be the most widely reported (indeed, this is reasonable only if consider the information from the works by Meinhold et al. [50] and Parnell [51]).

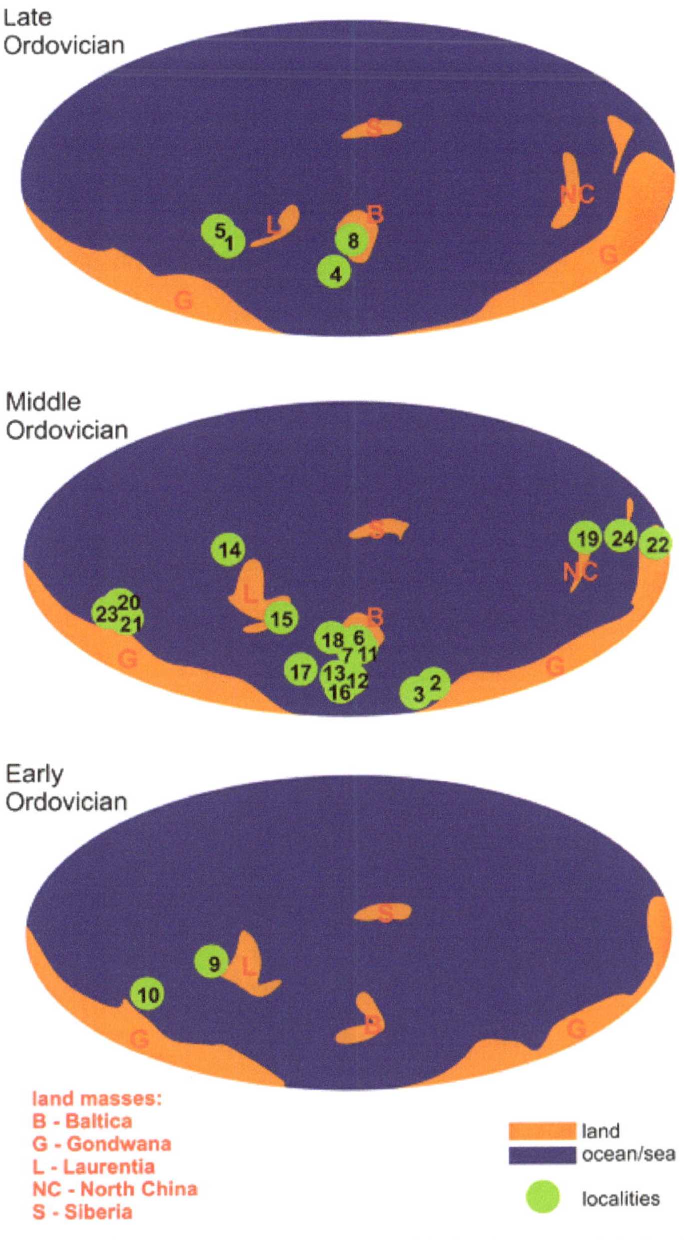

Figure 2. The distribution of the hypothesized Ordovician tsunamis in the global paleospace. See Table 1 for an explanation of the locality numbers. The paleogeographical contours are strongly generalized from [13].

4. Discussion

4.1. Representativeness of the Literature-Based Evidence

The present synthesis of the knowledge of Ordovician tsunamis raises the question of its completeness. The number of established localities is very limited (Table 1). Even if we suppose that some of them bear the evidence of two or more tsunami events, the total number of the presently known events is measured in the dozens. During the past

120 years, more than 1300 tsunamis were recorded [23]. Taking into account the length of the Ordovician Period [1,2], it is possible to calculate that the actual number of Ordovician tsunamis could exceed the number of the presently known events many millions of times (this may appear to be too much if even the geological records are always incomplete). In other words, we know no more than 0.00001% of them. The significant incompleteness of the knowledge of ancient tsunamis is established for the much younger periods [27]. As noted above, the sources of this incompleteness can be both objective and subjective [24–28]. Tsunami deposits are not necessarily preserved, and even when so, it is very difficult to identify and interpret them correctly (for instance, they can easily be misinterpreted as storm deposits or anything else). Moreover, the work by Parnell [51] gives a bold example of how massive the findings of the possible tsunami-related geological objects can be in the case of the special interest of any researcher(s). However, such interests appear only occasionally.

In regard to what is known about the common scarcity of information about pre-Quaternary tsunamis [24–28], one can conclude that Ordovician tsunamis are known more or less appropriately (at least, in regard to the present level of the development of geology). They were reported from all epochs of the period (Figure 1) and from many major palaeotectonic domains (Figure 2). The degree of certainty of the available information differs; although it is relatively high in a few cases, it is not too low in many other cases (Table 2). Moreover, the diversity of the hypothesized triggers (Table 1) seems to be even larger than found for the younger periods [27]. Taken together, these inferences mean that the knowledge of Ordovician tsunamis remains hypothetical and very incomplete, but its amount is expected in regard to the chances of documentation of pre-Quaternary tsunamis.

The present analysis indicates several biases. First, the stratigraphical bias reflects the insufficient (even absent) attention of the previous researchers to some geological time slices, especially the entire Early Ordovician Epoch and the Hirnantian Stage of the Late Ordovician Epoch (Figure 1). The latter was marked by the major glaciation [52–56], and it is known that glacial conditions and the related paleoenvironmental shifts may be favorable for tsunamis (in regard to icebergs and submarine landslides) [57–59]. The same bias is also determined using the unequal number of localities considered in the sources. For instance, Meinhold et al. [50] and Parnell [51] undertook a global-scale analysis, as a result of which many Middle Ordovician localities are known. In contrast, the other specialists focused on single localities. The situation is similar to that in paleontology when some time slice "deserved" monographic descriptions of fossils, and others do not. Second, the (paleo)geographical bias reflects the attention of the previous researchers on only selected regions and paleotectonic domains, especially in the Early and Late Ordovician (Figure 2). The distribution of the earthquake-generating subduction zones [14,15], which could source multiple tsunamis and other possible triggers, was wider. Therefore, one can expect tsunamis in many other places of the Ordovician world. Third, the causal bias reflects the unequal attention to the different triggers of tsunamis. Apparently, this bias is relatively weak in our case because the really different triggers of Ordovician tsunamis were discussed (Table 1). Many principal triggers of tsunamis [23,25] were considered by the previous researchers (it is necessary to stress that landslides were also taken into account for complex mechanisms). Apparently, meteorological tsunamis [60–62] were not reported from the Ordovician. The present-day sedimentological observations [63] can enhance deciphering such natural hazards in the Deep Past.

It is interesting to note that the collected sources were published since the end of the 1990s, and many of them appeared in the 2010s. However, the sources dated in the 2020s were not found. Indeed, such interruptions in the development of even so highly specific research directions limit the accumulation of knowledge. But the same interruptions may also be helpful to reconsider the previous experience critically and with "fresh eyes" in the course of the subsequent research developments.

4.2. Geoheritage Implications

Geoheritage comprises various unique (rare, peculiar) geological features, and its common manifestations are geosites. The latter is not only important to conserve the "fragile" elements of the global geological environment but also useful for scientific developments, educational activities, and tourism. In other words, geoheritage can be described in the words of valuable resources [64–66]. To facilitate its effective exploitation in a given area, these resources should be identified properly. Although this can be carried out in the course of territorial geological mapping and specialized research, a more systematic approach can be employed. Tsunami-related geoheritage has already been established [32–34,67,68], and it is reasonable to search for geosites representing tsunamis for all periods of the Phanerozoic, including the Ordovician. The bibliographical surveys may provide important lines for evidence for choosing the proper localities.

Contemporary geoheritage research pays much attention to the scientific value of geosites and their measurement [30,65,69–77]. Indeed, the scientific importance seems to be a fundamental quality of geosites, which stimulates their inventories and conservation efforts. In particular, these localities can provide matter for new discoveries, interpretations, and hypotheses, fix some standards, ideas, and theoretical concepts (e.g., stratigraphical boundaries or typical facies), and serve to show the extraordinary peculiarities of the geological environment. Therefore, the scientific reasons for the geosite selection are really important (nonetheless, there may be other reasons, including those educational and touristic), and the localities, which have already become important in the ancient tsunami studies, are promising for establishing geosites.

The information collected for the purposes of the present study indicates two dozen localities hypothetically representing Ordovician tsunamis. Principally, all of them can be linked to geoheritage. Nonetheless, the most valuable, world-class localities can be selected among them. It appears logical to prefer those that are distinguished with a high degree of certainty of the interpretations (scores 4 or 5), the clear trigger, and the undisputable age with full correspondence to the study time interval. From these requirements, the first one is of special importance because it determines the relatively higher probability of the ancient tsunami. More generally, the approach starts with the treatment of the evidence, continues with the selection of the proper localities, and culminates with the careful analysis of the published information when any stratigraphical section can be recommended as a potential geosite (Figure 3). Such an analysis permits judgments about the only availability of the highly valuable geoheritage resources in the particular area, i.e., the potential for the subsequent geosite establishment. This is so because the original publications do not necessarily bear the information about the potential geosites, and many technical details (for instance, accessibility, vulnerability, and aesthetic properties) were not reported in the "pure" geological works that dealt with tsunamis (or the published information needs updates). Some specialized geoheritage-focused field investigations are required to justify the information about the potential geosites and to formulate their official proposals. In other words, we can establish the presence of the previous geoheritage resources in particular areas and indicate the targets for future specialized studies, which can lead to the formal proposal of geosites.

From the localities of Ordovician tsunamis (Tables 1 and 2), two seem to be especially promising in regard to the geoheritage resources identified with the proposed approach (Figure 3). It should be stressed that the latter indicates where geosites can be found, but only the literature information does not allow formal geosite proposals and descriptions. Moreover, it cannot be excluded that the localities and their settings changed significantly since the time of the initial publication (e.g., due to human activities, slope collapses, etc.).

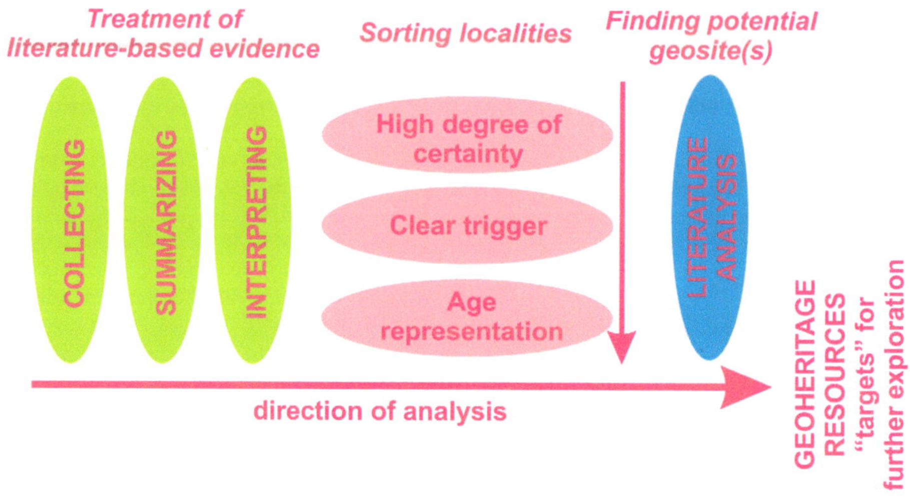

Figure 3. The proposed approach for the identification of the paleotsunami-related geoheritage resources.

The first locality is the Western Precordillera, where Pratt et al. [49] hypothesized Early Ordovician tsunamis. In western Argentina, near San Juan City, the La Silla Formation was studied in eight sections. It appears that the tsunami-related horizons of the Rio Blanco Member may exist in all of them. Although additional field investigations are necessary to propose one of these sections as a geosite, the Quebrada de la Flecha section can be preferred provisionally because it was used in the original source to demonstrate the tsunami-related facial features. This or any other neighboring section can be regarded as a world-class geosite representing the Ordovician tsunami (at least, bearing the evidence of hypothesizing the latter). As far as it is possible to deduce from the photographs available in the original source [49], the noted section may be relatively well-accessible and have a good scenic view. The rising interest in geoheritage and geotourism in Argentina and the related scientific debates [78–82] form sufficient premises for further field-based study in the noted locality, the formal geosite proposal, and its possible use in tourism.

The second promising locality is North Wales, where Kokelaar and Königer [43] hypothesized Late Ordovician tsunamis. In western Great Britain, the Pitts Head Tuff crops out. The relatively lengthy section examined at eight logs was described. The authors argued that the entry of the pyroclastic flow into the sea and the subsequent submarine sliding of the ignimbrite could trigger a tsunami. The entire section can be a world-class geosite representing the geological features, which allow for hypothesizing tsunamis. The original source does not contain any information, permitting us to judge its technical properties. The United Kingdom and, particularly, Wales are known for their traditions of geoconservation and geotourism and the availability of related, well-established institutions [83–87]. If so, the noted geosite (if established after additional, field-based examination) can be managed very effectively.

It is very important that the two potential geosites differ in age and tsunami mechanisms. Indeed, both already played a notable role in the studies of Ordovician tsunamis. This means that they are equally important as constituents of global geoheritage resources. Therefore, exploitation and sustainable management of the latter are reasonable tasks. Their solution will mark the shift from the realization of the geoheritage potential of the localities to full-scale geoconservation and geotouristic activities.

5. Conclusions

The present synthesis of the information about Ordovician tsunamis hypothesized in the geological literature allows for some general conclusions. First, evidence of tsunamis was found in different parts of the world and in all epochs of the Ordovician Period. Second, the triggers of the hypothesized Ordovician tsunamis were rather diverse, and they were not restricted to only earthquakes. Third, the degree of certainty of the interpretations of Ordovician tsunamis varies from very low to very high, but it is commonly moderate, which means that these tsunamis are chiefly neither too realistic nor too hypothetical. Fourth, despite the significant incompleteness and the stratigraphical and (paleo)geographical biases, the available knowledge of Ordovician tsunamis is not worse than one would expect. Fifth, this study contributes to the understanding of the geoheritage resources and indicates two potential world-class geosites with significant scientific value.

Such a bibliography-based study cannot avoid limitations. First, it is always possible to miss some lines of evidence (e.g., published in any local edition and/or in a language other than English). Second, the collected information is heterogeneous: when it is summarized, some generalizations and simplifications are necessary. Both limitations are principally unavoidable, and, supposedly, they cannot affect the outcomes of this paper strongly. The choice of the literature sources was careful, and the generalizations/simplifications were not so frequent and strong to question the analysis. The large incompleteness and the various biases leave the impression that the knowledge of Ordovician tsunamis is still too scarce to be reviewed. However, the present synthesis demonstrates that this knowledge is not so limited, and it is already found in many sources. If so, it deserves to be summarized and interpreted with certain cautions. Moreover, ancient tsunamis (not only Ordovician) are very difficult to document, and, most probably, the situation will not change in the near future. Therefore, scientists have nothing to do other than to try to understand these events on the basis of what is already known.

Two perspectives for future research can be outlined. First, this present study records several biases in the information of Ordovician tsunamis. Filling the related knowledge gaps is urgent. For instance, the researchers should pay attention to the sedimentary successions of the regions, which could be affected by major earthquakes, volcanism, landslides, and extraterrestrial impacts in Ordovician, but where tsunamis were not yet hypothesized. The present state of the palaeogeographical knowledge allows us to indicate such regions promising for tsunami searches. Second, this work demonstrates the importance of synthetic studies. Ancient tsunamis were reported from the different time slices, and it is very reasonable to summarize and interpret the information for each principal time slice. Subsequent comparisons of the outcomes of these studies may reveal some interesting long-term trends of the changes in tsunamis throughout the Phanerozoic (or, alternatively, this may indicate the absence of such changes). We expect that scientists can learn more about the appearance of tsunamis in geological history after the proper systematization of the multiple pieces of evidence.

Author Contributions: Conceptualization, D.A.R.; investigation, D.A.R. and N.N.Y.; writing—original draft preparation, D.A.R. and N.N.Y. All authors have read and agreed to the published version of the manuscript.

Funding: This research received no external funding.

Institutional Review Board Statement: Not applicable.

Informed Consent Statement: Not applicable.

Data Availability Statement: Not applicable.

Acknowledgments: We gratefully thank the academic editor and all anonymous reviewers for their thorough examination of our work and helpful suggestions.

Conflicts of Interest: The authors declare no conflict of interest.

References

1. Gradstein, F.M.; Ogg, J.G.; Schmitz, M.D.; Ogg, G.M. *Geologic Time Scale 2020*; Elsevier: Amsterdam, The Netherlands, 2020.
2. International Commission on Stratigraphy. Available online: Stratigraphy.org (accessed on 24 August 2023).
3. Sadler, P.M.; Cooper, R.A.; Melchin, M. High-resolution, early Paleozoic (Ordovician-Silurian) time scales. *Bull. Geol. Soc. Am.* **2009**, *121*, 887–906. [CrossRef]
4. Droser, M.L.; Finnegan, S. The Ordovician radiation: A follow-up to the Cambrian explosion? *Integr. Comp. Biol.* **2003**, *43*, 178–184. [CrossRef] [PubMed]
5. Ruban, D.A. Palaeoenvironmental setting (glaciations, sea level, and plate tectonics) of Palaeozoic major radiations in the marine realm. *Ann. De Paleontol.* **2010**, *96*, 143–158. [CrossRef]
6. Servais, T.; Cascales-Minana, B.; Harper, D.A.T.; Lefebvre, D.; Munnecke, A.; Wang, W.; Zhang, Y. No (Cambrian) explosion and no (Ordovician) event: A single long-term radiation in the early Paleozoic. *Palaeogeogr. Palaeoclimatol. Palaeoecol.* **2023**, *623*, 11592. [CrossRef]
7. Rasmussen, C.M.Ø.; Vandenbroucke, T.R.A.; Nogues-Bravo, D.; Finnegan, S. Was the Late Ordovician mass extinction truly exceptional? *Trends Ecol. Evol.* **2023**, *38*, 812–821. [CrossRef] [PubMed]
8. Sheehan, P.M. The late Ordovician mass-extinction. *Annu. Rev. Earth Planet. Sci.* **2001**, *29*, 331–364. [CrossRef]
9. Wang, G.; Zhan, R.; Percival, I.G. The end-Ordovician mass extinction: A single-pulse event? *Earth-Sci. Rev.* **2019**, *192*, 15–33. [CrossRef]
10. Hallam, A. Pre-Quaternary sea-level changes (Phanerozoic eustasy). *Annu. Rev. Earth Planet. Sci.* **1984**, *12*, 205–244. [CrossRef]
11. Haq, B.U.; Schutter, S.R. A chronology of Paleozoic sea-level changes. *Science* **2008**, *322*, 64–68. [CrossRef]
12. Marcilly, C.M.; Torsvik, T.H.; Conrad, C.P. Global Phanerozoic sea levels from paleogeographic flooding maps. *Gondwana Res.* **2022**, *110*, 128–142. [CrossRef]
13. Scotese, C.R. An atlas of Phanerozoic paleogeographic maps: The seas come in and the seas go out. *Annu. Rev. Earth Planet. Sci.* **2021**, *49*, 679–728. [CrossRef]
14. Torsvik, T.H.; Cocks, L.R.M. Gondwana from top to base in space and time. *Gondwana Res.* **2013**, *24*, 999–1030. [CrossRef]
15. von Raumer, J.F.; Stampfli, G.M. The birth of the Rheic Ocean-Early Palaeozoic subsidence patterns and subsequent tectonic plate scenarios. *Tectonophysics* **2008**, *461*, 9–20. [CrossRef]
16. Fortey, R.A.; Cocks, L.R.M. Late Ordovician global warming-The Boda event. *Geology* **2005**, *33*, 405–408. [CrossRef]
17. Ghienne, J.-F.; Desrochers, A.; Vandenbroucke, T.R.A.; Achab, A.; Asselin, E.; Dabard, M.-P.; Farley, C.; Loi, A.; Paris, F.; Wickson, S.; et al. A Cenozoic-style scenario for the end-Ordovician glaciation. *Nat. Commun.* **2014**, *5*, 4485. [CrossRef] [PubMed]
18. Grossman, E.L.; Joachimski, M.M. Ocean temperatures through the Phanerozoic reassessed. *Sci. Rep.* **2022**, *12*, 8938. [CrossRef] [PubMed]
19. Munnecke, A.; Calner, M.; Harper, D.A.T.; Servais, T. Ordovician and Silurian sea-water chemistry, sea level, and climate: A synopsis. *Palaeogeogr. Palaeoclimatol. Palaeoecol.* **2010**, *296*, 389–413. [CrossRef]
20. Scotese, C.R.; Song, H.; Mills, B.J.W.; van der Meer, D.G. Phanerozoic paleotemperatures: The Earth's changing climate during the last 540 million years. *Earth-Sci. Rev.* **2021**, *215*, 103503. [CrossRef]
21. Glikson, A.Y. An asteroid impact origin of the Hirnantian (end-Ordovician) glaciation and mass extinction. *Gondwana Res.* **2023**, *118*, 153–159. [CrossRef]
22. Schmitz, B.; Harper, D.A.T.; Peucker-Ehrenbrink, B.; Stouge, S.; Alwmark, C.; Cronholm, A.; Bergstrom, S.M.; Tassinari, M.; Xiaofeng, W. Asteroid breakup linked to the Great Ordovician Biodiversification Event. *Nat. Geosci.* **2008**, *1*, 49–53. [CrossRef]
23. Gusiakov, V.K. Global Occurrence of Large Tsunamis and Tsunami-Like Waves within the Last 120 Years (1900–2019). *Pure Appl. Geophys.* **2020**, *177*, 1261–1266. [CrossRef]
24. Bourgeois, J. Geologic effects and records of tsunamis. In *The Sea*; Robinson, A.R., Bernard, E.N., Eds.; Harvard University Press: Cambridge, MA, USA, 2009; Volume 15, pp. 53–91.
25. Dawson, A.G.; Stewart, I. Tsunami deposits in the geological record. *Sediment. Geol.* **2007**, *200*, 166–183. [CrossRef]
26. Luczyński, P. The tsunamites problem. Why are fossil tsunamites so rare? *Prz. Geol.* **2012**, *60*, 598–604.
27. Ruban, D.A. Tsunamis Struck Coasts of Triassic Oceans and Seas: Brief Summary of the Literary Evidence. *Water* **2023**, *15*, 1590. [CrossRef]
28. Shanmugam, G. Process-sedimentological challenges in distinguishing paleo-tsunami deposits. *Nat. Hazards* **2012**, *63*, 5–30. [CrossRef]
29. Pratt, B.R. Storms versus tsunamis: Dynamic interplay of sedimentary, diagenetic, and tectonic processes in the Cambrian of Montana. *Geology* **2002**, *30*, 423–426. [CrossRef]
30. Brilha, J. Inventory and Quantitative Assessment of Geosites and Geodiversity Sites: A Review. *Geoheritage* **2016**, *8*, 119–134. [CrossRef]
31. Prosser, C.; Murphy, M.; Larwood, J. *Geological Conservation: A Guide to Good Practice*; English Nature: Peterborough, UK, 2006.
32. Henriques, M.H. Broadening Frontiers in Geoconservation: The Concept of Intangible Geoheritage Represented by the 1755 Lisbon Earthquake. *Geoheritage* **2023**, *15*, 57. [CrossRef]
33. Roig-Munar, F.X.; Martín-Prieto, J.Á.; Rodríguez-Perea, A.; Gelabert, B.; Vilaplana, J.M. Proposed geosites for tsunamitic blocks in the rocky coasts of Formentera (Balearic Islands). *Rev. De La Soc. Geol. De Esp.* **2018**, *31*, 35–48.

34. Rusyana, A.; Nurhasanah; Astuti, S. Forecasting the number of visitors of Aceh state museum using decomposition method. *J. Phys. Conf. Ser.* **2018**, *1116*, 022041. [CrossRef]
35. Pranckutė, R. Web of science (WoS) and Scopus: The titans of bibliographic information in today's academic world. *Publications* **2021**, *9*, 12. [CrossRef]
36. Singh, V.K.; Singh, P.; Karmakar, M.; Leta, J.; Mayr, P. The journal coverage of Web of Science, Scopus and Dimensions: A comparative analysis. *Scientometrics* **2021**, *126*, 5113–5142. [CrossRef]
37. Takahashi, R.; Kaibe, K.; Suzuki, K.; Takahashi, S.; Takeda, K.; Hansen, M.; Yumoto, M. New concept of the affinity between research fields using academic journal data in Scopus. *Scientometrics* **2023**, *128*, 3507–3534. [CrossRef]
38. Aucoin, C.D.; Brett, C.E. Refined stratigraphy of the Late Ordovician (Katian; Richmondian) Waynesville Formation across the northeastern and northwestern margin of the Cincinnati Arch. *Stratigraphy* **2015**, *12*, 307–317.
39. Pope, M.C.; Read, J.F.; Bambach, R.; Hofmann, H.J. Late Middle to Late Ordovician seismites of Kentucky, southwest Ohio and Virginia: Sedimentary recorders of earthquakes in the Appalachian basin. *GSA Bull.* **1997**, *109*, 489–503. [CrossRef]
40. Dattilo, B.J.; Freeman, R.L.; Zubovic, Y.M.; Brett, C.E.; Straw, A.M.; Frauhiger, M.J.; Hartstein, A.R.; Shoemaker, L.M. Time-richness and phosphatic microsteinkern accumulation in the Cincinnatian (Katian) Ordovician, USA: An example of polycyclic phosphogenic condensation. *Palaeogeogr. Palaeoclimatol. Palaeoecol.* **2019**, *535*, 109362. [CrossRef]
41. Emig, C.C.; Gutiérrez-Marco, J.C. Lingulid beds at the upper limit of the Armorican Quartzite (Ordovician, Arenig, SW Europe). *Geobios* **1997**, *30*, 481–495. [CrossRef]
42. Kokelaar, P.; Königer, S. Marine emplacement of welded ignimbrite: The Ordovician Pitts Head Tuff, North Wales. *J. Geol. Soc.* **2000**, *157*, 517–536. [CrossRef]
43. Lam, A.R.; Stigall, A.R.; Matzke, N.J. Dispersal in the Ordovician: Speciation patterns and paleobiogeographic analyses of brachiopods and trilobites. *Palaeogeogr. Palaeoclimatol. Palaeoecol.* **2018**, *489*, 147–165. [CrossRef]
44. Monson, C.C.; Sweet, D.; Segvic, B.; Zanoni, G.; Balling, K.; Wittmer, J.M.; Ganis, G.R.; Cheng, G. The Late Ordovician (Sandbian) Glasford structure: A marine-target impact crater with a possible connection to the Ordovician meteorite event. *Meteorit. Planet. Sci.* **2019**, *54*, 2927–2950. [CrossRef]
45. Põldsaar, K.; Ainsaar, L. Extensive soft-sediment deformation structures in the early Darriwilian (Middle Ordovician) shallow marine siliciclastic sediments formed on the Baltoscandian carbonate ramp, northwestern Estonia. *Mar. Geol.* **2014**, *256*, 111–127. [CrossRef]
46. Suuroja, K.; Kirsimae, K.; Ainsaar, L.; Kohv, M.; Mahaney, V.C.; Suuroja, S. The Osmusaar Breccia in Northwestern Estonia–evidence of a 475 Ma earthquake of an impact? In *Impact Markers in the Stratigraphic Record*; Koeberl, C., Martinez-Ruiz, F.C., Eds.; Springer: Berlin/Heidelberg, Germany, 2003; pp. 333–347.
47. Põldsaar, K.; Ainsaar, L.; Nemliher, R.; Tinn, O.; Stinkulis, G. A siliciclastic shallow-marine turbidite on the carbonate shelf of the Ordovician Baltoscandian palaeobasin. *Est. J. Earth Sci.* **2019**, *68*, 1–14.
48. Suuroja, K.; Suuroja, S.; All, T.; Floden, T. Kärdla (Hiiumaa Island, Estonia)—The buried and well-preserved Ordovician marine impact structure. *Deep Sea Res. Part II Top. Stud. Oceanogr.* **2002**, *49*, 1121–1144. [CrossRef]
49. Pratt, B.R.; Raviolo, M.M.; Bordonaro, O.L. Carbonate platform dominated by peloidal sands: Lower Ordovician La Silla Formation of the eastern Precordillera, San Juan, Argentina. *Sedimentology* **2012**, *59*, 843–866. [CrossRef]
50. Meinhold, G.; Arslan, A.; Lehnert, O.; Stampfli, G. Global mass wasting during the Middle Ordovician: Meteoritic trigger or plate-tectonic environment? *Gondwana Res.* **2011**, *19*, 535–541. [CrossRef]
51. Parnell, J. Global mass wasting at continental margins during Ordovician high meteorite influx. *Nat. Geosci.* **2009**, *2*, 57–61. [CrossRef]
52. Delabroye, A.; Vecoli, M. The end-Ordovician glaciation and the Hirnantian Stage: A global review and questions about Late Ordovician event stratigraphy. *Earth-Sci. Rev.* **2010**, *98*, 269–282. [CrossRef]
53. Finlay, A.J.; Selby, D.; Gröcke, D.R. Tracking the Hirnantian glaciation using Os isotopes. *Earth Planet. Sci. Lett.* **2010**, *293*, 339–348. [CrossRef]
54. Villas, E.; Vennin, E.; Álvaro, J.J.; Hammann, V.; Herrera, Z.A.; Piovano, E.L. The late Ordovician carbonate sedimentation as a major triggering factor of the Hirnantian glaciation. *Bull. Soc. Geol. Fr.* **2002**, *173*, 569–578. [CrossRef]
55. Yin, L.-M.; Borjigin, T.; Ou, Z.-J.; Bian, L.-Z. Late Ordovician microphytoplankton in Southwest China: Recording paleogeographic variations across the Hirnantian glaciation. *Palaeoworld* **2023**, *32*, 79–92. [CrossRef]
56. Zhou, L.; Algeo, T.J.; Shen, J.; Hu, Z.F.; Gong, H.; Xie, S.; Huang, J.H.; Gao, S. Changes in marine productivity and redox conditions during the Late Ordovician Hirnantian glaciation. *Palaeogeogr. Palaeoclimatol. Palaeoecol.* **2015**, *420*, 223–234. [CrossRef]
57. Ke, H.; Ai, S.; Yan, B.; Zhou, C.; Wang, Z.; Yang, Y.; Liu, T.; An, J.; Chen, Y. Iceberg-Induced Tsunamis near Dålk Glacier, Antarctica. *J. Surv. Eng.* **2022**, *148*, 04021027. [CrossRef]
58. Lee, H.J. Timing of occurrence of large submarine landslides on the Atlantic Ocean margin. *Mar. Geol.* **2009**, *264*, 53–64. [CrossRef]
59. Lüthi, M.P.; Vieli, A. Multi-method observation and analysis of a tsunami caused by glacier calving. *Cryosphere* **2016**, *10*, 995–1002. [CrossRef]
60. Pattiaratchi, C.; Wijeratne, E.M.S. Observations of meteorological tsunamis along the south-west Australian coast. *Nat. Hazards* **2014**, *74*, 281–303. [CrossRef]
61. Rabinovich, A.B.; Vilibić, I.; Tinti, S. Meteorological tsunamis: Atmospherically induced destructive ocean waves in the tsunami frequency band. *Phys. Chem. Earth* **2009**, *34*, 891–893. [CrossRef]

62. Saito, T.; Kubota, T.; Chikasada, N.Y.; Tanaka, Y.; Sandanbata, O. Meteorological Tsunami Generation Due to Sea-Surface Pressure Change: Three-Dimensional Theory and Synthetics of Ocean-Bottom Pressure Change. *J. Geophys. Res. Ocean.* **2021**, *126*, 1–29. [CrossRef]
63. Alonso, G.; Dragani, W.; Péreza, I. The role of meteorological tsunamis on beach erosion in the buenos aires coast: Some numerical experiments. *Coast. Eng. J.* **2018**, *60*, 299–307. [CrossRef]
64. Bétard, F.; Hobléa, F.; Portal, C. Geoheritage as new territorial resource for local development. *Ann. Geogr.* **2017**, *717*, 523–543. [CrossRef]
65. Sallam, E.S.; Ruban, D.A.; Ermolaev, V.A. Geoheritage resources and new direction of infrastructural growth in Egypt: From geosite assessment to policy development. *Resour. Policy* **2022**, *79*, 103127. [CrossRef]
66. Santangelo, N.; Valente, E. Geoheritage and Geotourism resources. *Resources* **2020**, *9*, 80. [CrossRef]
67. Németh, K. Geoheritage and geodiversity aspects of catastrophic volcanic eruptions: Lessons from the 15th of January 2022 Hunga Tonga-Hunga Ha'apai eruption, SW Pacific. *Int. J. Geoheritage Parks* **2022**, *10*, 546–568. [CrossRef]
68. Sansò, P.; Margiotta, S.; Mastronuzzi, G.; Vitale, A. The geological heritage of Salento Leccese area (Apulia, southern Italy). *Geoheritage* **2015**, *7*, 85–101. [CrossRef]
69. Bruschi, V.M.; Cendrero, A. Geosite evaluation; can we measure intangible values? *Alp. Mediterr. Quat.* **2005**, *18*, 293–306.
70. Çetiner, Z.S.; Ertekin, C.; Yiğitbaş, E. Evaluating Scientific Value of Geodiversity for Natural Protected Sites: The Biga Peninsula, Northwestern Turkey. *Geoheritage* **2018**, *10*, 49–65. [CrossRef]
71. Corbí, H.; Fierro, I.; Aberasturi, A.; Sánchez Ferris, E.J. Potential Use of a Significant Scientific Geosite: The Messinian Coral Reef of Santa Pola (SE Spain). *Geoheritage* **2018**, *10*, 427–441. [CrossRef]
72. Dincă, I.; Keshavarz, S.R.; Almodaresi, S.A. Landscapes of the Yazd-Ardakan Plain (Iran) and the Assessment of Geotourism—Contribution to the Promotion and Practice of Geotourism and Ecotourism. *Land* **2023**, *12*, 858. [CrossRef]
73. Mucivuna, V.C.; Motta Garcia, M.D.G.; Reynard, E. Comparing quantitative methods on the evaluation of scientific value in geosites: Analysis from the Itatiaia National Park, Brazil. *Geomorphology* **2022**, *396*, 107988. [CrossRef]
74. Nazaruddin, D.A. Systematic Studies of Geoheritage in Jeli District, Kelantan, Malaysia. *Geoheritage* **2017**, *9*, 19–33. [CrossRef]
75. Reynard, E.; Perret, A.; Bussard, J.; Grangier, L.; Martin, S. Integrated Approach for the Inventory and Management of Geomorphological Heritage at the Regional Scale. *Geoheritage* **2016**, *8*, 43–60. [CrossRef]
76. Suzuki, D.A.; Takagi, H. Evaluation of Geosite for Sustainable Planning and Management in Geotourism. *Geoheritage* **2018**, *10*, 123–135. [CrossRef]
77. Tamang, L.; Mandal, U.K.; Karmakar, M.; Banerjee, M.; Ghosh, D. Geomorphosite evaluation for geotourism development using geosite assessment model (GAM): A study from a Proterozoic terrain in eastern India. *Int. J. Geoheritage Parks* **2023**, *11*, 82–99. [CrossRef]
78. Coronato, A.; Schwarz, S. Approaching geodiversity and geoconservation in Argentina. *Int. J. Geoheritage Parks* **2022**, *10*, 597–615. [CrossRef]
79. Lech, R.R.; Marcus, A.; Reinoso, J.R. Science and tourism together in the management for the preservation of the geological heritage. *Ser. Correl. Geol.* **2018**, *34*, 35–41.
80. Medina, W.M. Geo-environmental Evolutionary Interpretation of Geosites Analysed in La Quebrada de Humahuaca, Northwestern Argentina. *Geoheritage* **2023**, *15*, 51. [CrossRef]
81. Medina, W.; Vejsbjerg, L.; Aceñolaza, G. Legal framework for geoconservation. Presence of geology in protected areas of the Argentine Republic. *Rev. Del Mus. Argent. De Cienc. Nat. Nueva Ser.* **2016**, *18*, 53–64. [CrossRef]
82. Miranda, F.; Lema, H. Current situation of geologiacal heritage in Argentina. *Bol. Parana. Geosci.* **2013**, *70*, 87–102.
83. Brown, E.J.; Evans, D.H.; Larwood, J.G.; Prosser, C.D.; Townley, H.C. Geoconservation and geoscience in England: A mutually beneficial relationship. *Proc. Geol. Assoc.* **2018**, *129*, 492–504. [CrossRef]
84. Burek, C. The Role of LGAPs (Local Geodiversity Action Plans) and Welsh RIGS as Local Drivers for Geoconservation within Geotourism in Wales. *Geoheritage* **2012**, *4*, 45–63. [CrossRef]
85. Hose, T.A.; Vasiljević, D.A. Defining the Nature and Purpose of Modern Geotourism with Particular Reference to the United Kingdom and South-East Europe. *Geoheritage* **2012**, *4*, 25–43. [CrossRef]
86. Price, W.R.; Ronck, C.L. Quarrying for World Heritage Designation: Slate Tourism in North Wales. *Geoheritage* **2019**, *11*, 1839–1854. [CrossRef]
87. Prosser, C.D. The history of geoconservation in England: Legislative and policy milestones. *Geol. Soc. Spec. Publ.* **2008**, *300*, 113–122. [CrossRef]

Journal of
Marine Science and Engineering

MDPI

Article

The Biotoxic Effects of Ag Nanoparticles (AgNPs) on *Skeletonema costatum*, a Typical Bloom Alga Species in Coastal Areas

Ke Shi [1], Yuehong Yao [1], Jianliang Xue [1,2], Dongle Cheng [1,2] and Bo Wang [1,*]

1 College of Safety and Environmental Engineering, Shandong University of Science and Technology, Qingdao 266590, China; shike@sdust.edu.cn (K.S.); 202182200059@sdust.edu.cn (Y.Y.); ll-1382@163.com (J.X.); donglelecheng@gmail.com (D.C.)
2 Institute of Yellow River Delta Earth Surface Processes and Ecological Integrity, Shandong University of Science and Technology, Qingdao 266590, China
* Correspondence: wangbo@sdust.edu.cn

Abstract: With the rapid development of nanotechnology, nanomaterials have been widely utilized in many industries and daily life applications due to their unique properties. However, their potential release and the human health/environmental consequences have raised public concern greatly. In this study, we compared the toxic effects of AgNPs and $AgNO_3$ on *Skeletonema costatum* in 10, 100, and 500 $\mu g \cdot L^{-1}$ Ag treatments. In all the AgNP exposure experiments, cell membrane damage and growth inhibition occurred. However, the cellular damage only obviously appears on exposure to a high concentration of $AgNO_3$. The antioxidant enzyme (SOD and CAT) activities and lipid peroxidation in *Skeletonema costatum* were also induced significantly in the AgNP treatments. In addition, the percentage of Ag release in seawater increased with the increase in AgNP concentrations (13%, 32% for 100 and 500 $\mu g \cdot L^{-1}$ AgNPs). Thus, the biotoxic effects of AgNPs were found to be due to a combination of the solubilization of particles into toxic metal ions and the nature of the nanoparticles. It was worth noting that the induction of oxidative stress and damage to the cell membrane comprised the dominant mechanism of toxicity for AgNPs. Therefore, the behavior of nanometals in seawater affects the biotoxic effect on the phytoplankton. These results shed light on the biological toxicity of nanometals and their possible toxicity mechanism.

Keywords: biotoxic; AgNPs; soluble Ag release; oxidative stress; *Skeletonema costatum*

Citation: Shi, K.; Yao, Y.; Xue, J.; Cheng, D.; Wang, B. The Biotoxic Effects of Ag Nanoparticles (AgNPs) on *Skeletonema costatum*, a Typical Bloom Alga Species in Coastal Areas. *J. Mar. Sci. Eng.* **2023**, *11*, 1941. https://doi.org/10.3390/jmse11101941

Academic Editors: Assimina Antonarakou, George Kontakiotis and Dmitry A. Ruban

Received: 1 September 2023
Revised: 24 September 2023
Accepted: 29 September 2023
Published: 8 October 2023

1. Introduction

In recent years, silver nanoparticles (AgNPs) have received widespread attention [1–3]. Due to their stable physical and chemical properties, AgNPs show a better performance than silver in many aspects, and are widely used in textiles, food, water treatment, cosmetics, and other products [4,5]. In nature, many environmental factors affect the formation, aggregation, and dispersion of nanoparticles [6]. The colloidal organic matter is an important regulatory factor for the speciation of trace metals in seawater and even their biogeochemical cycling processes. These colloidal organic compounds can form nanoparticles by joining complexes with metals, thereby affecting the biogeochemical cycle of metals [2,7,8]. Dissolved organic matter in the aquatic environment can adsorb on the surface of nanometals and inhibit the aggregation of nanoparticles. The presence of organic matter affects the bioavailability and toxicity of nanometals [6,9,10], while ions in the water mass compress suspended nanometals and cause them to destabilize and precipitate [11]. Studies [12,13] have shown that a low concentration of humus can stabilize silver nanoparticles, while a high concentration can induce aggregation, thereby changing the bioavailability and biotoxicity of nanoparticles. For AgNPs, the ion release from the nanoparticle is significant (~34% of silver ions resolved from the nanoparticles), resulting in the absorption of metals by organisms exposed to AgNPs in both nano and ionic forms.

After entering the aquatic environment, nanoparticles usually aggregate under physical, chemical, biological, and other conditions, which may cause their aggregates to settle onto the surface of sediment, thereby posing an exposure toxicity risk to organisms living in benthic environments [14,15]. In addition, some active metal nanoparticles may also release metal ions into the water, posing an exposure toxicity risk to plankton that live in the aquatic environment. The physical and chemical properties of nanoparticles (including their size, shape, specific surface area, composition, aggregation, and dissolution) can affect their biotoxicity, especially some nanomaterials containing toxic metals. However, there is controversy as to whether "the biotoxicity is caused by the nanometal itself or the free ions released". It is generally believed that the biotoxic effect constitutes oxidative stress induced in organisms [16–20]. However, the toxicological mechanism studies in nanometals to date have ignored metal absorption and accumulation in organisms. Ringwood et al. [21] found that low concentrations of AgNPs (<0.16 $\mu g \cdot L^{-1}$) did not affect the normal hatching of fertilized eggs in oysters. However, when the exposure concentration exceeded 0.16 μg/L, the content of metallothionein (MT) in the fertilized eggs of oysters increased significantly and reached 80 times the MT contents in the control groups. Thus, it can be seen that the effects of AgNPs on different biological levels may be different [22]. Molecular markers are, undoubtedly, more sensitive indicators. Therefore, it is necessary to study the biological toxic effects of nanometals using different biomarkers in combination with their accumulation in organisms.

Coastal waters are usually considered as the ultimate sink for pollutants, including nanomaterials. In the process of human activities, unintentional and intentional discharge can lead to a large amount of wastewater entering the coastal environment [5,23,24]. Similarly, widely used nanomaterials will inevitably enter the marine environment and cause harm to the marine ecosystem [25,26]. Marine microalgae are highly abundant in coastal waters, so microalgae growth is likely to be affected by anthropogenic pollution. *Skeletonema costatum* (*S. costatum*) is a nearshore diatom which is distributed in waters with a wide range of temperatures and salinity levels. Among more than 40 species of bloom algae recorded in Jiaozhou Bay, a region of high urbanization in China, *S. costatum* is one of the dominant species [27]. As a typical bloom alga with low levels in the food chain, the absorption of nanoparticles in *S. costatum* directlys affect material transfer throughout the food chain. Therefore, the biotoxic effects of AgNPs on *S. costatum* may affect the entire coastal ecosystem. Thus, we selected *S. costatum* as the experimental object that has a certain ecological significance.

As it is an emerging pollutant, the widespread application of AgNPs may cause pollution to the water mass and also induce adverse effects on human health. However, it is difficult to evaluate the ecotoxicity risk of nanomaterials because their toxicity mechanism is still unclear. Understanding the biotoxicity mechanism of AgNPs on marine algae can help us to predict the impact of nanomaterials on the food web and the entire ecosystem in coastal waters. It is unclear how the particle size, shape, and surface properties of NPs are altered after entering the seawater, and how these variations could affect their biotoxicity to marine organisms. Therefore, it is necessary to conduct deeper research to explore the biotoxicity mechanism of AgNPs on marine organisms. Studying the toxic effect of AgNPs on coastal algae is of great scientific value in the protection of coastal ecosystems and provides a scientific basis for further revealing the biotoxic mechanism of nanoparticles.

2. Methods and Materials

2.1. Characteristics of AgNPs

AgNPs (<100 nm, 99.5% trace metal basis) were purchased from Sigma-Aldrich (https://www.sigmaaldrich.cn/CN/zh, accessed on 3 April 2014). The standard solution of Ag ($AgNO_3$) was purchased from China Standard Substance Network (https://www.reagent.com.cn/, accessed on 12 March 2014).

AgNP solutions were prepared with filtered seawater and Milli-Q water, respectively. The concentrations of 10, 100, and 500 $\mu g \cdot L^{-1}$ AgNPs/$AgNO_3$ were adapted in this study.

The 500 $\mu g \cdot L^{-1}$ AgNPs in seawater solution were observed using a scanning electron microscope (SEM, QUANTA200, FEI Company, Hillsboro, OR, USA).

After 24 h of solution preparation in seawater, the AgNP solution was subjected to ultrafiltration using the 3 KD Millipore ultrafiltration centrifuge tube (UFC500396). The Ag^{+} ion in filtrate, released from the AgNPs, was measured in acidic conditions using a flame atomic absorption spectroscope (MD2134, Thermo Fisher Scientific, Waltham, MA, USA).

2.2. Determination of Cytotoxic Effects

The f/2 medium was prepared with reference to the method in the study by Guillard et al. [28]. To avoid the complexation of Ag ions, Na_2EDTA was not added to the culture medium. Thus, it is called "modified f/2 medium".

The cytotoxic experiments were performed in three groups (control, AgNP exposure, and $AgNO_3$ exposure). In a previous study [29], the concentration of AgNPs in industrial discharge water was found to reach 1–6 $mg \cdot L^{-1}$. Considering the dilution effect of seawater, we set up 10, 100, and 500 $\mu g \cdot L^{-1}$ AgNPs in a seawater culture medium. The $AgNO_3$ exposure experiment also used the concentrations of 10, 100, and 500 $\mu g \cdot L^{-1}$ in the culture medium for comparison. Every Ag concentration gradient was performed in triplicate. The alga was cultivated in a constant-temperature incubator and manually shaken three times a day.

Cysteine can be used as a complexing agent to eliminate the role of Ag ions in AgNP toxicity. In our study, cysteine was added to the medium to complex Ag^{+}. In the 500 $\mu g \cdot L^{-1}$ $AgNO_3$ treatments, cysteine was added to the medium at a concentration of 500 $\mu g \cdot L^{-1}$. In order to avoid the differences in the impact of cysteine, equivalent cysteine also be added to the 500 $\mu g \cdot L^{-1}$ AgNP treatment. The Ag-cysteine formed could decrease the toxic effect induced by the Ag^{+} released from the nanoparticles.

2.2.1. The Relative Growth Rates

The density of algal cells was measured using the microscope at the exposure times of 0, 24, 48, 72, and 96 h, and the relative growth rate was calculated as follows.

$$\text{Relative growth rate} = (\ln N_1 - \ln N_0)/(T_1 - T_0)$$

where N_1, N_0 are the density of the algal cell at time T_1 and T_0, respectively.

2.2.2. Ag Accumulation in the Algal Cell

After 48 h of exposure, 100 mL of algal culture was centrifuged into algal mud and digested with HNO_3 and H_2SO_4 using the microwave digestion method. Metal ligands, like Na_2EDTA and cysteine were not added, to avoid the complexation of Ag. The silver content in the algal cell was detected in the digestion solution via flame atomic spectrophotometry.

2.2.3. The Damage to the Cell Membrane

According to Tommaso Pisani's method [30], the electrolyte flow out after the destruction of the algal cell membrane causes changes in the conductivity of the algal solution. The conductivity E1 and E2 of the algal solution before and after boiling were measured. The degree of the damage to the cell membrane was expressed as E1/E2. The larger the value, the greater the degree of damage.

2.2.4. Determination of the Chl-a

Referring to the method of Porra et al. [31], the absorbance of the supernatant at the wavelengths of 665 nm and 652 nm was measured using a visible spectrophotometer. The Chl-a was calculated as follows.

$$\text{Chlorophyll a } (mg \cdot L^{-1}) = 16.29 \times A_{665} - 8.54 \times A_{652}$$

where A_{665} and A_{652} represent the absorbance of the supernatant at the wavelengths of 665 nm and 652 nm.

2.3. *Determination of Antioxidant Stress*

2.3.1. Preparation of Protease Extract

Firstly, 20 mL algal solution was sampled to centrifuge at 5000 r/min for 10 min. The supernatant was removed. The precipitate was added to an appropriate amount of 0.05 $mol{\cdot}L^{-1}$ phosphate buffer (pH = 7.0), then crushed ultrasonically. The liquid was centrifuged at 4 °C, 6500 r/min for 20 min, and the supernatant was the protease extract.

The preparation of the protease extract and the subsequent determination of the protein content, SOD, CAT, and MDA were performed with reference to the instructions in the biological reagent kit purchased from Nanjing Jiancheng Biological Company (http://www.njjcbio.com/, accessed on 15 May 2014).

2.3.2. Determination of Protein Content, SOD, CAT, and MDA

The protein content in the protease extract was determined via the Coomassie brilliant blue method, and the absorbance of the protease extract was measured at the wavelength of 595 nm. The protein content was calculated as follows:

$$\frac{C_{prot.}}{(g/L)} = \frac{OD_m - OD_b}{OD_s - OD_b} \times 0.563\ g/L$$

where OD_m and OD_s are the optical density of the tested sample and standard sample, respectively. OD_b is the optical density of the blank samples. $C_{prot.}$ is the content of the protein, and 0.563 g/L is the protein content of the standard sample.

The SOD content in the protease extract was determined via the hydroxylamine method, and the absorbance of the protease extract was measured at the wavelength of 550 nm. The SOD content was calculated as follows:

$$\frac{SOD}{(U/mg\ prot.)} = \frac{OD_c - OD_m}{OD_c} \div 50\% \times \frac{V_t}{V_m} \div C_{prot.}$$

where OD_c and OD_m are the optical density of the control and measured sample, respectively. V_t and V_m are the volume of the total sample and measured sample, respectively, in the unit of mL. The protein content is in the unit of mg prot/mL. The SOD content is in the unit of U/mg prot, representing the SOD unit per mg protein.

The CAT content in the protease extract was determined via the visible light photometric method, and the absorbance of the protease extract was measured at the wavelength of 405 nm. The CAT content was calculated as follows:

$$\frac{CAT}{(U/mg\ prot.)} = (OD_c - OD_m) \times 271 \times \frac{1}{60 \times V_m} \div C_{prot.}$$

where OD_c and OD_m are the optical density of the control and measured sample, respectively. V_m is the volume of the measured sample, in the unit of mL. The protein content is in the unit of mg prot/mL. The CAT content is in the unit of U/mg prot, representing the CAT unit per mg protein.

The MDA content in the protease extract was determined via the spectrophotometry method, and the absorbance of the protease extract was measured at the wavelength of 532 nm. The MDA content was calculated as follows:

$$\frac{MDA}{(nmol/\ mg\ prot.)} = \frac{OD_m - OD_c}{OD_s - OD_{bl}} \times \frac{Con._s}{(10\ nmol/mL)} \div C_{prot.}$$

where OD_m, OD_c, OD_s, and OD_{bl} are the optical density of the measured, control, standard, and blank sample, respectively. $Con._s$ is the concentration of MDA in the standard sample

(10 nmol/mL). The protein content is in the unit of mg prot/mL. The MDA content is in the unit of U/mg prot, representing the MDA unit per mg protein.

2.4. Statistical Analyses

The data are expressed as mean ± SD. The difference analysis between the control and the exposure experiments is conducted using XLstat and one-way ANOVA. The significant difference is $p < 0.05$.

3. Results and Discussion

3.1. Characteristics of AgNPs

In an aquatic environment, the physical and chemical conditions could affect the agglomerate and deposition of nanoparticles, thereby altering their toxic effects on organisms. Therefore, it is necessary to consider the impact of the possible flocculation of nanoparticles on their toxicity mechanism.

Although the PVP coated on the surface of AgNPs plays a role in dispersion and stability, the SEM image (Figure 1) shows that AgNPs flocculated to an irregular granular shape in filtered seawater at different magnifications, and particles are produced with a size of micron level. Due to the limitation of the accuracy of the device, the precision of the images in Figure 1 is 10 and 5 μm, respectively. It can be seen from the SEM results that the AgNPs obviously flocculated in seawater, and the size mostly varied between 100 nm and 10 μm.

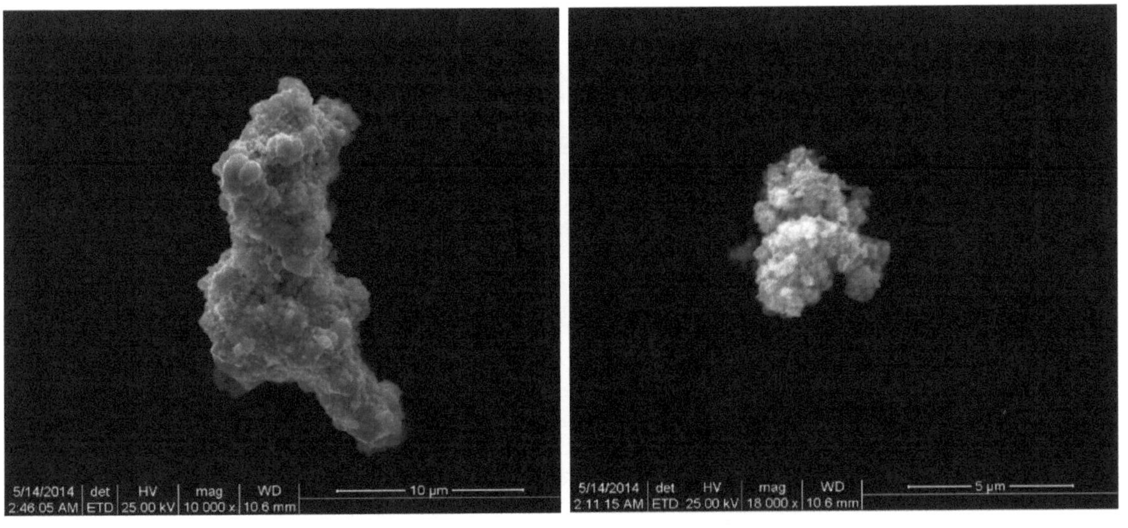

Figure 1. SEM images of AgNPs (500 $\mu g \cdot L^{-1}$, Sigma, <100 nm) dispersed in seawater.

After 24 h, the release of soluble Ag from Milli-Q water and seawater with different concentrations of AgNPs is shown in Figure 2. For 500 $\mu g \cdot L^{-1}$ AgNP solutions, ~32.43% Ag is released and exists in the filtered seawater medium as the soluble state (162.17 ± 67.14 $\mu g \cdot L^{-1}$), which is significantly higher than the release amount in Milli-Q water medium (11.9 ± 1.29 $\mu g \cdot L^{-1}$). However, for 100 $\mu g \cdot L^{-1}$ AgNP solutions, there is no significant difference in the released soluble Ag contents between the filtered seawater medium (13.01 ± 3.61 $\mu g \cdot L^{-1}$) and Milli-Q water medium (12.35 ± 4.17 $\mu g \cdot L^{-1}$). After 48 h, the content of soluble Ag released in both the filtered seawater and Milli-Q water medium were lower than the detection limit of AAS, indicating that the aggregation of AgNPs inhibited the release of soluble Ag from the nanoparticles.

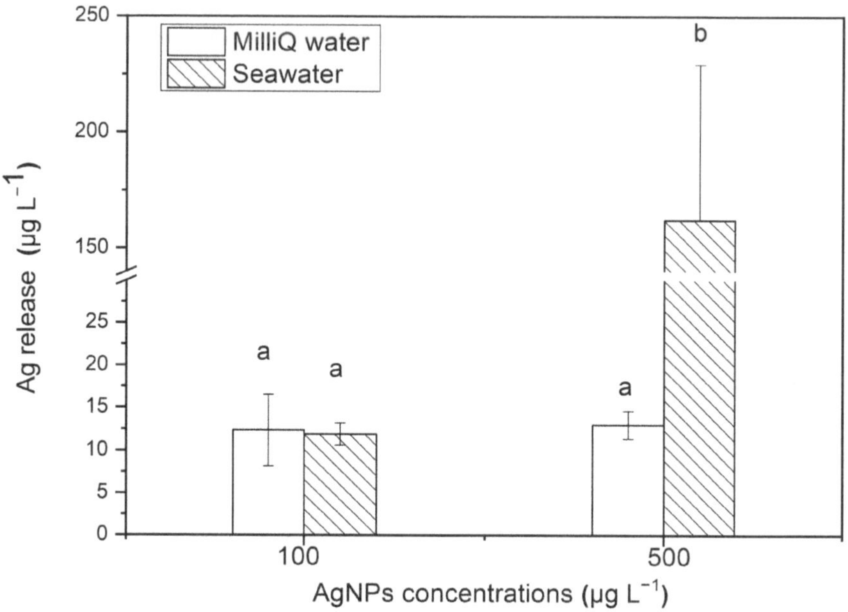

Figure 2. After 24 h, the release of soluble Ag from AgNPs in Milli-Q water and seawater. The same identification letter (e.g., a and a) represents no significant difference, while different identification letters (e.g., a and b) represent a significant difference.

Many studies have proven that AgNPs easily flocculate in seawater. Gomes et al. [32] showed that AgNPs with the initial particle size of 41.7 ± 9.6 nm flocculated to 144.2 ± 39.2 nm in a seawater suspension. Buffet et al. [13] demonstrated that the particle size of AgNPs would increase from 40 nm to 150–500 nm. The reasons for the change in particle size are summarized as the presence of inorganic ligands in seawater or the large specific surface area of nanoparticles, which may also react with organisms. Therefore, the environmental behavior of AgNPs and the reaction between AgNPs and biological cells are closely related [33,34].

The high ionic strength of seawater has a great influence on the aggregation and dissolution of nanoparticles. In this study, the percentage of released soluble Ag decreased with an increasing AgNP concentration in Milli-Q water (12% and 2% Ag released in 100 and 500 μg·L^{-1} AgNP solutions, respectively), while the results were the opposite in seawater (13% and 32%, respectively). In comparison, in the study by Zou et al. [35], 18% Ag was released from a 1500 μg·L^{-1} AgNP solution, which is higher than our results. The AgNPs in the study by Zou et al. [35] were synthesized by $AgNO_3$ and $NaBH_4$ in the presence of ATP-2Na. The research by Burchardt et al. [36] on the impact of AgNPs on diatoms showed that between 68% and 87% Ag was released from a 10 μg·L^{-1} AgNP solution in artificial seawater, and between 2% and 3% Ag was released in freshwater (BG11 medium). The content of Ag released in seawater was much higher than that in freshwater, which is the same as that in our study. In the study of ZnO NPs by Wang et al. [37], it was found that more than 50% Zn was released in the form of a dissolved state in 102 to 1700 μg·L^{-1} ZnO NPs, which is higher than the release of Ag from the AgNPs in our study. This showed that the release of ions from the nanoparticles—that is, the stability of the nanoparticle solution—would be affected by the surface wrapping, the synthetic method, the environmental media, the metal type and the environmental conditions (the organic content, pH, etc.). Therefore, understanding the possible morphological changes and dissolution in nanoparticles in different environments is of great significance in studying the toxic effects of nanoparticles on organisms [38,39].

3.2. Biological Effects of AgNPs on S. costatum

3.2.1. The Relative Growth Rate

From Figure 3, it can be seen that, after being exposed to different concentrations of AgNPs for 24 h, most of algae exhibited varying degrees of growth inhibition and, as the AgNP concentrations increased, the growth inhibition became apparent, except at 10 $\mu g \cdot L^{-1}$AgNP exposure, in which mild growth promotion occurred. In terms of ionic Ag exposure, *S. costatum* showed significant growth inhibition only at a high concentration (500 $\mu g \cdot L^{-1}$). Interestingly, at exposures with a low-concentration ionic Ag level (10 and 100 $\mu g \cdot L^{-1}$), the growth in *S. costatum* was promoted to a certain extent. This may be related to the "low toxicity stimulation" effect.

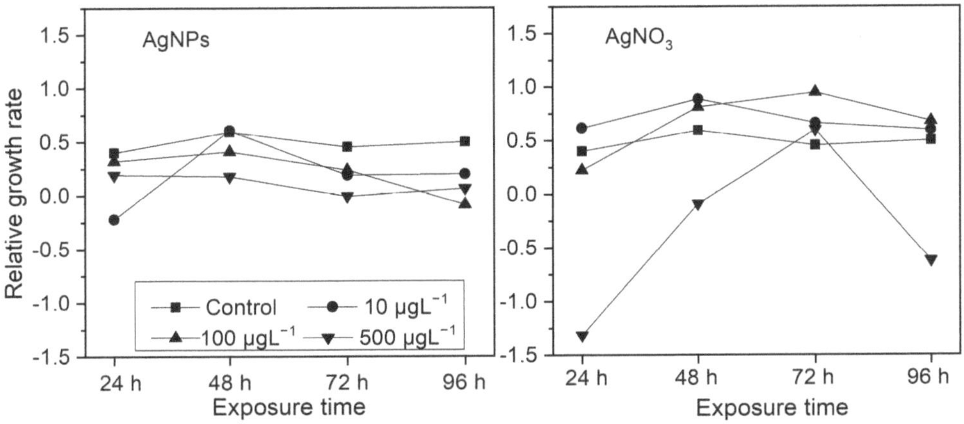

Figure 3. The effects of different concentrations of AgNPs and soluble Ag concentrations (10, 100, and 500 $\mu g \cdot L^{-1}$) on the relative growth rate of *S. costatum* at 24, 48, 72, and 96 h.

After the initial 24 h, the growth inhibition was more obvious in the 10 $\mu g \cdot L^{-1}$ AgNP treatment than in the 100 and 500 $\mu g \cdot L^{-1}$ AgNP treatments, while the growth inhibition of algae was more pronounced in the 500 $\mu g \cdot L^{-1}$ ionic Ag treatment than in the 10 and 100 $\mu g \cdot L^{-1}$ ionic Ag treatments. This may be due to the fact that nanoparticles with a higher concentration are more likely to form aggregates in an algal solution, which temporarily delays the growth inhibition of *S. costatum*, while a high concentration of ionic Ag has an immediate inhibitory effect on algal growth.

3.2.2. Ag Accumulation

The Ag accumulation in *S. costatum* at different exposures of AgNPs and ionic Ag is shown in Figure 4. As seen in Figure 4, the Ag accumulation in *S. costatum* at 48 h of exposure to ionic Ag (500 $\mu g \cdot L^{-1}$) was 3.70 times higher than that in the AgNP solution (500 $\mu g \cdot L^{-1}$). However, the Ag accumulation for the exposure with 100 $\mu g \cdot L^{-1}$ AgNPs was 1.64 times higher than that in the algae exposed to the ionic Ag solution. In addition, a higher concentration of Ag exposure induced a greater Ag accumulation. For example, the Ag accumulation in the algae exposed to 500 $\mu g \cdot L^{-1}$ AgNPs was 1.52 times that in the algae exposed to 100 $\mu g \cdot L^{-1}$ AgNPs. The Ag accumulation in the 500 $\mu g \cdot L^{-1}$ ionic Ag exposure was 3.35 times that in the 100 $\mu g \cdot L^{-1}$ ionic Ag exposure experiments. The Ag accumulation levels in the control and the 10 $\mu g \cdot L^{-1}$ Ag exposure experiments were all lower than the detection limit. The above results show that AgNPs can be absorbed in the form of particles. High concentrations of ionic Ag and low concentrations of AgNPs are more likely to accumulate in cells, which may be related to the aggregation of nanoparticles. There are more opportunities for high concentrations of AgNPs to form larger-particle-size aggregates through collisions between particles. The particle size of the aggregates exceeds the pore size of the cell membrane, hindering the entry of nanoparticles into the cell.

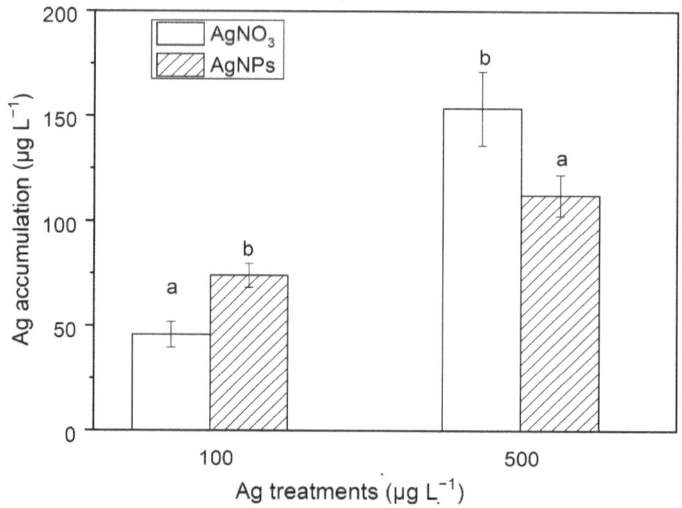

Figure 4. The Ag accumulation in algae exposed to soluble Ag and AgNPs for 48 h. The same identification letter (e.g., a and a) represents no significant difference, while different identification letters (e.g., a and b) represent a significant difference.

3.2.3. Destruction of Membrane Stability of *S. costatum* by AgNPs

The degree of damage of the cell membrane is represented in E1/E2. The larger the value, the greater the degree of damage. From Figure 5, it can be seen that, compared with the control, the AgNP exposure showed obvious damage to the algal cell membrane. With the ionic Ag exposure, only the 500 $\mu g \cdot L^{-1}$ levels showed obvious damage to the algal cell membrane. The results show that AgNPs can induce damage to the stability of algal cell membranes at a relatively low concentration level, which is related to the unique properties of nanoparticles.

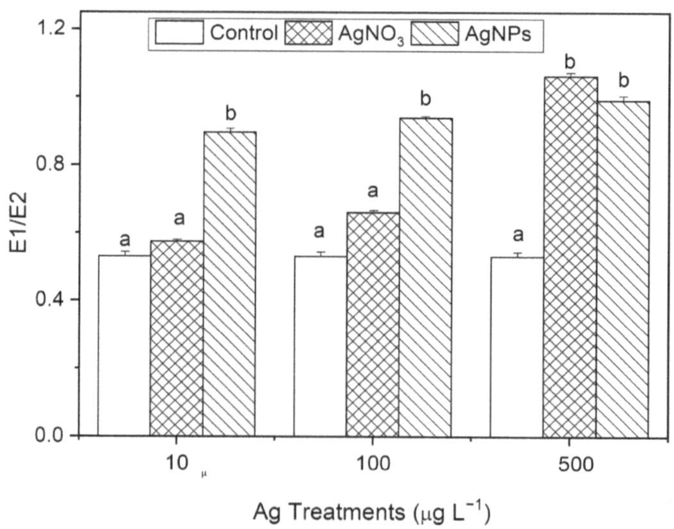

Figure 5. The stability changes in the cell membrane after *S. costatum* was exposed to different concentrations of soluble Ag and AgNPs for 48 h. The same identification letter (e.g., a and a) represents no significant difference, while different identification letters (e.g., a and b) represent a significant difference.

3.2.4. The Effect of AgNPs on Chl-a Contents in *S. costatum*

After 48 h of exposure to Ag solutions, the Chl-a contents in *S. costatum* under the AgNP treatments decreased slightly (from 4% to 22%) compared to the control ($p > 0.05$, Figure 6). Regarding the ionic Ag exposure, the Chl-a contents decreased significantly, by about 60%. This suggests that the ionic Ag was the main cause of the decrease in Chl-a in *S. costatum*. It is worth noting that, after the addition of cysteine, the Chl-a contents in *S. costatum* under the $AgNO_3$ treatments showed no significant difference from the control and AgNP exposures. This indicates that cysteine eliminates the inhibitory effect on the Chl-a production from complexing ionic Ag. Furthermore, it could be speculated that the inhibitory effect on Chl-a production was mainly caused by Ag ions, but the soluble Ag released in the AgNPs (100 to 500 $\mu g \cdot L^{-1}$) was not enough to induce an obvious inhibitory effect on the Chl-a production in *S. costatum*.

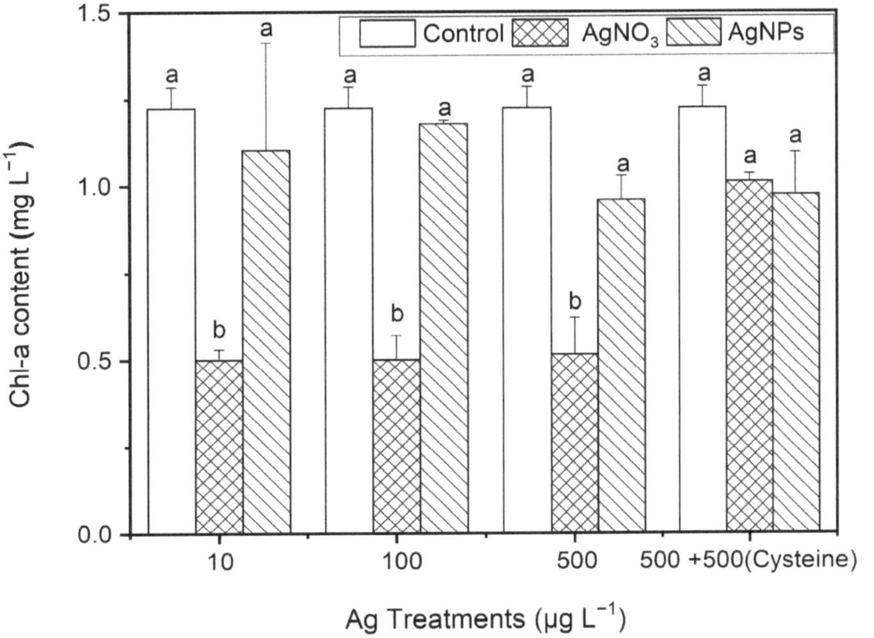

Figure 6. The evolution of the Chl-a content in algae exposed to different concentrations of ionic Ag and AgNPs for 48 h. The same identification letter (e.g., a and a) represents no significant difference, while different identification letters (e.g., a and b) represent a significant difference.

The adsorption of AgNPs on the surface of microalgae cells will induce obvious toxicity through direct or indirect physical reactions, such as the disintegration of cell walls [40], or the reduction in light required for algal growth ("shading effect") [41], and limit the utilization of nutrients [42,43]. In the study by Burchardt et al. [36], the same concentration of AgNPs also exhibited varying degrees of growth inhibition on *Pseudomonas aeruginosa* (inhibition percentage: from 40% to 60%) and *Polycystis cyanobacteria* (inhibition percentage: 100%). In this study, the growth inhibition percentage of *S. costatum* was 58%, 70%, and 86% for 10, 100, and 500 $\mu g \cdot L^{-1}$ AgNP exposure, respectively. Due to the limited research on the toxicity of AgNPs to algae, it can be inferred that the sensitivity of *S. costatum* to AgNPs is approximately between those of *Pseudomonas aeruginosa* and *Polycystis cyanobacteria*. In addition, low concentrations of AgNPs have a more obvious inhibitory effect on algal growth than relative high concentrations of AgNPs, while a high concentration of ionic Ag has an obvious algal inhibitory effect. This may be due to the fact that relative high concentrations of nanoparticles are more likely to form flocs in the algal medium, which temporarily delays the growth inhibition of *S. costatum*. The effect of ionic Ag on algal growth can be explained by the "Hormesis effect of toxins", which can be said to be an adaptive response of organisms to toxins (in the sense of, "What does not kill you makes you stronger"). That is to say, exposure to low-dose toxins will appropriately promote a certain function in the organism (referring to algal growth in this case), while exposure to high concentrations of toxins will have an inhibitory effect on that function.

The accumulation of metals in the organisms indicates that the toxicity of nanometals to organisms is, to some extent, based on the absorption of metals by organisms. Perreault et al. [44] emphasized, in their study on the toxicity of nano copper oxide to the green alga *Chlamydomonas reinhardtii*, that the exposure to polymer encapsulated nanoparticle solution could increase the absorption of nano copper oxide by algal cells. Hull et al. [45] studied the absorption and retention of nanometals by purple mussels, indicating that, after initial exposure for 8 h, more than 90% of nanometals were absorbed by purple mussels, and almost all nanometals were present in the digestive glands of mussels. In this study, after 48 h of exposure to 500 $\mu g \cdot L^{-1}$ AgNPs, approximately 22% of Ag was accumulated in the *S. costatum*. Therefore, the biological absorption of nanoparticles cannot be ignored when studying the biological toxicity effects of nanoparticles.

As is well known, the cell wall is the first reaction site between nanoparticles and cells, and serves as an important barrier to the absorption of nanoparticles. According to the relevant research [46–48], the pore size on the cell wall of diatoms is, approximately, from 3 to 50 nm. That is to say, if the particle size of the nanoparticles is smaller than the pore size, then they can enter the cell through endocytosis, diffusion, or reacting with carrier proteins [47,48]. Moreover, during the regeneration process, the permeability of the cell wall changes. The newly generated cell wall can make it easier for nanoparticles to enter [49]. In addition, the reaction between cells and nanoparticles induces a new pore size that is larger than the normal pore size, which increases the possibility of nanoparticles entering the cell through the cell wall [43,47].

As one of the important producers in the entire marine ecosystem, diatom photosynthesis plays a crucial role in energy flow and nutrient transfer in the food web. Therefore, studying the impact of nanoparticles on alga photosynthesis has a certain scientific significance. Oukarroum et al. [50] studied the inhibitory effect of AgNPs on freshwater algae (*Chlorella vulgaris*) and marine algae (*Dunaliella salina*), and the results showed that AgNPs caused a sharp decrease in chlorophyll production in these two algae. Chen et al. [51] studied the effect of nano titanium dioxide on the photosynthetic capacity (Fv/Fm) of *Chlamydomonas reinhardtii*. The results showed that, when the concentration of nano titanium dioxide was >1 $mg \cdot L^{-1}$, the photosynthetic capacity of *Chlamydomonas reinhardtii* decreased rapidly. Perreault et al. [44] studied the toxicity of nano copper oxide particles to the green alga *Chlamydomonas reinhardtii*, and indicated that the polymer-encapsulated nanoparticle solutions significantly reduced the electron transfer ability of *Chlamydomonas reinhardtii* photosynthetic system II. Miao et al. [52] studied the toxicity of nano zinc oxide to *Thalassiosira pseudonana*; it was found that, as the concentration of nanoparticles increased, the photosynthetic capacity (Fv/Fm) almost decreased to 0, and the Chl-a content decreased by 81%. Röhder et al. [53] showed that, when the concentration of nano tin oxide increased to 50 μM, the short-term toxicity of nano tin oxide particles was obviously promoted, and the photosynthesis yield decreased significantly. Therefore, the inhibition of photosynthesis is an important cause of the toxicity of nanoparticles to the aforementioned alga. However, in this study, although the decreases in Chl-a were induced by different concentrations of AgNPs, there was no significant difference compared to the control. This indicates that the inhibition of photosynthesis by AgNPs (within the concentration range from 10 to 500 $\mu g \cdot L^{-1}$) was not the main cause of toxicity to *S. costatum*.

3.3. The Effect of AgNPs on the Antioxidant System of S. costatum

3.3.1. SOD

From Figure 7, it can be seen that, after 48 h of exposure to Ag solutions with different concentration gradients, the SOD contents in the algae in the AgNP treatments were 1.17, 0.34, and 1.72 times higher than those in the ionic Ag treatments for 10, 100, and 500 $\mu g \cdot L^{-1}$, respectively. However, the SOD contents in *S. costatum* in the equal amount of cysteine addition treatments, there was no significant difference between the control and the two forms of Ag treatment.

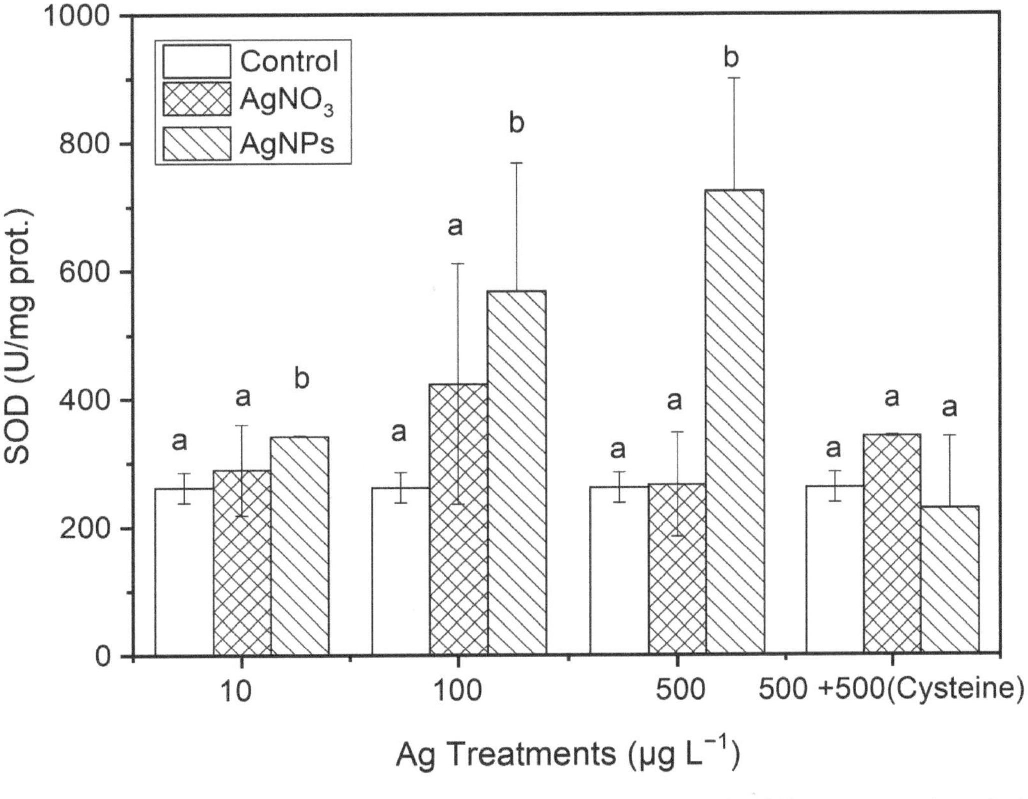

Figure 7. The evolution of SOD content in algae exposed to different concentrations of soluble Ag and AgNPs for 48 h. The same identification letter (e.g., a and a) represents no significant difference, while different identification letters (e.g., a and b) represent a significant difference.

3.3.2. CAT

As can be seen in Figure 8, the CAT contents in *S. costatum* under AgNP exposure were 2.37, 1.51, and 2.86 times higher than those in the control for the 10, 100, and 500 $\mu g \cdot L^{-1}$ treatments, respectively. Additionally, there was a significant difference in the CAT contents only between the 10 $\mu g \cdot L^{-1}$ AgNP and ionic Ag treatments. After the addition of an equal amount of cysteine to the 500 $\mu g \cdot L^{-1}$ Ag solutions, there was no significant difference in the CAT contents between the control and the two forms of Ag treatments, which was similar to the changes in SOD contents in Section 3.3.1.

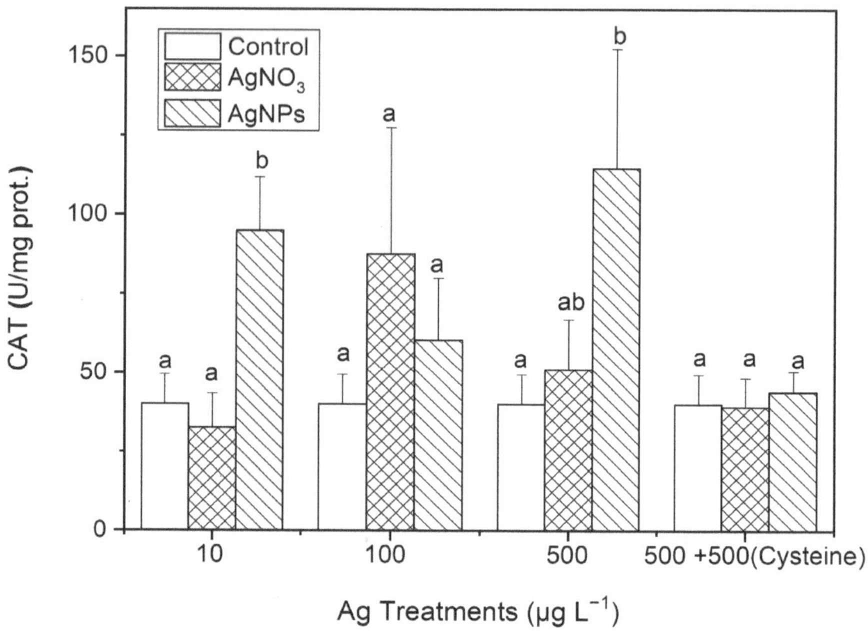

Figure 8. The evolution of the CAT content in algae exposed to different concentrations of soluble Ag and AgNPs for 48 h. The same identification letter (e.g., a and a) represents no significant difference, while different identification letters (e.g., a and b) represent a significant difference.

3.3.3. MDA

After organisms are subjected to oxidative stress, the production of MDA can effectively represent the degree of lipid peroxidation. From Figure 9, it can be seen that the MDA contents in *S. costatum* had also undergone varying degrees of change. In the 100 and 500 $\mu g \cdot L^{-1}$ AgNP treatments, the MDA contents were 3.42 and 20.67 times higher than those in the soluble Ag treatments, respectively. There was no significant difference in the MDA contents between the soluble Ag treatments and the control. After an equal amount of cysteine was added to the 500 $\mu g \cdot L^{-1}$ Ag solutions, no significant difference in MDA contents occurred between the control and the two forms of Ag treatments, which was similar to the changes in SOD and CAT contents. This indicates that AgNPs could induce an obvious oxidative stress on *S. costatum*.

The effect of AgNPs on oxidative stress and the antioxidant defense can alter the enzyme activity of SOD and CAT. The induction of these antioxidant enzymes is an effective detoxification mechanism for alga cells to resist exposure to nanoparticles. The production of lipid peroxide MDA indicates that the toxicity of AgNPs is related to oxidative stress. We speculate that, as the concentration of AgNPs continues to increase, the enzyme activity of SOD and CAT will decrease, and the MDA content will continue to increase, because the further increase in the degree of oxidative stress may exceed the ability of antioxidant enzymes to scavenge free radicals.

The study by Oukarroum et al. [50] showed that AgNPs induced lipid peroxidation in two green algae (*Chlorella vulgaris* and *Dunaliella salina*). Buffet et al. [54] studied the toxicity of AgNPs to two benthic organisms (the silkworm and bivalve), and the results showed that AgNP and ionic Ag had a significantly enhanced effect on CAT activity, and AgNPs also significantly induced the SOD activity in silkworms. Jiang et al. [55] found the toxic effects of AgNPs on the aquatic plant *Spirodela polyrhiza*, and that exposure to AgNPs significantly enhanced SOD and CAT activities in the duckweed, while particle Ag on the micrometer level had no obvious effect on SOD and CAT activities. Lin et al. [40] found that nano titanium dioxide could cause an increase in the levels of reactive oxygen species and MDA in *Chlorella vulgaris*. Chen et al. [51] showed that nano titanium dioxide also increased the MDA in *Chlamydomonas reinhardtii* to its maximum value after 8 h of exposure. Some studies [48,56] have also studied the toxic effect of AgNPs on large fleas using the ion complexation method with the addition of cysteine, confirming that the biological toxicity of nanoparticles was greatly reduced after cysteine complexed the released ionic Ag. Therefore, nanoparticles can cause oxidative stress in organisms such as *S. costatum*, leading to toxic effects on organisms [57,58].

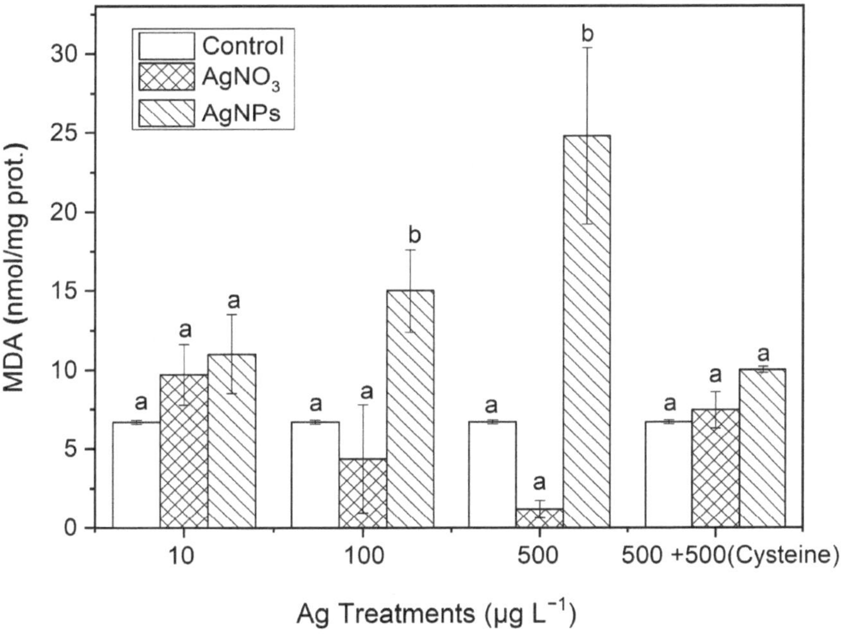

Figure 9. The evolution of the MDA content in algae exposed to different concentrations of soluble Ag and AgNPs for 48 h. The same identification letter (e.g., a and a) represents no significant difference, while different identification letters (e.g., a and b) represent a significant difference.

4. Conclusions

In this study, the biotoxic effects of AgNPs were found to comprise a combination of the solubilization of particles into toxic metal ions and the nature of the nanoparticles. The ion complexation with cysteine could reduce the biological toxicity of the released ionic Ag from the AgNPs. However, the induction of oxidative stress and the damage to the cell membrane comprised the dominant mechanism of toxicity for AgNPs. Therefore, the nature of nanometals in an aquatic environment is the key factor for understanding their biotoxic effect on phytoplankton. The study into the toxicity mechanism of nanoparticles in depth, and the possible morphological changes and dissolution of nanoparticles in different environments, is of great significance.

Author Contributions: Conceptualization, B.W.; Methodology, K.S. and B.W.; Software, K.S. and B.W.; Validation, D.C. and B.W.; Data curation, D.C., Y.Y. and B.W.; Writing—original draft preparation, K.S.; Writing—review and editing, B.W.; Visualization, Y.Y. and J.X.; Supervision, B.W.; Project administration, B.W.; Funding acquisition, J.X. and B.W. All authors have read and agreed to the published version of the manuscript.

Funding: This work was supported by the Natural Science Foundation of Shandong Province (ZR2021QD051); the scientific research fund project of the National Natural Science Foundation of China (52070123); and Project of Shandong Province Higher Educational Young Innovative Talent Introduction and Cultivation Team [Wastewater Treatment and Resource Innovation Team].

Institutional Review Board Statement: Not applicable.

Informed Consent Statement: Not applicable.

Data Availability Statement: The data that support the findings of this study are available from the corresponding author upon reasonable request.

Conflicts of Interest: The authors declare no conflict of interest.

References

1. Mahjoubian, M.; Naeemi, A.S.; Moradi-Shoeili, Z.; Tyler, C.R.; Mansouri, B. Toxicity of Silver Nanoparticles in the Presence of Zinc Oxide Nanoparticles Differs for Acute and Chronic Exposures in Zebrafish. *Arch. Environ. Con. Tox.* **2023**, *84*, 1–17. [CrossRef]
2. Bakr, Z.; Abdel-Wahab, M.; Thabet, A.A.; Hamed, M.; El-Aal, M.A.; Saad, E.; Faheem, M.; Sayed, A.E.-D.H. Toxicity of silver, copper oxide, and polyethylene nanoparticles on the earthworm *Allolobophora caliginosa* using multiple biomarkers. *Appl. Soil Ecol.* **2023**, *181*, 104681. [CrossRef]
3. Kose, O.; Mantecca, P.; Costa, A.; Carrière, M. Putative adverse outcome pathways for silver nanoparticle toxicity on mammalian male reproductive system: A literature review. *Part. Fibre Toxicol.* **2023**, *20*, 1. [CrossRef] [PubMed]
4. Lubick, N. Nanosilver toxicity: Ions, nanoparticles—Or both? *Environ. Sci. Technol.* **2008**, *42*, 8617. [CrossRef]
5. Arora, S.; Rajwade, J.M.; Paknikar, K.M. Nanotoxicology and in vitro studies: The need of the hour. *Toxicol. Appl. Pharm.* **2012**, *258*, 151–165.
6. Wang, L.-F.; Habibul, N.; He, D.-Q.; Li, W.-W.; Zhang, X.; Jiang, H.; Yu, H.-Q. Copper release from copper nanoparticles in the presence of natural organic matter. *Water Res.* **2015**, *68*, 12–23. [CrossRef] [PubMed]
7. Chen, M.; Dei, R.C.H.; Wang, W.-X.; Guo, L. Marine diatom uptake of iron bound with natural colloids of different origins. *Mar. Chem.* **2003**, *81*, 177–189. [CrossRef]
8. Chen, M.; Wang, W.X. bioavailability of natural colloid-bound iron to marine plankton: Influences of colloidal size and aging. *Limnol. Oceanogr.* **2001**, *46*, 1956–1967. [CrossRef]
9. Peyrot, C.; Wilkinson, K.J.; Desrosiers, M.; Sauve, S. Effects of silver nanoparticles on soil enzyme activities with and without added organic matter. *Environ. Toxicol. Chem.* **2014**, *33*, 115–125. [CrossRef]
10. Sharma, V.K.; Siskova, K.M.; Zboril, R.; Gardea-Torresdey, J.L. Organic-coated silver nanoparticles in biological and environmental conditions: Fate, stability and toxicity. *Adv. Colloid Interface Sci.* **2013**, *204*, 15–34. [CrossRef] [PubMed]
11. Saleh, N.; Kim, H.-J.; Phenrat, T.; Matyjaszewski, K.; Tilton, R.D.; Lowry, G.V. Ionic strength and composition affect the mobility of surface-modified Fe^0 nanoparticles in water-saturated sand columns. *Environ. Sci. Technol.* **2008**, *42*, 3349–3355. [CrossRef]
12. Buffet, P.-E.; Pan, J.-F.; Poirier, L.; Amiard-Triquet, C.; Amiard, J.-C.; Gaudin, P.; Faverney, C.R.-d.; Guibbolini, M.; Gilliland, D.; Valsami-Jones, E.; et al. Biochemical and behavioural responses of the endobenthic bivalve *Scrobicularia plana* to silver nanoparticles in seawater and microalgal food. *Ecotox. Environ. Safe* **2013**, *89*, 117–124. [CrossRef]
13. Buffet, P.E.; Zalouk-Vergnoux, A.; Chatel, A.; Berthet, B.; Metais, I.; Perrein-Ettajani, H.; Poirier, L.; Luna-Acosta, A.; Thomas-Guyon, H.; Risso-de Faverney, C.; et al. A marine mesocosm study on the environmental fate of silver nanoparticles and toxicity effects on two endobenthic species: The ragworm *Hediste diversicolor* and the bivalve mollusc *Scrobicularia plana*. *Sci. Total Environ.* **2014**, *470*, 1151–1159. [CrossRef] [PubMed]
14. Turan, N.B.; Erkan, H.S.; Engin, G.O.; Bilgili, M.S. Nanoparticles in the aquatic environment: Usage, properties, transformation and toxicity—A review. *Process Saf. Environ.* **2019**, *130*, 238–249. [CrossRef]
15. Garcés, M.; Cáceres, L.; Chiappetta, D.; Magnani, N.; Evelson, P. Current understanding of nanoparticle toxicity mechanisms and interactions with biological systems. *New J. Chem.* **2021**, *45*, 14328–14344. [CrossRef]
16. Miao, A.-J.; Schwehr, K.A.; Xu, C.; Zhang, S.-J.; Luo, Z.; Quigg, A.; Santschi, P.H. The algal toxicity of silver engineered nanoparticles and detoxification by exopolymeric substances. *Environ. Pollut.* **2009**, *157*, 3034–3041. [CrossRef] [PubMed]
17. Choi, O.; Hu, Z. Size dependent and reactive oxygen species related nanosilver toxicity to nitrifying bacteria. *Environ. Sci. Technol.* **2008**, *42*, 4583–4588. [CrossRef]
18. Newton, K.M.; Puppala, H.L.; Kitchens, C.L.; Colvin, V.L.; Klaine, S.J. Silver nanoparticle toxicity to Daphnia magna is a function of dissolved silver concentration. *Environ. Toxicol. Chem.* **2013**, *32*, 2356–2364. [CrossRef] [PubMed]
19. Lu, R.X.Z.; Radisic, M. Organ-on-a-chip platforms for evaluation of environmental nanoparticle toxicity. *Bioact. Mater.* **2021**, *6*, 2801–2819. [CrossRef]
20. Zhang, L.; Wang, W.-X. Silver nanoparticle toxicity to the larvae of oyster *Crassostrea angulata*: Contribution of in vivo dissolution. *Sci. Total Environ.* **2023**, *858*, 159965. [CrossRef]
21. Ringwood, A.H.; McCarthy, M.; Bates, T.C.; Carroll, D.L. The effects of silver nanoparticles on oyster embryos. *Mar. Environ. Res.* **2010**, *69*, S49–S51. [CrossRef] [PubMed]

22. Azadikhah, D.; Yalsuyi, A.M.; Saha, S.; Saha, N.C.; Faggio, C. Biochemical and Pathophysiological Responses in *Capoeta capoeta* under Lethal and Sub-Lethal Exposures of Silver Nanoparticles. *Water* **2023**, *15*, 585.
23. Liu, Z.; Malinowski, C.R.; Sepúlveda, M.S. Emerging trends in nanoparticle toxicity and the significance of using Daphnia as a model organism. *Chemosphere* **2022**, *291*, 132941. [CrossRef] [PubMed]
24. Padhye, L.P.; Jasemizad, T.; Bolan, S.; Tsyusko, O.V.; Unrine, J.M.; Biswal, B.K.; Balasubramanian, R.; Zhang, Y.; Zhang, T.; Zhao, J.; et al. Silver contamination and its toxicity and risk management in terrestrial and aquatic ecosystems. *Sci. Total Environ.* **2023**, *871*, 161926. [CrossRef] [PubMed]
25. Gao, J.; Powers, K.; Wang, Y.; Zhou, H.; Roberts, S.M.; Moudgil, B.M.; Koopman, B.; Barber, D.S. Influence of Suwannee River humic acid on particle properties and toxicity of silver nanoparticles. *Chemosphere* **2012**, *89*, 96–101. [CrossRef] [PubMed]
26. Dube, E.; Okuthe, G.E. Engineered nanoparticles in aquatic systems: Toxicity and mechanism of toxicity in fish. *Emerg. Contam.* **2023**, *9*, 100212. [CrossRef]
27. Liu, D.; Sun, J.; Zou, J.; Zhang, J. Phytoplankton succession during a red tide of *Skeletonema costatum* in Jiaozhou Bay of China. *Mar. Pollut. Bull.* **2005**, *50*, 91–94. [CrossRef]
28. Guillard, R.; Ryther, J. Studies of marine planktonic diatoms I: Cyclotella nanai Hustedt and Detonula confervacea Cleva. *Can. J. Microbiol.* **1962**, *8*, 229–239. [CrossRef]
29. He, Y.-L.; Bai, J.; Bai, X.-Y.; Li, K.-R. Effect of Nano-Silver on Planktonic-Bacterial Activity in Qingdao Coastal Waters. *P. Ocean Univ. China* **2020**, *50*, 76–82.
30. Pisani, T.; Munzi, S.; Paoli, L.; Bačkor, M.; Loppi, S. Physiological effects of arsenic in the lichen *Xanthoria parietina* (L.) Th. Fr. *Chemosphere* **2011**, *82*, 963–969. [CrossRef] [PubMed]
31. Porra, R.J.; Thompson, W.A.; Kriedemann, P.E. Determination of accurate extinction coefficients and simultaneous equations for assaying chlorophylls a and b extracted with four different solvents: Verification of the concentration of chlorophyll standards by atomic absorption spectroscopy. *Biochim. Biophys. Acta Bioenerg.* **1989**, *975*, 384–394. [CrossRef]
32. Gomes, T.; Pereira, C.G.; Cardoso, C.; Pinheiro, J.P.; Cancio, I.; Bebianno, M.J. Accumulation and toxicity of copper oxide nanoparticles in the digestive gland of Mytilus galloprovincialis. *Aquat. Toxicol.* **2012**, *118*, 72–79. [CrossRef] [PubMed]
33. Shin, S.W.; Song, I.H.; Um, S.H. Role of Physicochemical Properties in Nanoparticle Toxicity. *Nanomaterials* **2015**, *5*, 1351–1365. [CrossRef]
34. Sukhanova, A.; Bozrova, S.; Sokolov, P.; Berestovoy, M.; Karaulov, A.; Nabiev, I. Dependence of Nanoparticle Toxicity on Their Physical and Chemical Properties. *Nanoscale Res. Lett.* **2018**, *13*, 44. [CrossRef]
35. Zou, X.; Shi, J.; Zhang, H. Coexistence of silver and titanium dioxide nanoparticles: Enhancing or reducing environmental risks? *Aquat. Toxicol.* **2014**, *154*, 168–175. [CrossRef] [PubMed]
36. Burchardt, A.D.; Carvalho, R.N.; Valente, A.; Nativo, P.; Gilliland, D.; Garcìa, C.P.; Passarella, R.; Pedroni, V.; Rossi, F.; Lettieri, T. Effects of Silver Nanoparticles in Diatom *Thalassiosira pseudonana* and *Cyanobacterium Synechococcus* sp. *Environ. Sci. Technol.* **2012**, *46*, 11336–11344. [CrossRef] [PubMed]
37. Wang, D.; Gao, Y.; Lin, Z.; Yao, Z.; Zhang, W. The joint effects on *Photobacterium phosphoreum* of metal oxide nanoparticles and their most likely coexisting chemicals in the environment. *Aquat. Toxicol.* **2014**, *154*, 200–206. [CrossRef]
38. Seitz, F.; Rosenfeldt, R.R.; Storm, K.; Metreveli, G.; Schaumann, G.E.; Schulz, R.; Bundschuh, M. Effects of silver nanoparticle properties, media pH and dissolved organic matter on toxicity to *Daphnia magna*. *Ecotox Environ. Safe* **2015**, *111*, 263–270. [CrossRef]
39. Buchman, J.T.; Hudson-Smith, N.V.; Landy, K.M.; Haynes, C.L. Understanding Nanoparticle Toxicity Mechanisms to Inform. Redesign Strategies To Reduce Environmental Impact. *Acc. Chem. Res.* **2019**, *52*, 1632–1642. [CrossRef] [PubMed]
40. Lin, D.; Ji, J.; Long, Z.; Yang, K.; Wu, F. The influence of dissolved and surface-bound humic acid on the toxicity of TiO_2 nanoparticles to *Chlorella* sp. *Water Res.* **2012**, *46*, 4477–4487. [CrossRef]
41. Aruoja, V.; Dubourguier, H.-C.; Kasemets, K.; Kahru, A. Toxicity of nanoparticles of CuO, ZnO and TiO_2 to microalgae *Pseudokirchneriella subcapitata. Sci. Total Environ.* **2009**, *407*, 1461–1468. [CrossRef] [PubMed]
42. Hartmann, N.B.; Von der Kammer, F.; Hofmann, T.; Baalousha, M.; Ottofuelling, S.; Baun, A. Algal testing of titanium dioxide nanoparticles—Testing considerations, inhibitory effects and modification of cadmium bioavailability. *Toxicology* **2010**, *269*, 190–197. [CrossRef] [PubMed]
43. Xia, B.; Chen, B.; Sun, X.; Qu, K.; Ma, F.; Du, M. Interaction of TiO_2 nanoparticles with the marine microalga *Nitzschia closterium*: Growth inhibition, oxidative stress and internalization. *Sci. Total Environ.* **2015**, *508*, 525–533. [CrossRef] [PubMed]
44. Perreault, F.; Oukarroum, A.; Melegari, S.P.; Matias, W.G.; Popovic, R. Polymer coating of copper oxide nanoparticles increases nanoparticles uptake and toxicity in the green alga *Chlamydomonas reinhardtii. Chemosphere* **2012**, *87*, 1388–1394. [CrossRef]
45. Hull, M.S.; Vikesland, P.J.; Schultz, I.R. Uptake and retention of metallic nanoparticles in the Mediterranean mussel *(Mytilus galloprovincialis). Aquat. Toxicol.* **2013**, *140–141*, 89–97. [CrossRef]
46. Vrieling, E.G.; Beelen, T.P.M.; van Santen, R.A.; Gieskes, W.W.C. Diatom silicon biomineralization as an inspirational source of new approaches to silica production. *J. Biotechnol.* **1999**, *70*, 39–51. [CrossRef]
47. Navarro, E.; Baun, A.; Behra, R.; Hartmann, N.B.; Filser, J.; Miao, A.J.; Quigg, A.; Santschi, P.H.; Sigg, L. Environmental behavior and ecotoxicity of engineered nanoparticles to algae, plants, and fungi. *Ecotoxicology* **2008**, *17*, 372–386. [CrossRef]
48. Navarro, E.; Piccapietra, F.; Wagner, B.; Marconi, F.; Kaegi, R.; Odzak, N.; Sigg, L.; Behra, R. Toxicity of Silver Nanoparticles to *Chlamydomonas reinhardtii. Environ. Sci. Technol.* **2008**, *42*, 8959–8964. [CrossRef]

49. Ovečka, M.; Lang, I.; Baluška, F.; Ismail, A.; Illeš, P.; Lichtscheidl, I.K. Endocytosis and vesicle trafficking during tip growth of root hairs. *Protoplasma* **2005**, *226*, 39–54. [CrossRef]
50. Oukarroum, A.; Bras, S.; Perreault, F.; Popovic, R. Inhibitory effects of silver nanoparticles in two green algae, *Chlorella vulgaris* and *Dunaliella tertiolecta*. *Ecotox. Environ. Safe* **2012**, *78*, 80–85. [CrossRef]
51. Chen, L.; Zhou, L.; Liu, Y.; Deng, S.; Wu, H.; Wang, G. Toxicological effects of nanometer titanium dioxide (nano-TiO_2) on *Chlamydomonas reinhardtii*. *Ecotox. Environ. Safe* **2012**, *84*, 155–162. [CrossRef] [PubMed]
52. Miao, A.J.; Zhang, X.Y.; Luo, Z.; Chen, C.S.; Chin, W.C.; Santschi, P.H.; Quigg, A. Zinc oxide–engineered nanoparticles: Dissolution and toxicity to marine phytoplankton. *Environ. Toxicol. Chem.* **2010**, *29*, 2814–2822. [CrossRef]
53. Röhder, L.A.; Brandt, T.; Sigg, L.; Behra, R. Influence of agglomeration of cerium oxide nanoparticles and speciation of cerium (III) on short term effects to the green algae *Chlamydomonas reinhardtii*. *Aquat. Toxicol.* **2014**, *152*, 121–130. [CrossRef] [PubMed]
54. Buffet, P.-E.; Poirier, L.; Zalouk-Vergnoux, A.; Lopes, C.; Amiard, J.-C.; Gaudin, P.; Risso-de Faverney, C.; Guibbolini, M.; Gilliland, D.; Perrein-Ettajani, H.; et al. Biochemical and behavioural responses of the marine polychaete Hediste diversicolor to cadmium sulfide quantum dots (CdS QDs): Waterborne and dietary exposure. *Chemosphere* **2014**, *100*, 63–70. [CrossRef]
55. Jiang, H.-S.; Qiu, X.-N.; Li, G.-B.; Li, W.; Yin, L.-Y. Silver nanoparticles induced accumulation of reactive oxygen species and alteration of antioxidant systems in the aquatic plant *Spirodela polyrhiza*. *Environ. Toxicol. Chem.* **2014**, *33*, 1398–1405. [CrossRef] [PubMed]
56. Zhao, C.M.; Wang, W.X. Comparison of acute and chronic toxicity of silver nanoparticles and silver nitrate to *Daphnia magna*. *Environ. Toxicol. Chem.* **2011**, *30*, 885–892. [CrossRef]
57. Yu, Z.; Li, Q.; Wang, J.; Yu, Y.; Wang, Y.; Zhou, Q.; Li, P. Reactive Oxygen Species-Related Nanoparticle Toxicity in the Biomedical Field. *Nanoscale Res. Lett.* **2020**, *15*, 115. [CrossRef]
58. Horie, M.; Tabei, Y. Role of oxidative stress in nanoparticles toxicity. *Free Radic. Res.* **2021**, *55*, 331–342. [CrossRef] [PubMed]

Journal of
Marine Science and Engineering

MDPI

Article

Features of Seismological Observations in the Arctic Seas

Artem A. Krylov [1,2,3,*], Mikhail A. Novikov [1], Sergey A. Kovachev [1], Konstantin A. Roginskiy [1], Dmitry A. Ilinsky [1], Oleg Yu. Ganzha [1], Vladimir N. Ivanov [1], Georgy K. Timashkevich [1], Olga S. Samylina [4], Leopold I. Lobkovsky [1,2,3] and Igor P. Semiletov [2,3]

1 Shirshov Institute of Oceanology, Russian Academy of Sciences, Nakhimovskiy Prospekt 36, 117997 Moscow, Russia; mihail.novikow0@gmail.com (M.A.N.); kovachev@ocean.ru (S.A.K.); roginskiy@list.ru (K.A.R.); dilinskiy61@mail.ru (D.A.I.); ganzhaoy@mail.ru (O.Y.G.); ivanov.vl-mir@yandex.ru (V.N.I.); tim@ocean.ru (G.K.T.); llobkovsky@ocean.ru (L.I.L.)

2 V.I. Il'ichev Pacific Oceanological Institute, Far Eastern Branch of the Russian Academy of Sciences, Baltijskaya St. 43, 690041 Vladivostok, Russia; ipsemiletov@gmail.com

3 Tomsk State University, Lenina Prospekt 36, 634050 Tomsk, Russia

4 Winogradsky Institute of Microbiology, Research Center of Biotechnology, Russian Academy of Sciences, Prospect 60-Letiya Oktyabrya 7/2, 117312 Moscow, Russia; olga.samylina@gmail.com

* Correspondence: artyomkrlv@ocean.ru

Abstract: This paper is devoted to the features of seismological observations in the Arctic seas, which are complicated by harsh climatic conditions, the presence of ice cover, stamukhi and icebergs, and limited navigation. Despite the high risk of losing expensive equipment, the deployment of local networks of bottom seismographs or stations installed on ice is still necessary for studying the seismotectonic characteristics and geodynamic processes of the region under consideration, the deep structure of the crust and upper mantle, seismic hazards, and other marine geohazards. Various types of seismic stations used for long-term and short-term deployments in the Russian sector of the Arctic Ocean, as well as various schemes and workflows for their deployment/recovery, are described. The characteristics of seafloor seismic noise and their features are also considered. The results of deployments demonstrate that the characteristics of the stations make it possible to reliably record earthquake signals and seismic noise. Based on the experience gained, it was concluded that the preferred schemes for deploying ocean-bottom seismographs are those in which their subsequent recovery does not depend on their power resources. Usually, such schemes allow for the possibility of dismantling stations via trawling and are suitable for the shelf depths of the sea. The advantages of such schemes include the possibility of installing additional hydrophysical and hydrobiological equipment. When using pop-up ocean-bottom seismographs, special attention should be paid to the careful planning of the recovery because its success depends on the possibility of a passage to the deployment site, which is not always possible due to changing meteorological and ice conditions. Seismic records obtained on the seafloor are characterized by a high noise level, especially during periods of time when there is no ice cover. Therefore, it is recommended to install bottom stations for periods of time when ice cover is present. The frequency range of the prevailing noise significantly overlaps with the frequency range of earthquake signals that must be taken into account when processing bottom seismic records.

Keywords: marine seismology; submarine earthquake; marine geohazards; ice cover; ocean-bottom seismograph; deployment scheme; seafloor seismic noise; microseism; horizontal-to-vertical spectral ratio

Citation: Krylov, A.A.; Novikov, M.A.; Kovachev, S.A.; Roginskiy, K.A.; Ilinsky, D.A.; Ganzha, O.Y.; Ivanov, V.N.; Timashkevich, G.K.; Samylina, O.S.; Lobkovsky, L.I.; et al. Features of Seismological Observations in the Arctic Seas. *J. Mar. Sci. Eng.* **2023**, *11*, 2221. https://doi.org/10.3390/jmse11122221

Academic Editors: Dmitry A. Ruban, George Kontakiotis and Assimina Antonarakou

Received: 18 October 2023
Revised: 18 November 2023
Accepted: 21 November 2023
Published: 23 November 2023

1. Introduction

For the effective seismological monitoring of a tectonically active area, it is necessary to deploy a seismic station network as close as possible to the observed seismogenic structures. Most of the divergent, convergent, and transform boundaries of lithospheric plates, to which the most seismically active regions of the Earth are confined, are located under the bottom of the seas and oceans. This leads to the need to develop means of recording so-called submarine earthquakes.

Currently, the main way to record earthquakes, including submarine ones, is the deployment of a network of land-based seismographs. Despite the constant development of land-based seismological equipment, the accuracy of determining the main parameters of earthquakes would be much higher if seismographs were installed on the bottom of seas and oceans, closer to the observed tectonic structures. In addition, the objects of study can be not only earthquakes but also seafloor seismic noise, processes of interaction between the lithosphere and the hydrosphere, including tsunami generation, the deep structure of the Earth under the oceans, etc.

Nowadays, for marine observations, either autonomous pop-up ocean-bottom seismographs (OBSs) are used if observations are carried out far from the coast or cabled OBSs are used if there is a possibility of communication between the observed structures and the coast. There are many types of OBSs; their design and characteristics vary significantly. An example of a large project involving OBS deployments in many areas is the OBSIP (Ocean Bottom Seismograph Instrument Pool) observation series [1,2]. Through the SAGE electronic resource [3], information is available about experiments throughout the world ocean conducted from 2001 to 2019: the location of OBSs, recording periods, descriptions and the technical characteristics of the equipment and recording channels, reports on scientific cruises, geological characteristics of the study area, etc.

The studies of the OBSIP project [4–6], like many others [7–9], are primarily focused on determining the deep structure of the crust and upper mantle, rather than studying the seismic regime, which requires a longer recording period. For this purpose, permanent cabled OBS networks located in seismically active zones are more suitable. Examples of such systems are DONET (Dense Oceanfloor Network) and S-net, which are part of the Japanese earthquake and tsunami warning system [10,11], or the Canadian network Ocean Networks Canada (ONC) located in the Cascadia region [12,13].

In addition to fundamental research, marine seismological observations find practical application due to the active industrial development of water areas, and they include marine seismic exploration and seismological monitoring. Marine seismic exploration plays a key role in the search for hydrocarbons on the shelf and continental slope. The design and construction of offshore oil and gas production facilities, such as oil platforms, marine terminals, and underwater pipelines, require accurate seismic hazard assessment, including seismological monitoring to study the natural seismic regime of the region [14] and seismic microzoning to assess the contribution of local conditions to the level of possible seismic impact [15]. For critical objects, seismological monitoring is also necessary during their operation, such as for oil and gas production facilities, since, due to the extraction of large volumes of oil and gas, the natural tectonic balance in the Earth's crust is disrupted, which can lead to the activation of seismic processes [16–18].

Many earthquakes also occur in the water area of the Arctic Ocean [19]. The Arctic is one of the key regions influencing global processes on Earth that causes great scientific interest in its complex geodynamics and tectonics. This is especially true for the Laptev Sea region since it is located, on the one hand, on the border of the Eurasian and North American plates and, on the other hand, in the region of transition from the spreading axis of the Gakkel Ridge to continental rifting on the shelf. Therefore, in recent years, a number of marine scientific cruises have been carried out in the Laptev Sea, including a seismological program of work [20–23].

Active tectonic processes and their manifestations in the form of earthquakes are interconnected with such indirect hazardous phenomena as liquefaction and the subsidence of soils, tsunamis, underwater landslides, and the massive release of bubbled methane from marine sediments [24]. Studying possible geohazards is necessary for the economic development of the Arctic region, including the development of both extraction and transport infrastructure.

Marine seismological observations are complicated primarily by the inaccessibility of study areas and the high cost of marine scientific expeditions. In addition, seafloor seismic records are characterized by a high level of natural noise and distortion associated with placing the device on the surface of soft marine soil [7,21,22,25,26]—currently, it is rather rare to have the opportunity to provide good coupling to stiff soil or installation in a borehole.

When researching in the Arctic seas, additional difficulties arise due to the presence of ice cover in most of the Arctic Ocean. In this case, it is necessary either to ensure the long-term operation of OBSs in order to wait for the next season for recovery, when the study area is again free of ice, to deploy OBSs through a hole in the ice, or to deploy seismographs on the ice surface. In the first decade of the 2000s, the fundamental possibility of recording seismic signals from earthquakes using a network of seismographs (seismic array) located on drifting ice was demonstrated: the Arctic Mid-Ocean Ridge Expedition (AMORE2001) project [27–29] and the Arctic Gakkel Vents (AGAVE) project [30,31]. During these experiments, signals from hundreds of microearthquakes associated with volcanic activity on the Gakkel Ridge were recorded.

This paper describes the design of a number of seismographs that have been used or are planned to be used for seismological research in the seas of the Russian Arctic, as well as a demonstration of some observational results. The article also aims to discuss the features of seismographs' deployment/recovery schemes and workflows, as well as the features of seafloor seismic records, the consideration of which is necessary when processing them.

2. Types of Seismographs for Use in the Arctic Seas

Depending on the tasks, geography, season, and duration of observations in the Arctic seas, we use various seismometric equipment. The stations were mainly developed at the Shirshov Institute of Oceanology of the Russian Academy of Sciences (IO RAS) and by the individual entrepreneur Dmitry Ilinsky (IP Ilinsky D.A.). If the scientific aim of the work involves recording seismic signals in a broad frequency band that is usually necessary for recording not only local earthquakes but also remote ones, as well as seismic noise, then seismic stations based on broadband molecular–electronic transfer (MET) seismometers are used [32]. This choice is due to the optimal price/quality ratio. Hydrophones are also usually used for seismological observations at the seabed of water areas since hydrophone records usually have a higher quality P-wave arrival than vertical seismometer records. But they should have a low-frequency "seismological" amplitude–frequency characteristic.

At shallow shelf depths of up to 100 m, we use heavy, non-pop-up OBS versions of MPSSR (a Russian abbreviation for marine bottom station for seismoacoustic survey) and Typhoon (Figure 1). They have a similar design, including a block of MET seismic sensors, a hydrophone, a recorder, and a battery pack. MPSSR stations are distinguished by their larger dimensions, the lower-frequency characteristics of their main set of seismometers, the presence of an additional set of high-frequency geophones, and an electronic compass module. Both OBS versions are rigidly attached to heavy, bowl-shaped concrete ballast. The dismantling of such stations is ensured either via trawling or via the functionality of an external buoy-acoustic release-ballast system.

At depths greater than 100 m and up to 6000 m, we use pop-up OBS versions, GNS (an abbreviation for geonode station) and GNS-C (Figure 2). They differ primarily in the type and characteristics of the main block of seismic sensors (GNS-C stations have a lower-frequency one), as well as in the dimensions and possible installation period—GNS stations are used for observations of up to 2–3 months, and GNS-C stations can be installed for a year.

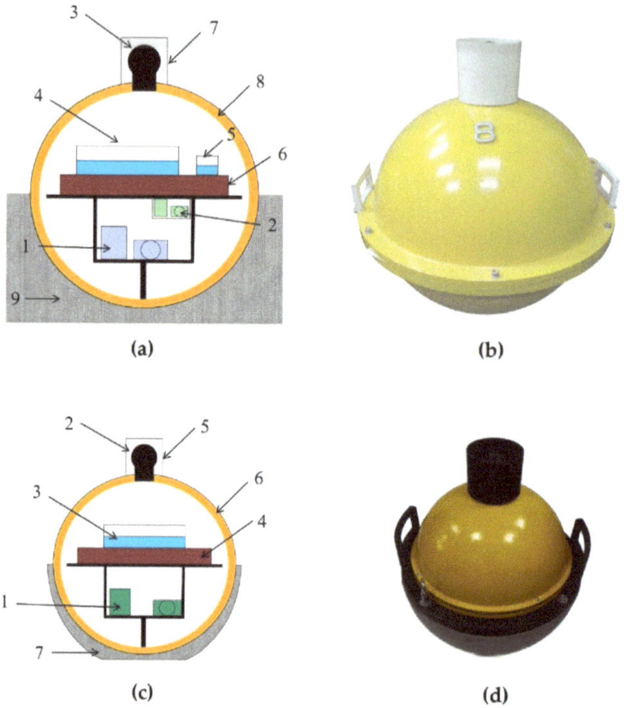

Figure 1. (**a**) Design of the MPSSR ocean-bottom seismographs: 1—three-component broadband MET seismometer CME-4311 [32], 2—three-component short-period electrodynamic seismometer (SV-10 and SG-10) placed in gimbal, 3—hydrophone 5007 m, 4—seismic recorder URS-S, 5—digital compass module, 6—battery pack, 7—protective half-cover for hydrophone, 8—duralumin sphere, 9—concrete ballast. (**b**) External view of the MPSSR. (**c**) Design of the Typhoon ocean-bottom seismographs: 1—three-component short-period MET seismometer CME-3311 [32], 2—hydrophone 5007 m, 3—seismic recorder URS-S, 4—battery pack, 5—protective half-cover for the hydrophone, 6—duralumin sphere, 7—concrete ballast. (**d**) External view of the Typhoon.

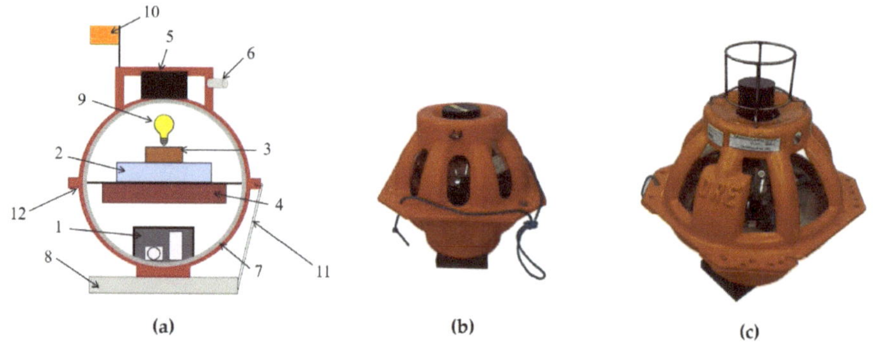

Figure 2. (**a**) Design of the GNS and GNS-C ocean-bottom seismographs: 1—three-component geophone SM-6 for GNS or MET seismometer CME-4111 [32] for GNS-C, 2—seismic recorder GNS, 3—acoustic modem, 4—battery (or accumulator) pack, 5—acoustic hydrophone, 6—seismic hydrophone (HTI-94-SSQ for GNS or EDBOE RAS for GNS-C), 7—glass spherical housing, 8—ballast, 9—lamp, 10—flag, 11—anchor release, 12—plastic case. (**b**) External view of the GNS. (**c**) External view of the GNS-C.

To carry out short-term measurements of seismic noise at shallow depths of up to 30 m through holes in fast ice, a corresponding modification of the Typhoon station was developed (Figure 3). The sensors, except the hydrophone, and the recorder are the same. The housing, ballast, internal layout of the station, and size of the battery pack differ. The dimensions of the plastic case and ballast make it possible to lower the device on a rope through a hole in the ice. The diameter of the hole is 300 mm, and it corresponds to the standard diameter of an ice drill.

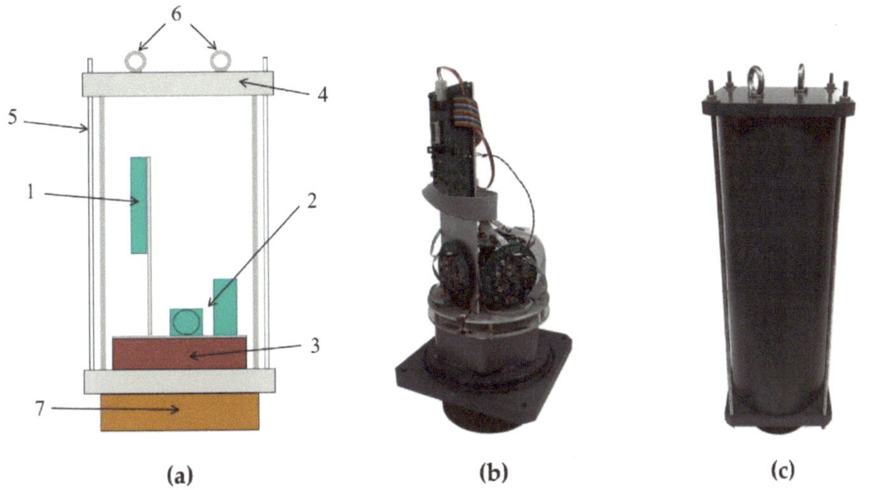

Figure 3. (**a**) Design of OBS for installation through holes in ice: 1—seismic recorder URS-S, 2—three-component short-period MET seismometer CME-3311 [32], 3—battery pack, 4—plastic housing, 5—fastening pins, 6—eyes for fastening a rope or cable, 7—ballast. (**b**) View of the internal layout of the OBS. (**c**) External view of the OBS.

To conduct long-term seismological monitoring on drifting ice, a specific seismic station was developed (Figure 4). Metal housing that is frozen into ice contains a three-component seismometer, a seismic recorder, and a digital compass/inclinometer capable of operating at low temperatures [33]. The station is equipped with a hydrophone on a cable, which is lowered into a hole in the ice to a depth of 20 m.

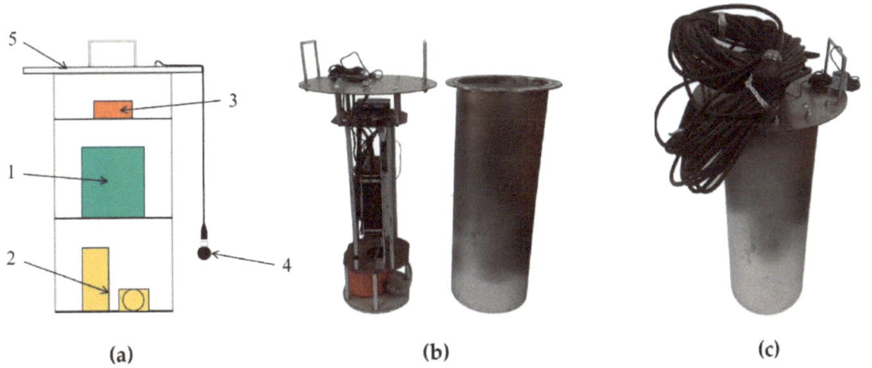

Figure 4. (**a**) Design of seismic stations for installation on ice: 1—seismic recorder DELTA-03M [33], 2—three-component electrodynamic seismometer SPV-3K [33], 3—block of additional sensors (inclinometer, thermometer, electronic compass, and battery voltage indicator), 4—hydrophone 5007 m, 5—metal housing. (**b**) View of the internal layout of the station. (**c**) External view of station with the hydrophone on a cable.

Table 1 shows the main characteristics of the seismic stations described above. Most of them (MPSSR, Typhoon, GNS, and GNS-C) have already been repeatedly used in the Arctic seas. Seismographs designed for installation on ice or through a hole in the ice are currently planned for use in the fast ice zone of the Laptev Sea.

Table 1. Main parameters of seismic stations for use in Arctic seas.

Type	MPSSR	Typhoon	GNS	GNS-C	OBS for Installation through Ice Holes	Seismograph for Installation on Ice
Developer	IO RAS	IO RAS	IP Ilinsky D.A.	IO RAS/IP Ilinsky D.A.	IO RAS	IO RAS
Dimensions	44 cm (diam.)	37 cm (diam.)	33 cm (diam.)	43 cm (diam.)	21 × 21 × 68 cm	276 cm (diam.) 75 cm (height)
Maximum depth (housing)	3000 m	2000 m	6000 m	6000 m	30 m	–
Sensors	Three-component seismometer CME-4311, three-component geophone SH/SV-10, hydrophone 5007 m	Three-component seismometer CME-3311, hydrophone 5007 m	Three-component seismometer SM-6, hydrophone HTI-94-SSQ	Three-component seismometer CME-4111/4311, hydrophone EDBOE RAS	Three-component seismometer CME-3311	Three-component seismometer SPV-3K, hydrophone 5007 m
Number of channels	7	4	4	4	3	4/8
Frequency band	0.0167–50 Hz (CME-4311), 10–250 Hz (SH/SV-10), 0.04–2500 Hz (5007 m)	1–50 Hz (CME-3311), 0.04–2500 Hz (5007 m)	4.5–140 Hz (SM-6), 2–30,000 Hz (HTI-94-SSQ)	0.0083–50 Hz (CME-4111), 0.067–30,000 Hz (EDBOE RAS)	1–50 Hz (CME-3311)	0.5–65 Hz (SPV-3K), 0.04–2500 Hz (5007 m)
Sensitivity	2000 V/m/s (CME-4311), 28 V/m/s (SH/SV-10), 7.2 ± 0.5 mV/Pa (5007 m)	2000 V/m/s (CME-3311), 7.2 ± 0.5 mV/Pa (5007 m)	28.8 V/m/s (SM-6), 12.6 V/Bar (HTI-94-SSQ without preamp)	4000 V/m/s (CME-4111/4311), 200 V/bar (EDBOE RAS)	2000 V/m/s (CME-3311)	500 V/m/s (SPV-3K)
Dynamic range	122 dB (CME-4311), 100 dB (5007 m)	118 dB (CME-3311), 100 dB (5007 m)	140 dB (SM-6), 198 dB (HTI-94-SSQ without preamp)	122 dB (CME-4111), 120 dB (EDBOE RAS)	118 dB (CME-3311)	120 dB (SPV-3K), 100 dB (5007 m)
Sample rates, Hz	20, 25, 40, 50, 80, 100, 160, 200, 400, 800	20, 25, 40, 50, 80, 100, 160, 200, 400, 800	62.5, 125, 250, 500, 1000, 2000, 4000	62.5, 125, 250, 500, 1000, 2000, 4000	20, 25, 40, 50, 80, 100, 160, 200, 400, 800	31.25, 62.5, 125, 250, 500, 1000
Time synchronization	GPS	GPS	GPS GLONASS	GPS GLONASS	GPS	GPS
Temperature stability of the quartz generator	$\pm 5 \times 10^{-9}$	$\pm 5 \times 10^{-9}$	$\pm 5 \times 10^{-9}$	$\pm 5 \times 10^{-9}$	$\pm 5 \times 10^{-9}$	10^{-7} (basic) 10^{-8} (optional)
Memory	SD card up to 64 Gb	SD card up to 64 Gb	SD card up to 128 Gb	SD card up to 128 Gb	SD card up to 64 Gb	SD card, 32 Gb
Allowed installation tilt angle	±15°	±15°	±20°	±15°	±15°	±15°
Temperature range (sensors)	−12...+55 °C (basic), −40...+55 °C (optional)	−12...+55 °C (basic), −40...+55 °C (optional)	−40...+100 °C	−12...+55 °C (basic), −40...+55 °C (optional)	−12...+55 °C (basic), −40...+55 °C (optional)	−30. . .+55 °C

3. Some Observation Results Obtained from the Arctic Seas

A study of the relationship between the phenomenon of the massive release of methane from the seafloor of the Arctic seas and various other natural processes, including seismotectonic processes, requires the involvement of specialists from many disciplines, as well as significant resources. Nevertheless, the relevance and scientific significance of such work is determined by the possible contribution of the released gas to climate change. In addition, the gas seepage from marine sediments, of course, refers to geohazards that must be assessed when developing the Arctic waters.

Regular scientific marine expeditions to the Arctic seas of Russia have an extensive multidisciplinary program of work, including geological, biological, geophysical, and chemical research [24,34–36]. For several years, OBSs were deployed for both a long recording period (several months) and a short one (several days). Figure 5 shows the locations of OBSs in the Barents Sea, Kara Sea, Laptev Sea, and East Siberian Sea. Table 2 provides information about the type of devices, coordinates, and recording periods in the experiments performed.

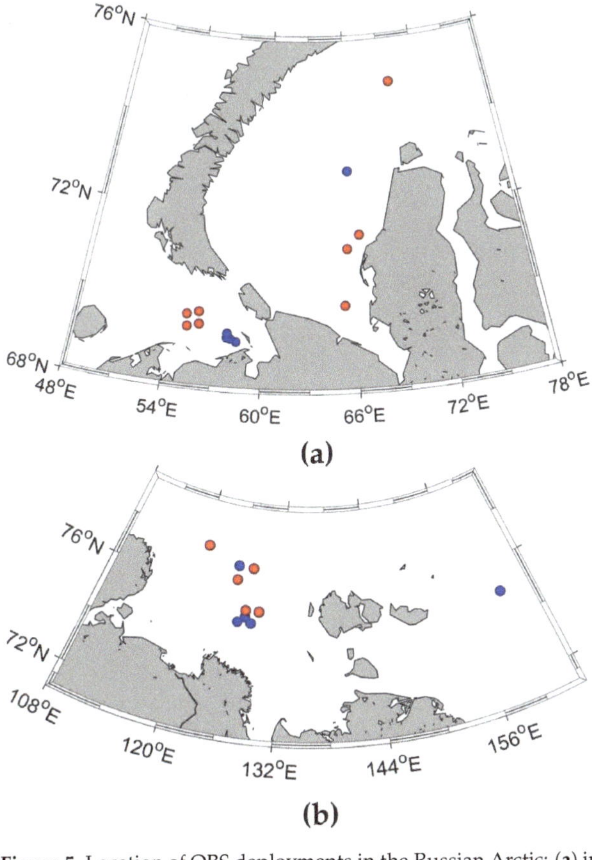

Figure 5. Location of OBS deployments in the Russian Arctic: (**a**) in the Barents and Kara seas, and (**b**) in the Laptev Sea and the East Siberian Sea. Long-term deployments (3–7 months)—red circles, short-term deployments (1 h–15 days)—blue circles.

Table 2. Locations, types of OBSs, and their operation periods in the Arctic seas.

Type	Latitude, ° N	Longitude, ° E	Depth, m	Water Area	Operation Period
			Long-term deployments		
MPSSR	75.42	127.39	42	Laptev Sea	October 2018–February 2019
MPSSR	75.43	129.13	40	Laptev Sea	October 2018–March 2019
GNS-C	77.31	120.61	350	Laptev Sea	October 2018–May 2019
MPSSR	69.67	55.18	39	Barents Sea	Aug 2018–November 2019
MPSSR	69.40	55.26	29	Barents Sea	Aug 2018–November 2019
MPSSR	69.48	56.01	29	Barents Sea	Aug 2018–November 2019
MPSSR	69.75	55.93	44	Barents Sea	Aug 2018–November 2019
MPSSR	76.39	125.66	51	Laptev Sea	October 2019–January 2020
Typhoon	76.83	127.69	61	Laptev Sea	October 2019–February 2020
Typhoon	71.54	66.47	46	Kara Sea	October 2021–January 2022
Typhoon	71.24	65.60	42	Kara Sea	October 2021–March 2022
Typhoon	69.97	65.30	41	Kara Sea	October 2021–February 2022
MPSSR	74.90	69.72	42	Kara Sea	October 2021–March 2022
			Short-term deployments		
GNS-C	75.42	129.13	40	Laptev Sea	30 September 2018–6 October 2018
GNS-C	76.86	125.57	75	Laptev Sea	28 September 2018–13 October 2018
MPSSR	75.01	126.52	37	Laptev Sea	6 October 2018–9 October 2018
MPSSR	75.01	128.26	36	Laptev Sea	5 October 2018–12 October 2018
MPSSR	75.20	127.40	40	Laptev Sea	6 October 2018–8 October 2018
MPSSR	74.94	160.52	45	East Siberian Sea	30 September 2019–3 October 2019
Typhoon	72.98	65.87	82	Kara Sea	4 November 2022 (~1 h)
Typhoon	69.13	58.42	17	Barents Sea	10 November 2022 (~1 h)
Typhoon	69.20	58.03	21	Barents Sea	10 November 2022 (~1 h)
Typhoon	69.22	57.81	22	Barents Sea	11 November 2022 (~1 h)
Typhoon	69.32	57.82	22	Barents Sea	11 November 2022 (~1 h)

The main purpose of OBS deployments was to study the relationship between gas seeps and tectonic processes. This required the registration of local and regional earthquakes, as well as the determination of their magnitudes and spatial distribution. During the observation period, several hundred local earthquakes were recorded—Figure 6a shows examples of waveforms of recorded signals, as well as their spectra. In particular, the results of OBS observations, along with catalogs of global and regional networks, revealed a number of new patterns in the distribution of earthquake epicenters on the shelf of the Laptev Sea, the most seismically active sea in the Russian sector of the Arctic Ocean [23]. Figure 7a shows Guttenberg–Richter curves for the Laptev Sea shelf—demonstrating an improvement in the completeness of the earthquake catalog in the range of magnitudes of less than 4 after taking into account seismic events recorded by OBSs, but this is still not enough to make the recurrence curve linear.

The use of broadband, highly sensitive sensors made it possible to record signals from remote earthquakes that occurred at a distance of several thousand kilometers. An example of such a signal is shown in Figure 6b. Records of remote earthquakes can be useful for determining the deep structure of the crust and upper mantle in the vicinity of the location of OBSs. This is important in particular for studying the deep geological "roots" of gas seeps. Figure 7b shows the correspondence between the travel times of seismic waves from recorded remote earthquakes and travel–time curves constructed using the ak135 model [37]. The figure demonstrates the correspondence of real arrivals of P-waves from earthquakes to the ak135 travel–time curves for different focal depths. It can also be seen that events with epicentral distances of up to 100° were recorded.

In the Arctic seas, tectonic processes are interconnected both with the release of geofluids from marine sediments and with the state of underwater permafrost. Thus, increased heat flow in the vicinity of active faults can lead to the thawing of a continuous layer of permafrost, releasing the gas contained in it, as well as creating pathways for the penetration of deep gas to the surface of the seabed. By analyzing seismic noise recorded via OBSs, it is possible to determine whether there is a contrast boundary under the station location, which can be associated with the top of underwater permafrost [22,38]. Figure 8 shows an example of HVSR (horizontal-to-vertical spectral ratio) curves calculated via the workflow [22] using the records obtained in the Laptev Sea. There are resonance peaks, indicating the presence of a reflecting boundary under the seabed.

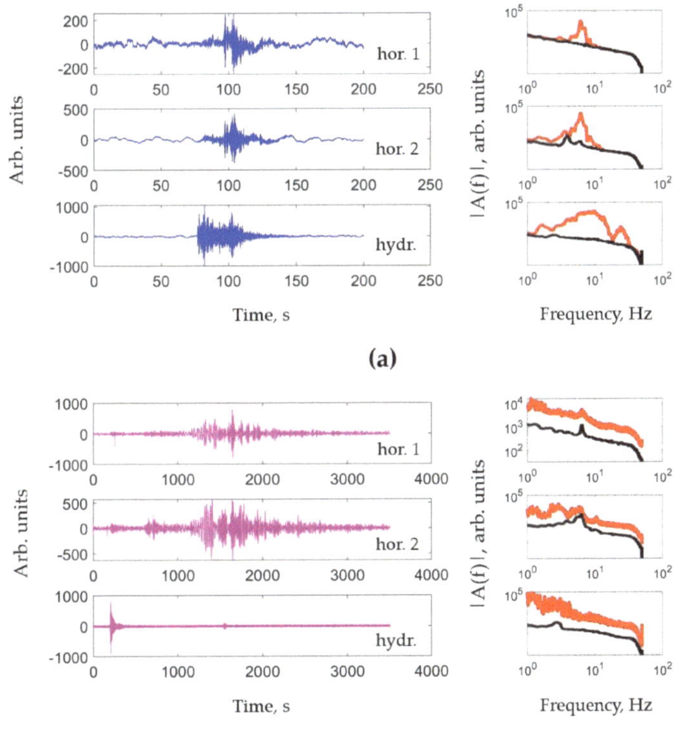

Figure 6. Examples of waveforms and their Fourier spectra from earthquakes recorded in the Laptev Sea: (**a**) a local earthquake (M = 2.6) in the Laptev Sea that occurred on 3 November 2018, 18:01:51, at a distance of 154 km and (**b**) a remote earthquake (M = 6.3) that occurred in Japan on 8 January 2019, 12:39:31, at a distance of approximately 5000 km. On the Fourier spectra plots, the black lines indicate the noise spectra before the arrival of the P-wave from the earthquake, and the red line indicates the spectra after its arrival.

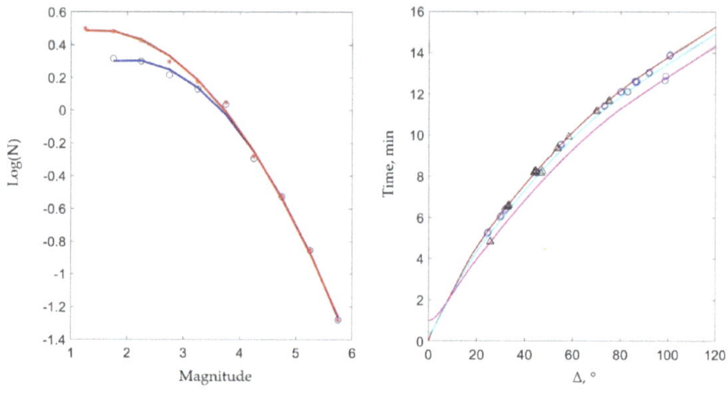

(a) (b)

Figure 7. (**a**) Cumulative recurrence curves for the Laptev Sea shelf: blue circles and fitting line—from the combined ISC regional catalog [39], US Geological Survey [40], and the Russian Earthquakes database [41]; red circles and fitting line—the same with the addition of events recorded via OBSs in 2018–2019. (**b**) P-wave travel–time curves constructed using the ak135 model [37] for depths of 0 km (the red line), 150 km (the blue line), and 500 km (the purple line) with plotted symbols corresponding to the arrivals of the P-waves of remote earthquakes recorded in the Laptev Sea (circles—in the 2018–2019 season, triangles—in the 2019–2020 season).

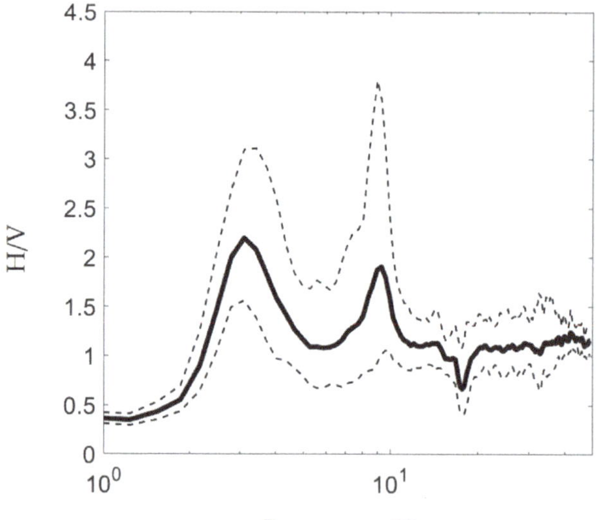

Figure 8. Example of HVSR curves calculated for OBS records of 6–8 h duration obtained on the Laptev Sea shelf on 25 October 2019. The thick solid line is the curve of the average values, while the thin dashed lines are the curves of the maximum and minimum values.

4. Features of Seismographs' Deployment in the Arctic Seas and Seafloor Seismic Records

4.1. Deployment Schemes and Workflows

In order to ensure the effectiveness and quality of experiments with OBSs in the Arctic seas and reduce the likelihood of instrument loss, it is necessary to take into account climatic conditions, bathymetry features, the ice conditions at the proposed location of the station, as well as the assigned tasks and the desired recording duration. If it is necessary to ensure the operation of an OBS network for a long time (several months or longer), then special attention should be paid to the dynamics of the ice cover.

In many seas of the Arctic Ocean, such as the Laptev Sea, the East Siberian Sea, or the Chukchi Sea, navigation is only possible for 1–2 months a year. Sometimes some straits are blocked with ice all year round—for example, this often happens with the Vilkitsky Strait. Thus, a situation is possible when it will be possible to get to the place of OBS deployment by ship only in a year or even several years. In this case, completely relying on the acoustic release mechanism to dismantle the OBS is impractical due to the limited power resources in the stations.

At shallow shelf depths of up to 100 m, this problem can be solved by providing the possibility to recover OBSs via trawling. Figure 9 shows examples of the corresponding deployment schemes that we used in the Arctic seas. These schemes assumed the presence of a piece of rope laid out on the bottom between the OBS and the ballast. The ballast could also be accompanied by an acoustic system for its release and popping up of buoys with the end of a rope that is also tied to the OBS (Figure 9a). One of the advantages of such schemes is the permissibility of a large weight of the station since it is not expected to float. This allows for equipping the station with a large number of batteries for long-term operation, as well as a rigid bounding with heavy ballast to improve the coupling of the OBS to the sea soil. This coupling improves the quality of seismic recordings. Another advantage of such schemes is the ability to install additional devices, such as ADCPs (acoustic Doppler current profilers), CTD (conductivity, temperature, and depth), wave recorders, thermistors, and plexiglass plates for biofouling. This significantly expands the range of solving scientific problems. For example, the presence of wave recorders makes it possible to study in detail the influence of sea waves on the characteristics of seafloor seismic noise [21,22]. Biofouling plates make it possible to conduct studies of bottom fouling that are unique to the Arctic seas [42].

With all the advantages of the described schemes, there are some significant nuances. Thus, for deploying the OBS, it is advisable to choose the flattest seabed areas. Sandy soils are preferable to clay and silt soils because devices can become bogged down in them. One should very carefully save the installation coordinates of the OBS and the ballast, preferably with a handheld GPS navigator directly above the place and at the time of installation. Without all this, the successful recovery of stations via trawling is impossible. It is also recommended to choose areas with a sea depth of at least 40 m since, at the seabed, up to several tens of meters deep, a large number of plow marks left from stamukhi were found [43,44]. At depths of more than 80–100 m, the probability of successful trawling also decreases. When installing according to the presented schemes, several joints of individual elements occur, so attention should be paid to avoiding increased galvanic corrosion in places where different metals are attached. Such corrosion, for example, can occur when fastenings made of galvanized and stainless steel come into contact.

The installation of equipment on the seabed in accordance with the schemes presented in Figure 9 is carried out while a ship is drifting. Figure 10 shows three stages of the deployment process: first, the OBS is carefully lowered on a rope to the seafloor (the first fixation of the exact coordinates). Then, the rope is laid out in the drift, and at the final stage, the ballast is dropped (the second fixation of the exact coordinates). Recovery via trawling is carried out carefully in several stages (Figure 11): First, the vessel lowers a system of several grapnels to the bottom on a rope that does not reach 100–200 m before the intersection of the rope line between the OBS and the ballast. Then, laying out the cable with the grapnels, at a low speed of 3–5 knots, the ship crosses the rope line perpendicularly. After this, the vessel stops and is fixed in one position. Then, the cable with the grapnels is pulled out with a winch. If unsuccessful, the procedure is repeated on another tack.

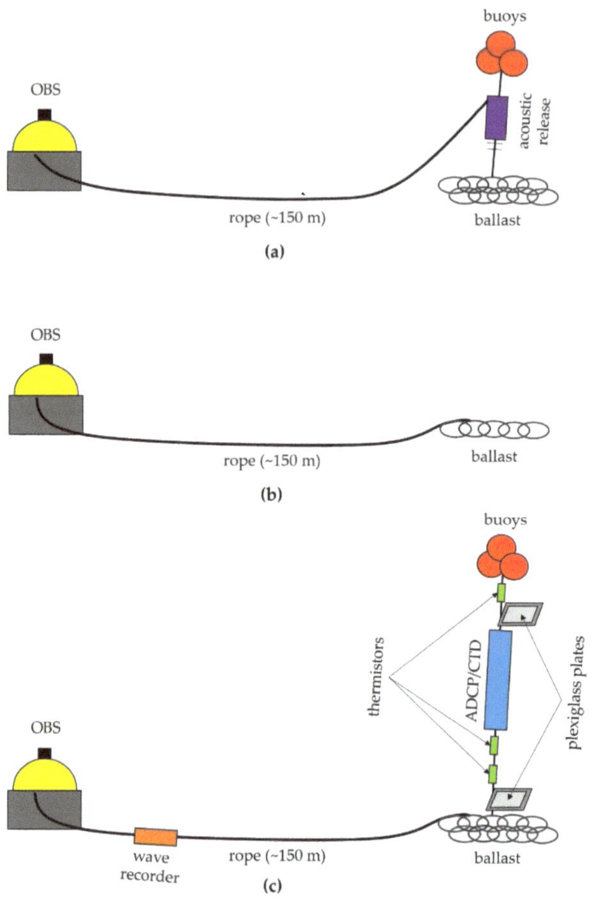

Figure 9. Deployment schemes used for non-pop-up OBSs on the shelf to depths of 100 m: (**a**) using submerged buoys and an external acoustic release, (**b**) without using submerged buoys and an external acoustic release, and (**c**) using additional measuring devices.

At sea depths of more than 100 m, only pop-up OBSs are suitable for observation. In order to use them, it is necessary to carefully select the installation site, making sure that researchers can certainly approach the installation site on a ship before the power supply to the station runs out. Figure 12 shows a workflow of the installation of pop-up OBSs using a mechanical or automatic pelican hook at a low vessel speed of 3–5 knots and a workflow of the dismantling of the OBSs via releasing ballast after an acoustic signal.

For the short-term recording of seafloor seismic noise and the subsequent application of the HVSR (horizontal-to-vertical spectral ratio) method, a seismic station must be installed for a period of from several minutes to several hours, depending on the frequency response of the equipment [45]. For such observations at sea, it is more convenient to use a pop-up OBS, moving the ship away from the deployment site to minimize noise from the ship. If only a non-pop-up OBS is available, then it is enough to lower it to the bottom in accordance with the scheme shown in Figure 13, controlling it with a rope throughout the registration period. In this case, it is better to use a long rope and intermediate ballast between the OBS and the vessel, thus minimizing the parasitic noise associated with the influence of the vessel on the rope. The ship must be held precisely at one point to avoid pulling on the rope and preventing the rope from getting caught in the propellers. Processing seismic records obtained in this way requires the mandatory consideration of hydroacoustic noise from the vessel, the characteristics of which depend primarily on the size, number of blades, and rotation speed of the ship's propellers [46].

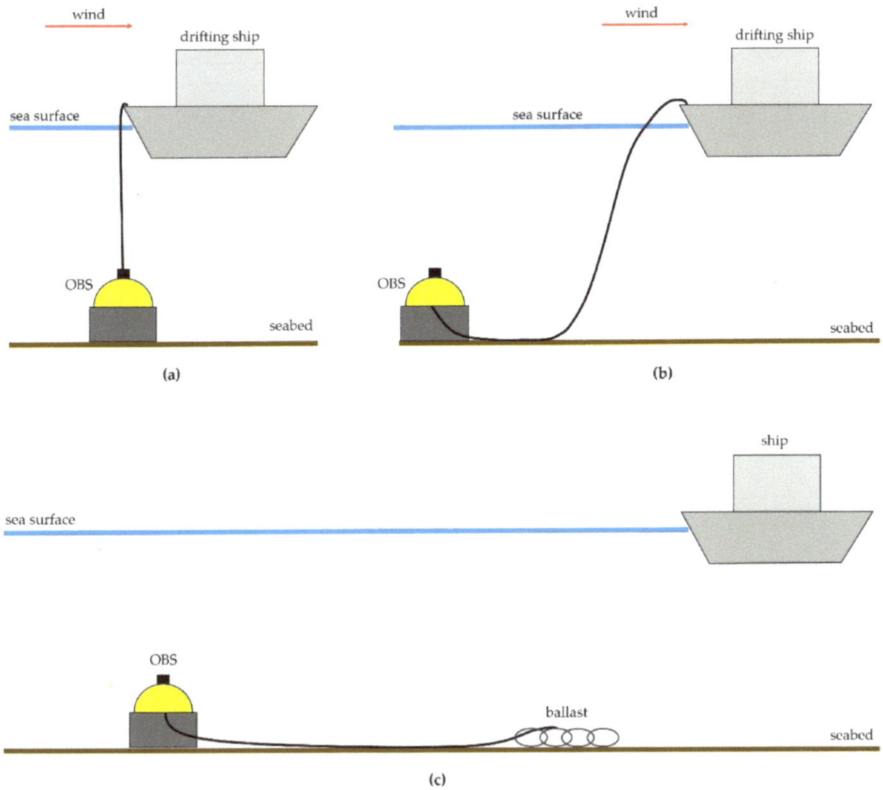

Figure 10. The workflow of deploying non-pop-up OBSs on the shelf to depths of 100 m: (**a**) lowering the station to the seafloor, (**b**) laying out a rope on the seafloor in a drift, and (**c**) dropping the ballast.

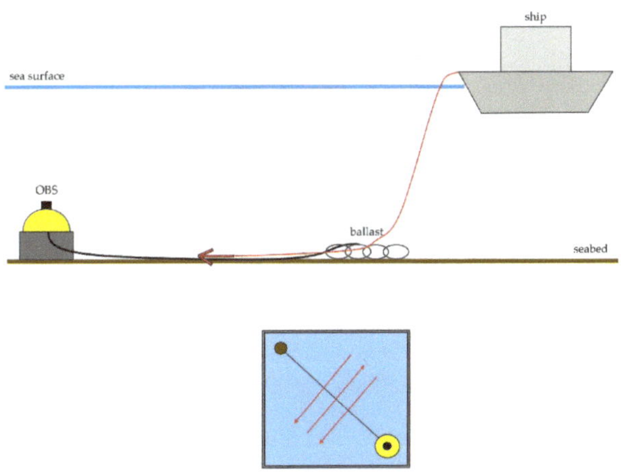

Figure 11. The workflow for dismantling non-pop-up OBSs on the shelf using the trawling method.

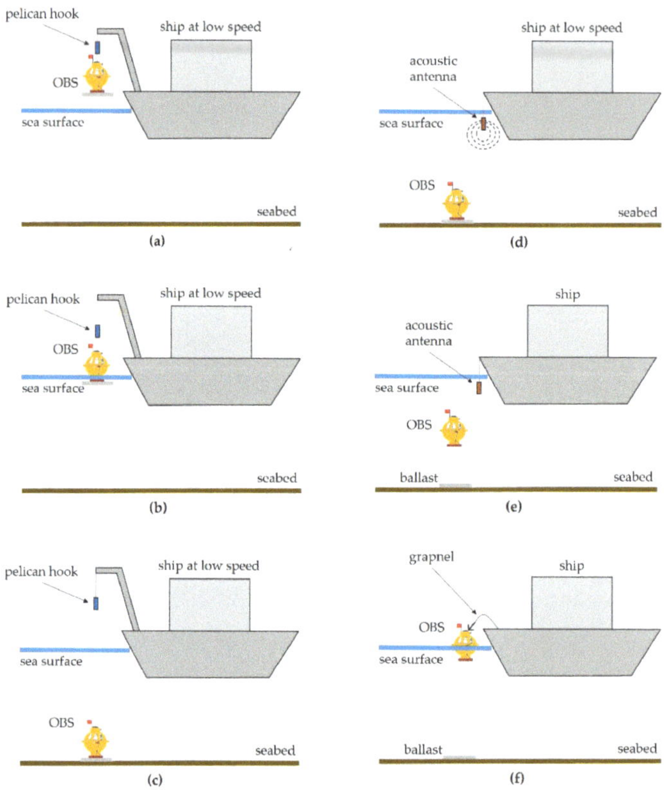

Figure 12. The workflow for deploying (**a–c**) and dismantling (**d–f**) pop-up OBSs.

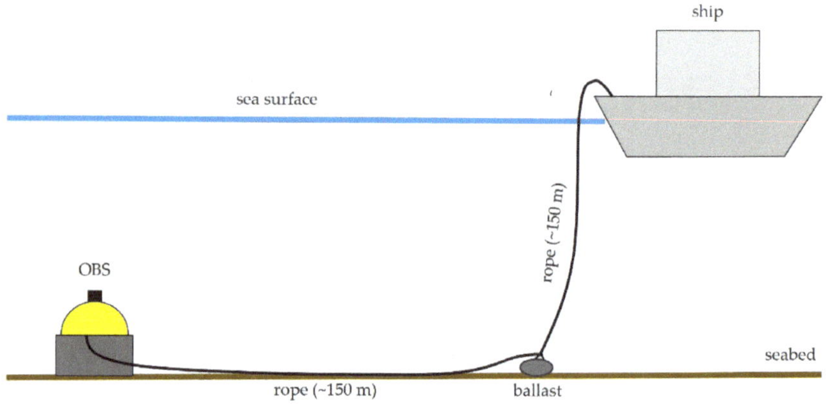

Figure 13. A scheme for the short-term installation of non-pop-up OBSs on the shelf in order to measure seafloor seismic noise using the HVSR method.

Such seismic noise measurements can be obtained from ice (Figure 14). Obviously, it is only possible from fast ice that does not experience large displacements. The position of the OBS plays an important role—if it is tilted or turned over, the registration results will not be indicative.

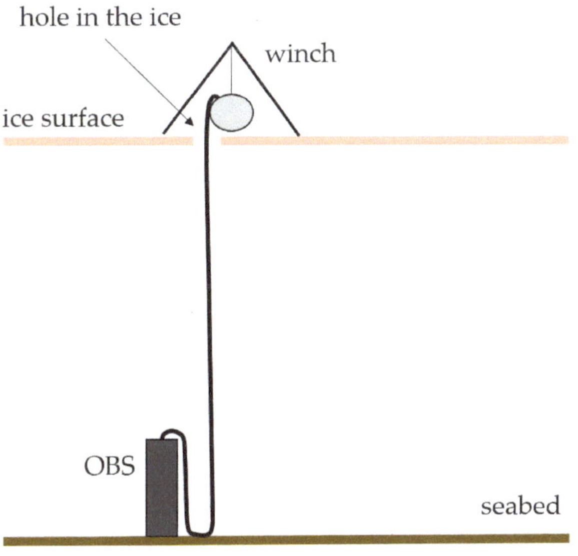

Figure 14. A scheme of the short-term installation of OBSs on the shelf through holes in the ice.

The scheme presented in Figure 14 is poorly suited for long-term seismological monitoring. A specialized seismic station (Figure 4) has been designed for freezing into the ice while a hydrophone on a cable is lowered into a hole in the ice to a depth of 20 m. Deepening the housing makes it possible to create a relatively stable operating temperature inside the container that does not fall below −30 °C [47]. The durable housing protects the equipment from water and avoids possible deformation due to ice and exposure to animals. The container is also accompanied by a replaceable battery pack, a GPS antenna for synchronizing the internal clock of the recorder and determining current coordinates, as well as a wireless data transmission modem with a sectoral antenna for organizing a wireless bridge, together with a circular antenna located in a camp on ice or a frozen icebreaker at a distance up to 5 km or more (Figure 15).

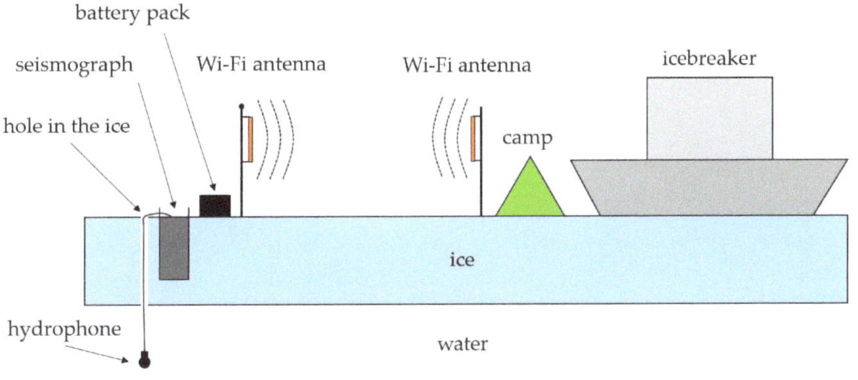

Figure 15. A scheme for deploying a seismograph on ice.

4.2. Characteristics of Seafloor Seismic Noise

In order to process seismic records for earthquake searching or determining the reflecting boundary in the upper part of a geological section, it is necessary to know the characteristics of natural seismic noise at the OBS deployment site. Seismic noise on the seabed, especially in the Arctic seas, has a number of specific characteristics. Figure 16 shows noise power–spectral density (PSD) curves obtained on the Laptev Sea shelf for two recording periods: 15 October 2018–31 October 2018, when the sea was still free of ice, and 1 November 2018–15 November 2018, when, for a short period of time, solid ice formed above the installation site. Figure 17 shows the corresponding time dependence of ice concentration obtained from the reanalysis data [48]. Figure 16 shows a sharp decrease in the noise level after ice formation in almost the entire frequency range of the seismograph, especially in the range below 6 Hz, apparently due to the smoothing of wind-induced gravitational sea waves via ice. This observation led to the conclusion that seismological monitoring using OBS on the shelf of the Arctic seas is preferably carried out during a period of time when there is ice cover above the stations.

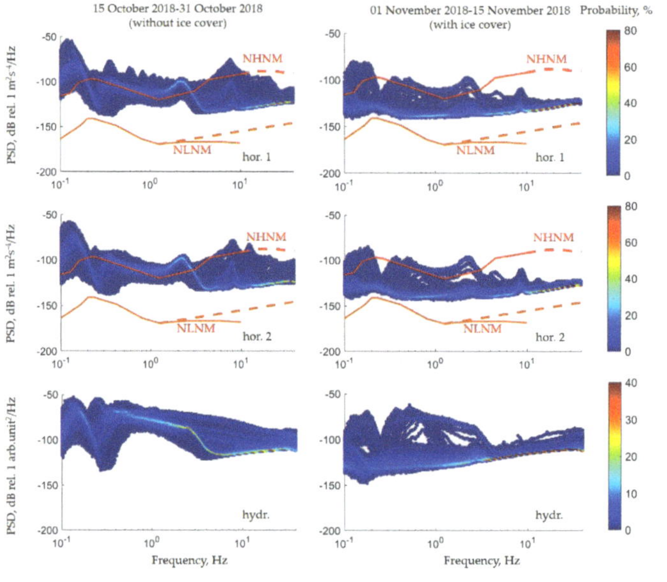

Figure 16. Seismic noise PSD curves recorded on the Laptev Sea shelf during the periods of 15 October 2018–31 October 2018 (**left**) and 1 November 2018–15 November 2018 (**right**). The solid red lines indicate the curves of the NHNM (new high-noise model) and NLNM (new low-noise model) [49], while the red dotted lines indicate the extension of the NHNM and NLNM according to [50]. The color indicates the probability of finding the noise curve in a certain area of the graph.

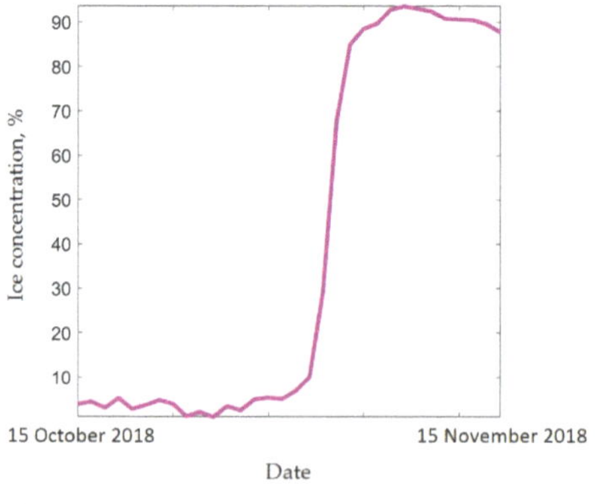

Figure 17. Ice concentration curve according to EUMETSAT reanalysis (mesh size: 25 km) [48] for the period of 15 October 2018–15 November 2018 for the closest node to the location of the OBS that corresponded to the PSD curves shown in Figure 16.

In general, the spectral characteristics of seabed seismic noise show that records obtained on the shelf are very noisy, which significantly complicates their processing. In the frequency range from 0.1 to 2 Hz, the noise associated with primary and secondary microseisms dominated [25,51,52], as well as their overtones. The presence of a series of additional maxima, which are especially prominent in the upper left graph in Figure 16, indicates a resonance effect that enhanced the tilting of the OBS under the influence of wind gravity waves. It is interesting that the noise associated with wind waves decreased significantly with the appearance of ice but was still present—this can also be seen in Figure 16. However, the dominant source of seafloor seismic noise during the period of time when ice was present became longer-period (30–300 s) infragravity waves [22], which are apparently capable of deforming extensive ice fields.

Figure 16 also shows a stable maximum in the vicinity of 9–10 Hz, in some cases also with additional overtones. The frequency range of 5–10 Hz is often associated with the so-called coupling effect—parasitic oscillations that occur in the station–ballast–soft sediment system [25]. Moreover, these parasitic oscillations intensify at high wind speed values—Figure 18 shows a spectrogram of a fragment of a seafloor seismic record and its comparison with the corresponding time dependence of wind speed obtained from reanalysis data [53].

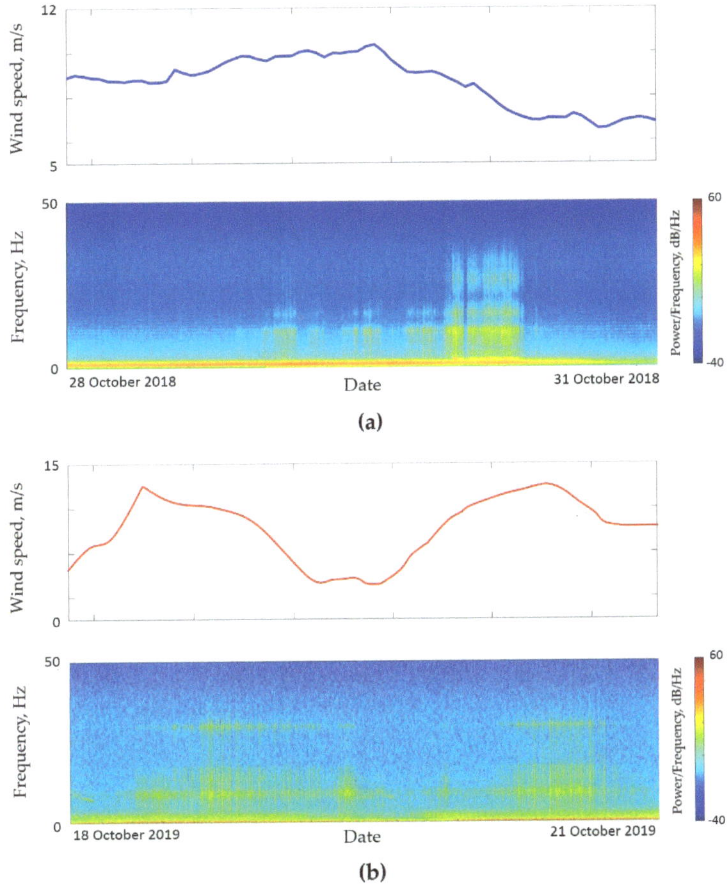

Figure 18. Time dependence of wind speed for the location of the OBS in the Laptev Sea obtained as a result of the ERA5 reanalysis [53] and its correlation with the seafloor seismic noise level at the same point shown in the spectrogram—two record fragments (**a**,**b**).

5. Conclusions

This paper reveals the features and difficulties of seismological observations in the Arctic seas. The following main results of the work and the following conclusions can be formulated:

1. The characteristics of ocean-bottom pop-up and non-pop-up seismographs, as well as stations for deployment on ice, were described in detail. The results of the deployments demonstrated that the characteristics of the stations make it possible to reliably record both high-frequency signals from local earthquakes and low-frequency signals from distant earthquakes on the shelf and continental slope of the Arctic seas.
2. Various schemes for seismic stations' deployment were described. It was concluded that the preferred schemes for deploying OBSs are those in which their subsequent dismantling does not depend on their power resources. Usually, such schemes provide for the possibility of dismantling stations via trawling and are suitable for shallow sea depths of up to 100 m. The advantages of such schemes include the possibility of installing additional hydrophysical and hydrobiological equipment, such as ADCP, CTD, thermistors, wave recorders, and biofouling plates.
3. The nuances of offshore work on the installation and recovery of equipment were outlined. It was concluded that particular attention should be paid to planning the recovery of seismic stations due to possible difficulties associated with the passage of a vessel to the deployment site due to unfavorable ice conditions. When deploying an OBS, it is advisable to choose the flattest seabed areas. Sandy soils are preferable to clay and silt soils because devices can become bogged down in them. Attention should be paid to avoiding increased galvanic corrosion in places where different metals are attached.
4. The features of seabed seismic records in the Arctic seas were demonstrated. It turned out that seabed seismic records are characterized by a high level of noise, especially during periods of time when there is no ice cover. Therefore, it is recommended to deploy OBSs for periods of time when ice cover is present. Seismic noise is caused by wind gravity waves, infragravity waves, and the coupling effect, and it also strongly depends on meteorological conditions, primarily on wind speed. The frequency range of the prevailing noise significantly overlaps with the frequency range of signals from both weak local earthquakes and strong distant ones. This must be taken into account when searching for and processing signals from earthquakes obtained in the Arctic seas.

Author Contributions: Conceptualization, A.A.K.; methodology, A.A.K.; validation, A.A.K.; formal analysis, A.A.K.; investigation, A.A.K., S.A.K., D.A.I., O.Y.G. and M.A.N.; resources, A.A.K., S.A.K., D.A.I., O.Y.G., V.N.I. and G.K.T.; data curation, A.A.K.; writing—original draft preparation, A.A.K.; writing—review and editing, all authors; visualization, A.A.K., M.A.N. and D.A.I.; supervision, A.A.K., L.I.L. and I.P.S.; project administration, A.A.K., I.P.S. and K.A.R.; funding acquisition, A.A.K., I.P.S., O.S.S. and K.A.R. All authors have read and agreed to the published version of the manuscript.

Funding: This work was financially supported by the following initiatives: the Russian Science Foundation, project no. 23-17-00125 (analysis and description of seismicity in the Laptev Sea region) and project no. 21-77-30001 (partially organizing expeditions to the Eastern Arctic Seas for the investigation of the permafrost state); the Ministry of Science and High Education of the Russian Federation, project "Priority-2030", Tomsk State University (partially organizing expeditions to the Eastern Arctic Seas for the investigation of methane seepage); and the Russian State Assignment, project no. FMWE-2021-0011 (the design and description of a pop-up OBS and seismograph for installation on ice).

Institutional Review Board Statement: Not applicable.

Informed Consent Statement: Not applicable.

Data Availability Statement: Data are contained within the article.

Conflicts of Interest: The authors declare no conflict of interest.

References

1. OBSIP Experiment Archive. Available online: https://obsic.whoi.edu/experiments/obsip-experiment-archive/ (accessed on 1 October 2023).
2. Aderhold, K.; Woodward, R.; Frassetto, A. *Ocean Bottom Seismograph Instrument Pool*; Final Report; National Science Foundation: Alexandria, VA, USA, 2019; 80p.
3. SAGE. Seismological Facility for the Advancement of Geoscience. Available online: http://ds.iris.edu/mda/_OBSIP/ (accessed on 1 October 2023).
4. Cai, C.; Wiens, D.A.; Shen, W.; Eimer, M. Water input into the Mariana subduction zone estimated from ocean-bottom seismic data. *Nature* **2018**, *563*, 389–392. [CrossRef] [PubMed]
5. Isse, T.; Kawakatsu, H.; Yoshizawa, K.; Takeo, A.; Shiobara, H.; Sugioka, H.; Ito, A.; Suetsugu, D.; Reymond, D. Surface wave tomography for the Pacific Ocean incorporating seafloor seismic observations and plate thermal evolution. *Earth Planet. Sci. Lett.* **2019**, *510*, 116–130. [CrossRef]
6. Mark, H.F.; Lizarralde, D.; Collins, J.A.; Miller, N.C.; Hirth, G.; Gaherty, J.B.; Evans, R.L. Azimuthal seismic anisotropy of 70 Ma Pacific-plate upper mantle. *J. Geophys. Res. Solid Earth* **2019**, *124*, 1889–1909. [CrossRef]
7. Corela, C.; Loureiro, A.; Duarte, J.L.; Matias, L.; Rebelo, T.; Bartolomeu, T. The effect of deep ocean currents on ocean- bottom seismometers records. *Nat. Hazards Earth Syst. Sci.* **2023**, *23*, 1433–1451. [CrossRef]
8. Zali, Z.; Rein, T.; Krüger, F.; Ohrnberger, M.; Scherbaum, F. Ocean bottom seismometer (OBS) noise reduction from horizontal and vertical components using harmonic–percussive separation algorithms. *Solid Earth* **2023**, *14*, 181–195. [CrossRef]
9. Liu, D.; Yang, T.; Wang, Y.; Wu, Y.; Huang, X. Pankun: A New Generation of Broadband Ocean Bottom Seismograph. *Sensors* **2023**, *23*, 4995. [CrossRef] [PubMed]
10. Nakano, M.; Nakamura, T.; Kamiya, S.-I.; Kaneda, Y. Seismic activity beneath the Nankai trough revealed by DONET ocean-bottom observations. *Mar. Geophys. Res.* **2014**, *35*, 271–284. [CrossRef]
11. Suzuki, K.; Nakano, M.; Takahashi, N.; Hori, T.; Kamiya, S.; Araki, E.; Nakata, R.; Kaneda, Y. Synchronous changes in the seismicity rate and ocean-bottom hydrostatic pressures along the Nankai trough: A possible slow slip event detected by the Dense Oceanfloor Network system for Earthquakes and Tsunamis (DONET). *Tectonophysics* **2016**, *680*, 90–98. [CrossRef]
12. Barnes, C.R.; Best, M.M.R.; Zielinski, A. The NEPTUNE Canada Regional Cabled Ocean Observatory. *Sea Technol.* **2008**, *49*, 10–14.
13. Barnes, C.R.; Best, M.M.R.; Johnson, F.R.; Pirenne, B. Final installation and initial operation of the world's first regional cabled ocean observatory (NEPTUNE Canada). *CMOS Bull. SCMO* **2010**, *38*, 2–9.
14. Lobkovsky, L.I.; Kuzin, I.P.; Kovachev, S.A.; Krylov, A.A. Seismicity of the Central Kuril Islands before and after the catastrophic M = 8.3 (November 15, 2006) and M = 8.1 (January 13, 2007) earthquakes. *Dokl. Earth Sci.* **2015**, *464*, 1101–1105. [CrossRef]
15. Kuzin, I.P.; Kovachev, S.A.; Lobkovskii, L.I. Seismic microzonation and assessment of earthquake hazard for the construction sites of sea-based facilities in low seismicity water areas. *J. Volcanol. Seism.* **2009**, *3*, 131–143. [CrossRef]
16. Simpson, D.W.; Leith, W. The 1976 and 1984 Gazli, USSR, earthquakes—Were they induced? *Bull. Seismol. Soc. Am.* **1985**, *75*, 1465–1468.
17. Vlek, C. Induced Earthquakes from Long-Term Gas Extraction in Groningen, the Netherlands: Statistical Analysis and Prognosis for Acceptable-Risk Regulation. *Risk Anal.* **2018**, *38*, 1455–1473. [CrossRef]
18. Kovachev, S.A.; Krylov, A.A. Results of Seismological Monitoring in the Baltic Sea and Western Part of the Kaliningrad Oblast Using Bottom Seismographs. *Izv. Phys. Solid Earth* **2023**, *59*, 190–208. [CrossRef]
19. Piskarev, A.L. *Arctic Basin: Geology and Morphology*; VNIIOkeangeologiya: St. Petersburg, Russia, 2016; 291p. (in Russian)
20. Krylov, A.A.; Ivashchenko, A.I.; Kovachev, S.A.; Tsukanov, N.V.; Kulikov, M.E.; Medvedev, I.P.; Ilinskiy, D.A.; Shakhova, N.E. The Seismotectonics and Seismicity of the Laptev Sea Region: The Current Situation and a First Experience in a Year-Long Installation of Ocean Bottom Seismometers on the Shelf. *J. Volcanol. Seism.* **2020**, *14*, 379–393. [CrossRef]
21. Krylov, A.A.; Egorov, I.V.; Kovachev, S.A.; Ilinskiy, D.A.; Ganzha, O.Y.; Timashkevich, G.K.; Roginskiy, K.A.; Kulikov, M.E.; Novikov, M.A.; Ivanov, V.N.; et al. Ocean-Bottom Seismographs Based on Broadband MET Sensors: Architecture and Deployment Case Study in the Arctic. *Sensors* **2021**, *21*, 3979. [CrossRef]
22. Krylov, A.A.; Kulikov, M.E.; Kovachev, S.A.; Medvedev, I.P.; Lobkovsky, L.I.; Semiletov, I.P. Peculiarities of the HVSR Method Application to Seismic Records Obtained by Ocean-Bottom Seismographs in the Arctic. *Appl. Sci.* **2022**, *12*, 9576. [CrossRef]
23. Krylov, A.A.; Lobkovsky, L.I.; Rukavishnikova, D.D.; Baranov, B.V.; Kovachev, S.A.; Dozorova, K.A.; Tsukanov, N.V.; Semiletov, I.P. New Data on Seismotectonics of the Laptev Sea from Observations by Ocean Bottom Seismographs. *Dokl. Earth Sci.* **2022**, *507*, 936–940. [CrossRef]
24. Krylov, A.A.; Ananiev, R.A.; Chernykh, D.V.; Alekseev, D.A.; Balikhin, E.I.; Dmitrevsky, N.N.; Novikov, M.A.; Radiuk, E.A.; Domaniuk, A.V.; Kovachev, S.A.; et al. A Complex of Marine Geophysical Methods for Studying Gas Emission Process on the Arctic Shelf. *Sensors* **2023**, *23*, 3872. [CrossRef]
25. Webb, S.C. Broadband seismology and noise under the ocean. *Rev. Geophys.* **1998**, *36*, 105–142. [CrossRef]
26. Janiszewski, H.A.; Eilon, Z.; Russell, J.B.; Brunsvik, B.; Gaherty, J.B.; Mosher, S.G.; Hawley, W.B.; Coats, S. Broad-band ocean bottom seismometer noise properties. *Geophys. J. Int.* **2023**, *233*, 297–315. [CrossRef]
27. Schlindwein, V.; Müller, C.; Jokat, W. Seismoacoustic evidence for volcanic activity on the ultraslow spreading Gakkel Ridge, Arctic Ocean. *Geophys. Res. Lett.* **2005**, *32*, L18306. [CrossRef]

28. Schlindwein, V.; Müller, C.; Jokat, W. Microseismicity of the ultraslow spreading Gakkel ridge, Arctic Ocean: A pilot study. *Geophys. J. Int.* **2007**, *169*, 100–112. [CrossRef]
29. Schlindwein, V.; Schmid, F. Mid-ocean-ridge seismicity reveals extreme types of ocean lithosphere. *Nature* **2016**, *535*, 276–279. [CrossRef]
30. Schlindwein, V.; Riedel, C. Location and source mechanism of sound signals at Gakkel ridge, Arctic Ocean: Submarine Strombolian activity in the 1999–2001 volcanic episode. *Geochem. Geophys. Geosyst.* **2010**, *10*, Q01002. [CrossRef]
31. Koulakov, I.; Schlindwein, V.; Liu, M.; Gerya, T.; Jakovlev, A.; Ivanov, A. Low-degree mantle melting controls the deep seismicity and explosive volcanism of the Gakkel Ridge. *Nat. Commun.* **2022**, *13*, 3122. [CrossRef] [PubMed]
32. R-Sensors. Seismic Instruments for Science and Engineering. Available online: https://r-sensors.ru/en/ (accessed on 1 October 2023).
33. Logys. Available online: https://logsys.ru/ (accessed on 1 October 2023).
34. Shakhova, N.; Semiletov, I.; Sergienko, V.; Lobkovsky, L.; Yusupov, V.; Salyuk, A.; Salomatin, A.; Chernykh, D.; Kosmach, D.; Panteleev, G.; et al. The East Siberian Arctic Shelf: Towards further assessment of permafrost-related methane fluxes and role of sea ice. *Philos. Trans. R. Soc. A* **2015**, *373*, 20140451. [CrossRef]
35. Shakhova, N.; Semiletov, I.; Gustafsson, Ö.; Sergienko, V.; Lobkovsky, L.; Dudarev, O.; Tumskoy, V.; Grigoriev, M.; Mazurov, A.; Salyuk, A.; et al. Current rates and mechanisms of subsea permafrost degradation in the East Siberian Arctic Shelf. *Nat. Commun.* **2017**, *8*, 15872. [CrossRef]
36. Shakhova, N.; Semiletov, I.; Chuvilin, E. Understanding the Permafrost–Hydrate System and Associated Methane Releases in the East Siberian Arctic Shelf. *Geosciences* **2019**, *9*, 251. [CrossRef]
37. Kennett, B.L.N. *Seismological Tables: ak135*; Research School of Earth Sciences, The Australian National University: Canberra, Australia, 2005; 289p.
38. Overduin, P.P.; Haberland, C.; Ryberg, T.; Kneier, F.; Jacobi, T.; Grigoriev, M.N.; Ohrnberger, M. Submarine permafrost depth from ambient seismic noise. *Geophys. Res. Lett.* **2015**, *42*, 7581–7588. [CrossRef]
39. International Seismological Centre. Available online: http://www.isc.ac.uk/iscbulletin/search/ (accessed on 1 October 2023).
40. U.S. Geological Survey. Available online: https://earthquake.usgs.gov/earthquakes/search/ (accessed on 1 October 2023).
41. "Earthquakes of Russia" Database. Geophysical Survey of the Russian Academy of Sciences. Available online: http://eqru.gsras.ru/ (accessed on 1 October 2023).
42. Chava, A.; Gebruk, A.; Kolbasova, G.; Krylov, A.; Tanurkov, A.; Gorbuskin, A.; Konovalova, O.; Migali, D.; Ermilova, E.; Shabalin, N.; et al. At the Interface of Marine Disciplines: Use of Autonomous Seafloor Equipment for Studies of Biofouling Below the Shallow-Water Zone. *Oceanography* **2021**, *34*, 61–70. [CrossRef]
43. Ananyev, R.; Dmitrevskiy, N.; Jakobsson, M.; Lobkovsky, L.; Nikiforov, S.; Roslyakov, A.; Semiletov, I. Sea-ice ploughmarks in the eastern Laptev Sea, East Siberian Arctic shelf. *Geol. Soc. Lond. Mem.* **2016**, *46*, 301–302. [CrossRef]
44. Nikiforov, S.L.; Ananiev, R.A.; Libina, N.V.; Dmitrevskiy, N.N.; Lobkovskii, L.I. Ice Gouging on Russia's Arctic Shelf. *Oceanology* **2019**, *59*, 422–424. [CrossRef]
45. *Guidelines for the Implementation of the H/V Spectral Ration Technique on Ambient Vibrations Measurements, Processing and Interpretation*; SESAME European research project WP12—Deliverable D23.12, European Commission—Research General Directorate Project No. EVG1-CT-2000-00026 SESAME; European Comission: Geneva, Switzerland, 2004; 62p.
46. Kuznetsov, M.Y.; Shevtsov, V.I.; Poljanichko, V.I. Inderwater noise characteristics of TINRO-Center's reseach vessels. *Izv. TINRO* **2014**, *177*, 235–256. (In Russian)
47. Pisarev, S.V. Modern drifting robotic devices for contact measurements of the physical characteristics of the Arctic basin. *Oceanol. Res.* **2019**, *47*, 5–31. (In Russian) [CrossRef]
48. OSI SAF Global Sea Ice Concentration Climate Data Record v2.0—Multimission, EUMETSAT SAF on Ocean and Sea Ice, 2017. Available online: https://navigator.eumetsat.int/product/EO:EUM:DAT:MULT:OSI-450 (accessed on 1 October 2023).
49. Peterson, J. *Observation and Modeling of Seismic Background Noise*; U.S. Geological Survey Open-File Report; US Geological Survey: Reston, VA, USA, 1993; pp. 93–322.
50. Wolin, E.; McNamara, D.E. Establishing High-Frequency Noise Baselines to 100 Hz Based on Millions of Power Spectra from IRIS MUSTANG. *Bull. Seism. Soc. Am.* **2020**, *110*, 270–278. [CrossRef]
51. Longuet-Higgins, M.S. A theory for the generation of microseisms. *Philos. Trans. R. Soc. A* **1950**, *243*, 1–35.
52. Lepore, S.; Grad, M. Analysis of the primary and secondary microseisms in the wavefield of the ambient noise recorded in northern Poland. *Acta Geophys.* **2018**, *66*, 915–929. [CrossRef]
53. Hersbach, H.; de Rosnay, P.; Bell, B.; Schepers, D.; Simmons, A.J.; Soci, C.; Abdalla, S.; Balmaseda, M.A.; Balsamo, G.; Bechtold, P.; et al. *Operational Global Reanalysis: Progress, Future Directions and Synergies with NWP*; ERA Report Series no. 27; European Centre for Medium Range Weather Forecasts: Reading, UK, 2018.

Journal of
Marine Science and Engineering

MDPI

Article

Net Transport Patterns of Surficial Marine Sediments in the North Aegean Sea, Greece

Ioannis Vakalas [1,2,*] and Irene Zananiri [1]

1 Hellenic Survey of Geology & Mineral Exploration (HSGME), 13677 Acharnae, Greece; izananiri@eagme.gr
2 Department of Geological Sciences, School of Mining and Metallurgical Engineering, National Technical University of Athens, Zografou Campus, 15773 Athens, Greece
* Correspondence: ivakalas@metal.ntua.gr; Tel.: +30-2107723970

Abstract: The spatial distribution of sediments on the seafloor reflects the various dynamic processes involved in the marine realm. To analyze sediment transport patterns in the North Aegean Sea, 323 surficial samples were obtained and studied. The granulometry data revealed a diverse range of grain sizes of surficial sediments, ranging from purely sandy to clay. The predominant size classes were silt and muddy sand, followed by sandy silt and mud. However, there were very few samples that fell within the clay classes. The sorting coefficient ranged from 0.21 to 5.48, while skewness ranged from −1.09 to 1.29. The sediment transport patterns were analyzed based on the grain-size parameters (mean, sorting, and skewness). The results showed the variability of flow parameters involved in sediment distribution. River influx and longshore drift near the shoreline are the most significant factors affecting sediment transport. At the open sea, sediment distribution is mainly controlled by general water circulation patterns, especially by the outflow of low-salinity waters from the Black Sea through the Dardanelles and the Marmara Sea. The heterogeneity of sediment textural parameters across the study area suggests that seafloor sediments are further reworked in areas where water masses are highly energetic. It can be concluded that open sea water circulation controls sediment distribution patterns at the open shelf, while close to the coast, river discharge plays a key role.

Keywords: sediment transport; Aegean Sea; grain size parameters

Citation: Vakalas, I.; Zananiri, I. Net Transport Patterns of Surficial Marine Sediments in the North Aegean Sea, Greece. *J. Mar. Sci. Eng.* **2024**, *12*, 512. https://doi.org/10.3390/jmse12030512

Academic Editor: Daniel C. Conley

Received: 8 February 2024
Revised: 12 March 2024
Accepted: 15 March 2024
Published: 20 March 2024

1. Introduction

Sediment distribution in the marine realm is the result of various processes described by the general term, sediment dynamics. Various factors are involved in the process of sediment transport and deposition including sediment supply sources, coastal processes (waves, prevailing winds, storm surges, etc.) water depth, seafloor topography, marine water circulation, physical and chemical properties of marine water (salinity, temperature), biogenic alteration, etc. The dynamic conditions of the system result in a continuous reworking of sediments with cycles of remobilization and redeposition. In terms of hydrodynamics, various mechanisms have been proposed. In the shoreline winds, storms and waves [1] can set sediment into a moving state in various ways (collisions between individual sediment grains, and effect of fluid stresses on individual sediment grains), as proposed by Bagnold [2]. Sediment in suspension can remain in motion for long periods due to various processes governed by fluid mechanics laws (diffusion, advection, convection). In the deeper parts of the basin floor, where coastal processes are less intense, gravity currents can also control sediment distribution. These currents can be the result of salinity contrasts, temperature differences, and sediment load differences and are associated with gravity flow deposits [3]. Other factors that affect sediment transport are the biochemical processes that can play a key role in the behavior and textural characteristics of grains. As described by Lee et al. and Cieśla et al. [4,5], the organic matter composed by living organisms can be absorbed by suspended sediments resulting in the stabilization or flocculation of sediments. This affects the grain-size properties of sediment by producing aggregates of fine material.

The aforementioned processes have a direct imprint on the textural parameters of sediments. In this context, various researchers have tried to uncover sediment transport processes by proposing ways to quantitively identify the spatial variations in the grain-size parameters (mean size, sorting, skewness). The first attempts were based in the assumption that mean grain size is reduced in the direction of sediment transport [6] (1972) but this approach was finally assessed as weak. McLaren [7] combined mean size, sorting, and skewness and proposed that in the direction of sediment transport will have a better sorting and a positive skewness while it is possible to become finer or coarser. McLaren and Bowles [8] presented a modified approach of the previous work, concluding that successive deposits in the direction of sediment transport will become either finer, well sorted and more negatively skewed or coarser, or well sorted and more positively skewed. Gao and Collins [9] identified that these are not the only possible cases and proposed a modified model. The model assumes that the trend analyzed has a higher possibility of occurrence in the direction of net transport than in other directions. It identifies the possible net transport paths by comparing two neighboring sampling sites and employing the three most commonly used grain-size parameters (m, s, and Sk) for trend analysis. Through the comparison of grain-size parameters of the two neighboring sampling sites, a dimensionless trend vector is defined, which has a unit length and an orientation that points from the station with the higher sorting coefficient to that with the lower sorting value. The grain-size trend vectors are generated for each sampling station and its neighboring stations, and a single resultant vector is produced after summing the identified trend vectors. The produced vector depicts the trend of sediment transport at the area that has been sampled. A smoothing process is finally applied in order to reduce background noise. This process averages the produced vector of the previous stage, taking into consideration the neighboring sampling sites, and finally provides a residual vector. Net sediment transport vectors methodology have maximized the utilization of grain size parameters and have been used by various researchers to unlock sediment distribution patterns. In this context, in the present study the methodology of sediment transport vectors has been applied in the North Aegean region of the Aegean Sea. The available dataset is part of the Hellenic Survey of Geology & Mineral Exploration (HSGME) of Greece database. The Hellenic Survey of Geology & Mineral Exploration (HSGME) of Greece (the former Institute of Geology and Mineral Exploration—IGME) has to exhibit a long period of activity in the field of marine geological research. The HSGME marine geology geodatabase [10] encompasses up to 25,000 km of bathymetry and seismic profiles, 3500 surface grab samples (from the upper ~15 cm of the seafloor), and 1200 cores (up to 3.5 m long) analyzed at the laboratories of HSGME, with a focus on sedimentology, and specifically, the distribution of grain size and related statistical parameters [11]. Through the interpretation of this data, sediment distribution maps (scale 1:200,000) have been constructed for the Aegean region [12–14].

2. Materials and Methods

2.1. Study Area

The North Aegean (Figure 1) is a tectonically complex area dominated by the presence of major structures, such as the North Aegean Trough, as a physical continuation of the North Anatolian Fault [15], as well as by many other active faults; this complexity favoring the accumulation of geological reserves (raw materials, hydrocarbons). Through this active geological framework, the geomorphological features are in a dynamic state, as revealed by various studies [16] depicting the significant modification during the Quaternary. These processes have resulted in a variable configuration of the marine realm consisting of linear and extended peninsulas (e.g., Chalkidiki), isolated gulfs (Thermaikos), and well-preserved continental plateaus incised by canyons and gullies that represent the past subaerial river pathways of fluvial systems like Strymonikos, Nestos [17], and Evros.

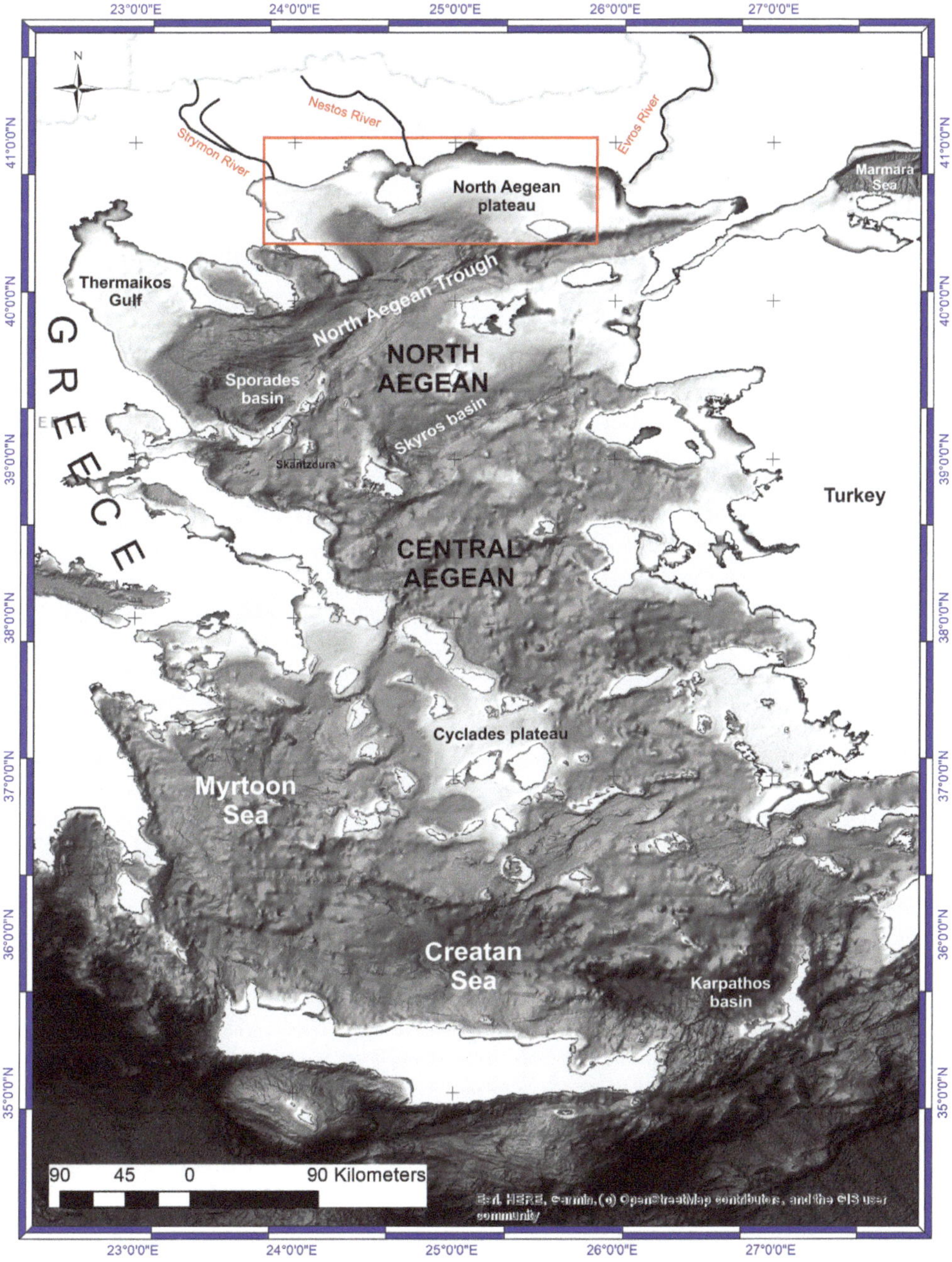

Figure 1. Reference map where the red rectangle marks the study area. The bathymetry hillshade is from the EMODNET database [18].

Regarding the composition of the sediments accumulating in the basin through the aforementioned fluvial systems (Strymonas, Nestos, and Evros), as expected, the controlling factor lies in the subaerial exposures of Macedonia and Thraki districts geological formations, providing through erosion a variety of materials consisting mainly of post-Alpine sediments, Eocene and Miocene granitoids, Triassic marbles, and Paleozoic gneisses. The various processes involved in sediment transport are imprinted in the variety of grain-size textures identified in the basin floor, where the entire range of grain sizes (gravely sands to clays) have been identified [14,19]. Close to the coastal area, coarser fractions are dominant (gravel and sand), while at the deeper parts, muddy deposits cover the seafloor.

In the western edge of the North Aegean area, corresponding to the Thermaikos Gulf (Figure 1), the geographical features have contributed to the formation of a relatively isolated gulf, protected from external influences. As a result of freshwater influx carrying fine-grained material, there is a prevalence of fine-grained clasts such as silts and clays in the area. Additionally, sandy material in the vicinity of the continental shelf has been identified as relict pre-Holocene deposits through previous research [20–22]. The extent of the continental shelf allows for the deposition of most of the material, while only a small fraction travels to the deeper parts of the North Aegean basin. This fraction is transported through the Sporades basin, an elongated basin that represents the western margin of the North Aegean Trough.

The North Aegean continental shelf, located to the east of the Chalkidiki Peninsula, is a relatively flat plateau with limited variations in its morphology. However, the sediment types present in this region vary significantly, with sand fractions exceeding 40% of the analyzed sediment, thus being of particular interest for raw material exploitation, according to sources [13,23,24] (Figure 2). The distribution of sediment in this region is influenced by hydrodynamic conditions, with fine-grained sediments dominating in low-energy areas and significant sand content in other regions. The highest sand content, exceeding 90%, is observed near the Evros River, Samothraki Island, and between the mainland and Thassos Island. Conversely, the Strymonikos and Kavala Gulfs, the deepest areas of the shelf, and the eastward region of the Athos Peninsula are characterized by mud types and fine-grained sediments. The northeastern to southwestern parts of the North Aegean Sea is dominated by fine-grained muddy sediments, the Sporades Basin is mostly covered by silt, while sandy clays are found in the Skyros Basin.

Given the variations in sediment distribution in this region, it is crucial to estimate sediment transport patterns using granulometry data from surficial samples. This is not only important for rare earth element mining but also for the placement of various infrastructures on the seafloor, such as cables and anchored offshore windmills. Therefore, the goal of the present study is to provide a detailed analysis of sediment transport patterns in the North Aegean Sea, enabling better planning and utilization of the region's resources.

2.2. Sampling Effort and Laboratory Testing

In order to identify sediment distribution patterns, a dataset of 323 point granulometry data, from samples collected during previous field campaigns [14,25], was used. The study area, comprising the northern coastal and open sea of Aegean (Figure 2), covered the region between the Chalkidiki Peninsula and Samothraki Island. The marine environment in this area is variable, ranging from the coastal zone to the shelf areas of Strymonikos and North Aegean plateaus, reaching the shelf break. The sampling depths varied between 0.5 and 693 m. This area is tectonically active, being adjacent to the north Aegean Trough, and exhibits several recent faults and various geomorphological features, such as submerged valleys, slopes, plains, canyons, submarine fans, erosional features, and landslide areas. The aforementioned geomorphological configuration, in combination with marine water circulation, controls sediment transport, reworking, and redeposition and significantly influences the seabed substrate.

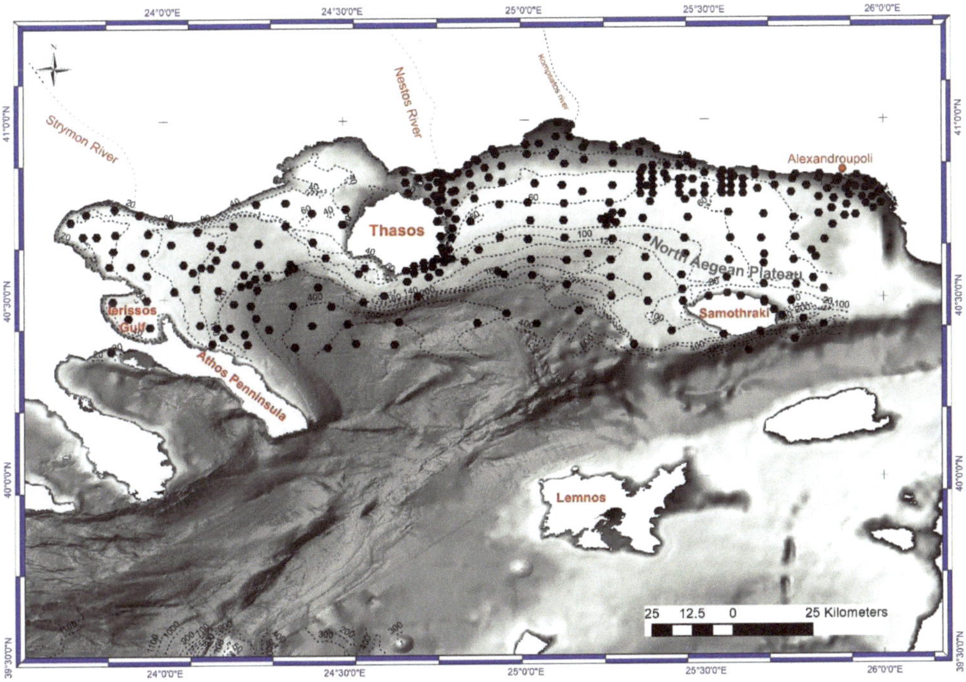

Figure 2. Sampling stations across the study area. Bathymetric contour lines are also shown. The bathymetry hillshade is from the EMODNET database [18].

The deposition of coarse-grained sediments is favored in high-energy environments, such as low-relief shallow areas. On the other hand, finer sediments are more likely to accumulate in low-energy habitats where sediments can remain in suspension and reach the most distant coastal areas. The continental shelf's paleogeographic evolution is shown in the two well-defined paleo-riverbeds of Strymon and Nestos, as identified by detailed bathymetry and seismic profiling studies [17,26]. The onshore geology mainly consists of post-Alpine sediments, Eocene and Miocene granitoids, Triassic marbles, and Paleozoic gneisses [27], whose erosion products are transported offshore via the Strymon River in the NW and a few streams elsewhere, controlling the sediment supply. The grain-size data used for this paper (323 data points) are stored in the Marine Geology Database of HSGME and include data collected from several expeditions carried out between 1979 and 2015 for national mapping and other research projects [10]. The sampling strategy followed a grid based on regular spacing, bathymetry, and acoustic data. The sampling effort varied from a few hundred meters in the shallower part to several kilometers in the deep (Figure 2), with an average of 14.25 points per 100 km^2, guided by expert knowledge according to area complexity. The collection of physical samples (grab samples) followed the HSGME Marine Geology Laboratory protocol, and all samples were analyzed in the laboratory according to international protocols [28]. The textural parameters (mean grain size, sorting, and asymmetry) were calculated using the graphical method of Folk and Ward [29]. The results of the textural analysis (mean size, sorting, and skewness) were then used to construct a two-dimensional sediment transport model that reflects sediment transport pathways [7,8]. Net sediment transport patterns were calculated using Gao's and Collins' [9,30] grain-size trend analysis. Similar studies from various aquatic realms address the applicability of grain-size trend analysis [31–34]. This study utilized a two-dimensional model developed by [30] to calculate grain-size trends. The model assumes that the trend analyzed has a higher possibility of occurrence in the direction of net transport than in other directions. It identifies the possible net transport paths by comparing two neighboring sampling sites and employing the three most commonly used grain-size parameters (m, s, and Sk) for trend analysis. The comparison of grain-size parameters of the two neighboring sampling sites, a dimensionless trend vector is defined, which has a unit length and an orientation that points from the station with the higher sorting coefficient to that with the lower sorting value. The grain-size trend vectors are generated for each sampling station and its neighboring stations, and a single resultant vector is produced after summing the identified trend vectors. The produced vector depicts the trend of sediment transport at the area that has been sampled. A smoothing process is finally applied in order to reduce background noise. This process averages the produced vector of the previous stage taking into consideration and the neighboring sampling sites and finally provides a residual vector. The transport vector maps were produced using MATLAB's GSTA model [35], designed based on Gao's FORTRAN program [36]. As input parameters were set the mean size, sorting, and skewness of the 323 samples as presented in Table A1 of the Appendix A. The scaling factor A, considering that the used projection is UTM 34N, was set to 1 [36] and the characteristic distance DC was set to 16,000 m.

3. Results

3.1. Grain Size Analyses

The surface sediments were analyzed and classified using the Folk classification method (Figure 3) [11]. The classification results were then plotted on a ternary diagram. The spatial distribution of the sediment classes was estimated using the methodology developed by [14]. Most samples lacked gravel components and were thus plotted onto a Sand/Silt/Clay ternary diagram.

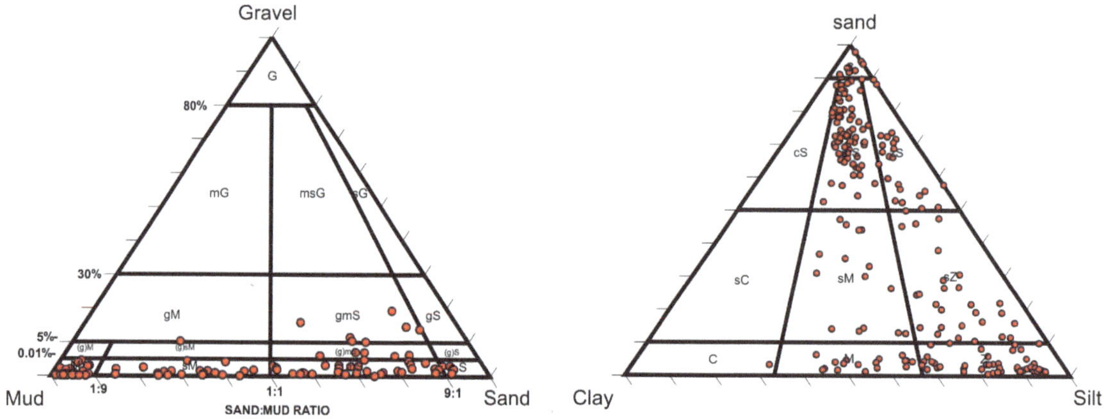

Figure 3. Ternary diagrams representing sample classification [11].

As shown in Figure 4, the eastern part of this study area has a more dominant sand component, reaching up to 80%. This is due to residual sands previously identified in the area during scientific expeditions by HSGME. Increased sand percentages also occur northwest of Thasos Island and in patches near the Athos Peninsula coastline. At depths below 120 m, the sand component decreases rapidly, consisting mainly of benthic and planktonic calcareous skeletal debris. The silt component accumulates in a semicircular depression extending from the east margin of Athos Peninsula to the southwest outer shelf of Thasos Island. A similar trend, with slightly decreased silt component percentages, characterizes the area between the Thasos and Samothraki islands. Silt is less than 30–40% in the rest of the study area, while in sand-dominated areas, it does not exceed values above 20%. The clay fraction is generally underrepresented in the study area, not exceeding 30–40%; however, in the vicinity of river mouths such as Strymon, Nestos, and Kompsatos, the clay fraction reaches up to 50%.

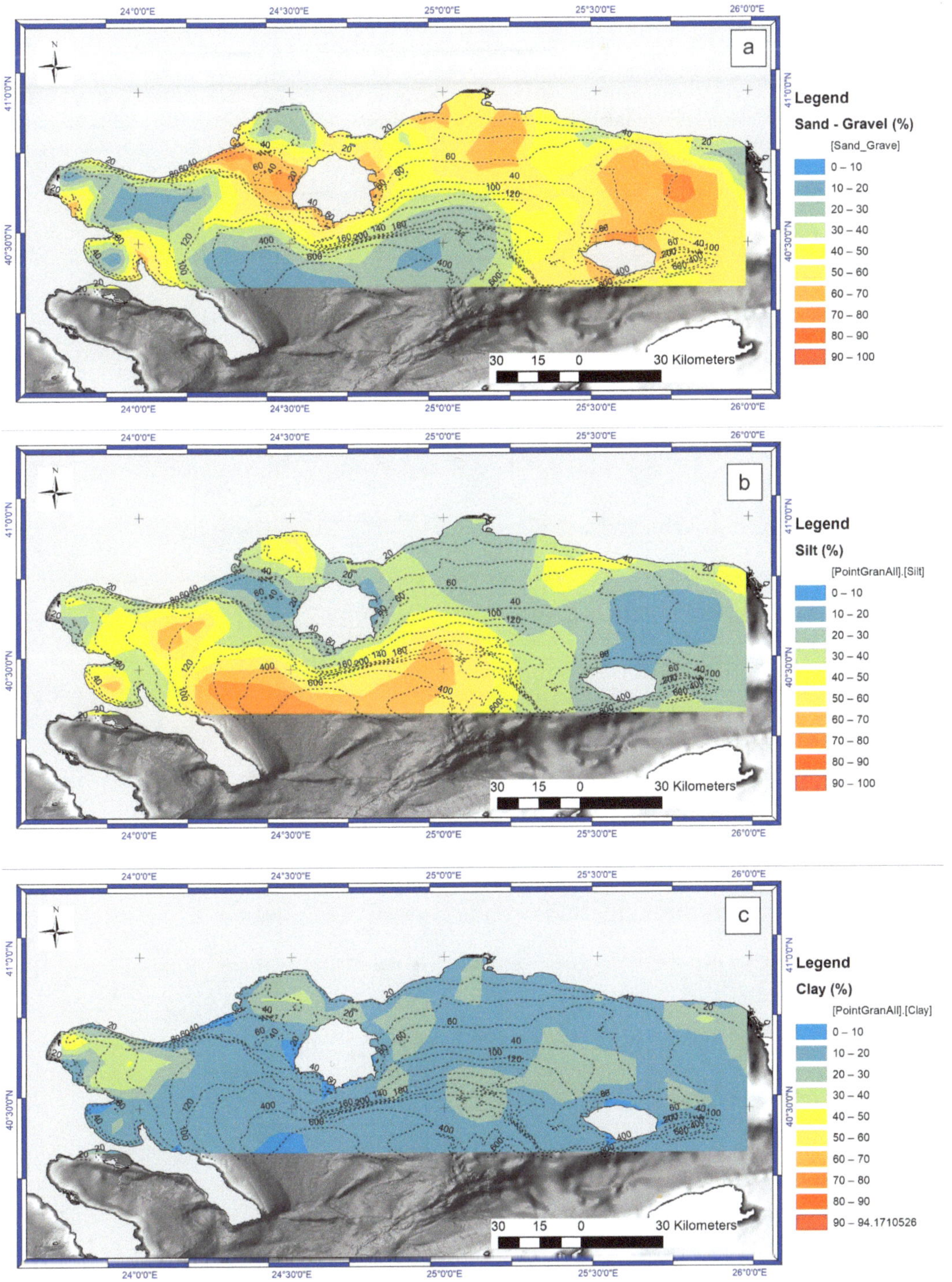

Figure 4. Spatial distribution of gravel–sand, silt, and clay ((**a**–**c**) respectively).

3.2. Grain Size Parameters

The spatial distribution of the mean diameter of the surface sediments reveals two trends (Figure 5a). Firstly, there is a trend of fining towards the deeper parts of the study area, with a slight variation close to Stymonikos Gulf. Secondly, there is a trend of coarsening towards the northeast. The mean diameter of the surface sediments ranges from -0.85φ to 8.31φ.

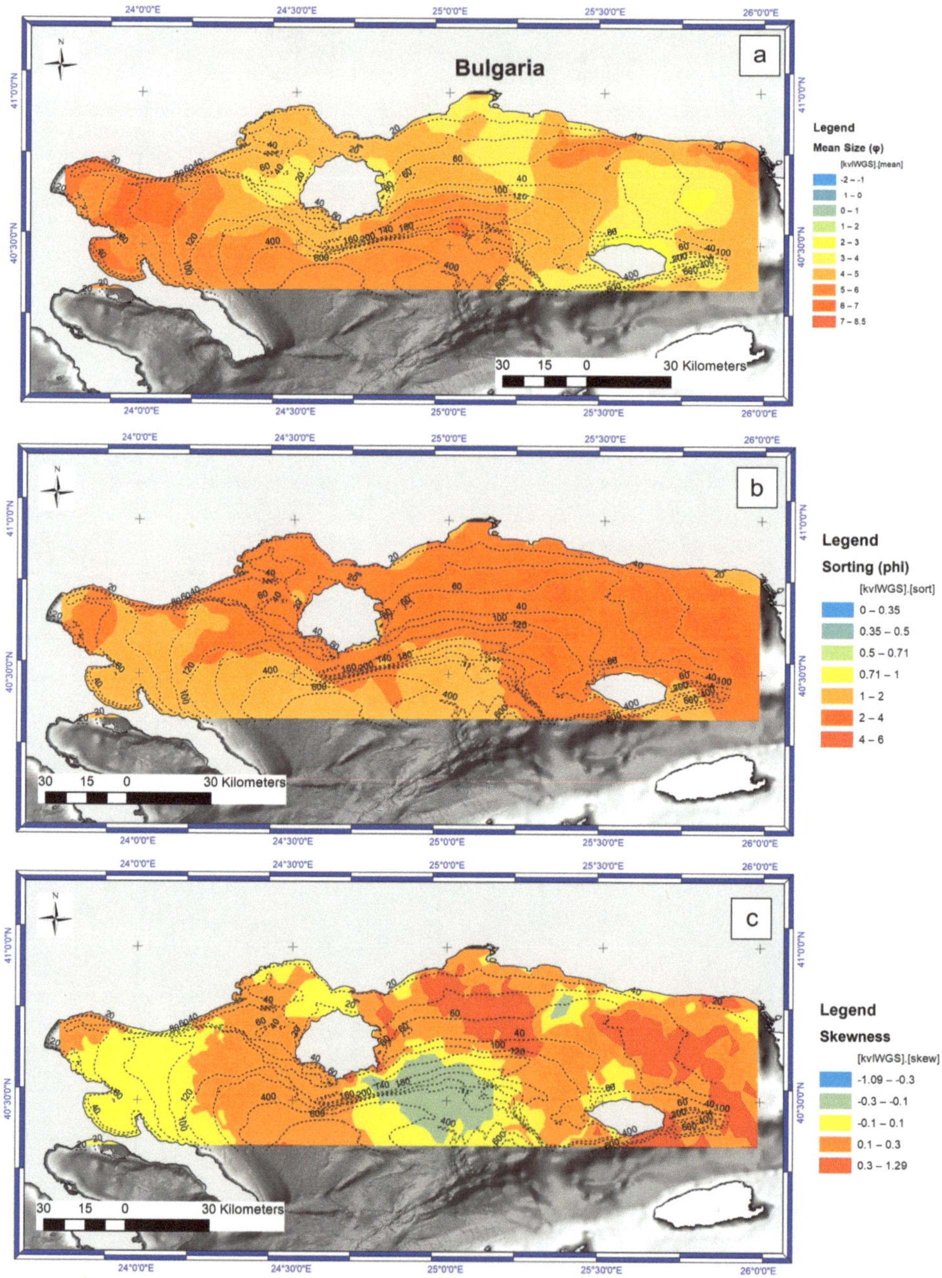

Figure 5. Statistical parameters (mean, sorting, skewness) spatial distribution ((**a**–**c**) respectively).

In the north of Samothraki Island, the mean diameters range from 2 to 4φ, corresponding to medium to very-fine sand, respectively. Finer sediments are found in the perimeter of this area, with mean size values ranging from 4φ to 8φ (very fine sands and silts). As expected, in the depression between Athos and Samothraki, which is generally marked by the 120 to 140-m bathymetric contours, finer sediments have been accumulated (5φ to 8φ). The sorting coefficients of surface sediment (Figure 5b) are generally grouped into two classes—poorly sorted (1–2φ) and very poorly sorted (2–4φ). Poorly sorted sediments are spatially distributed in deeper parts of the basin and partially in the Strymonikos Gulf area. On the other hand, very poorly sorted sediments occur in a wide area, from Kavala and Thasos Island towards the eastern margins of the study area. Few outliers (less than 15 samples) of moderately to well-sorted sediments occur generally close to the shoreline. The skewness coefficients present a more complex pattern (Figure 5c). In the Strymonikos Gulf area, sediments are generally symmetrical or fine-skewed, moving closer to the northernmost coastline. Two trends were identified between Athos and Samothraki Island. At the westernmost area, the sediments are generally fine skewed to symmetrical, while eastwards, symmetrical to coarse-skewed sediments are deposited. Around Samothraki and towards the north, fine-skewed to strongly fine-skewed trends dominate. Few patches of symmetrical distributions occur.

Figure 6 displays cross plots of the sorting coefficient, skewness, and mean diameter. Data were averaged within a range of one φ bin to identify any trends in these plots. The results demonstrate that the sorting coefficient decreases from −1φ to 5φ. Samples with mean grain sizes between −1φ and 0φ are moderately to very well sorted, while those ranging from 0φ to 5φ are poorly to very poorly sorted. Conversely, a better sorting trend is observed from 5φ to finer average values, resulting in a gradual improvement in the sorting of fine-grained sediments close to 1.5φ (poorly sorted). Furthermore, skewness exhibits a correlation with the mean grain diameter. Coarse-grained sediments (up to 0φ) are generally symmetrical, with a few outliers (2 samples) exhibiting strong negative or positive skewness. Up to 5φ, skewness shows a clear positive trend, as the examined samples generally have values greater than 0.3. However, skewness values decrease from 5φ to finer sediments, resulting in a coarse to strongly coarse skewed trend, as the samples exhibit finer average grain sizes.

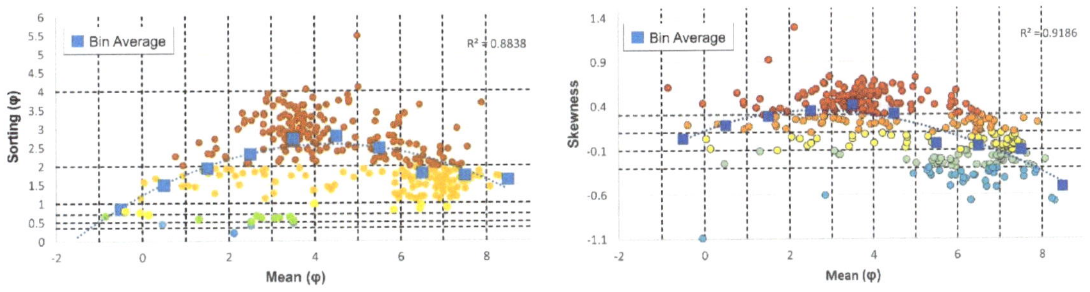

Figure 6. Statistical parameters correlations. The coloring of the various groups of Y-Axis is consistent with the classification presented in Figure 5.

3.3. Sediment Transport Patterns

Following Gao's [36] methodology, a sediment transport map was designed with transport vectors in each sampling site. Before analyzing the vector transport directions, a statistical test for an assessment of the significance of the produced vectors was conducted in 246 out of 323 samples (covering the region Thasos—Alexandroupoli), by comparing the characteristic vector length of the analysis with the averages of randomly distributed samples in the study area. The characteristic vector length is practically the average length of the vectors that have been calculated for each sampling site and for the examined area, has a value of L = 1.9613. The process was based on the following steps: (a) using Excel Vba coding, each grain-size analysis was randomly reallocated in a different sampling site by generating a sequence of random numbers (1 to 246). If, for example, a sample numbered as 1 was located in a site with coordinates x: 610,969 and y: 4,529,442, by randomly assigning the number 63 it was reallocated in a site with coordinates x: 656,838 and y: 507,446. This process was repeated 100 times producing 100 randomly distributed datasets. (b) For each dataset transport vectors were produced and the average for each dataset was calculated. (c) The frequency of the 100 datasets (Figure 7a) was used to generate a normal distribution curve in order to calculate the critical values L90, L95, and L99 (Figure 7b,c,d respectively). The real dataset characteristic length was then statistically tested for significance with aforementioned critical values. According to the statistical test (Z test), the estimated transport vectors are significant in a 0.1 and 0.05 level (for 90% CI $z = 4.13 > 1.645$, 95% CI $z = 2.69 > 1.96$) and it is not significant for 0.01 level (for 99% CI $z = 0.05 < 2.58$). According to Davis [37] a confidence interval of 95% is acceptable for testing the significance of a dataset.

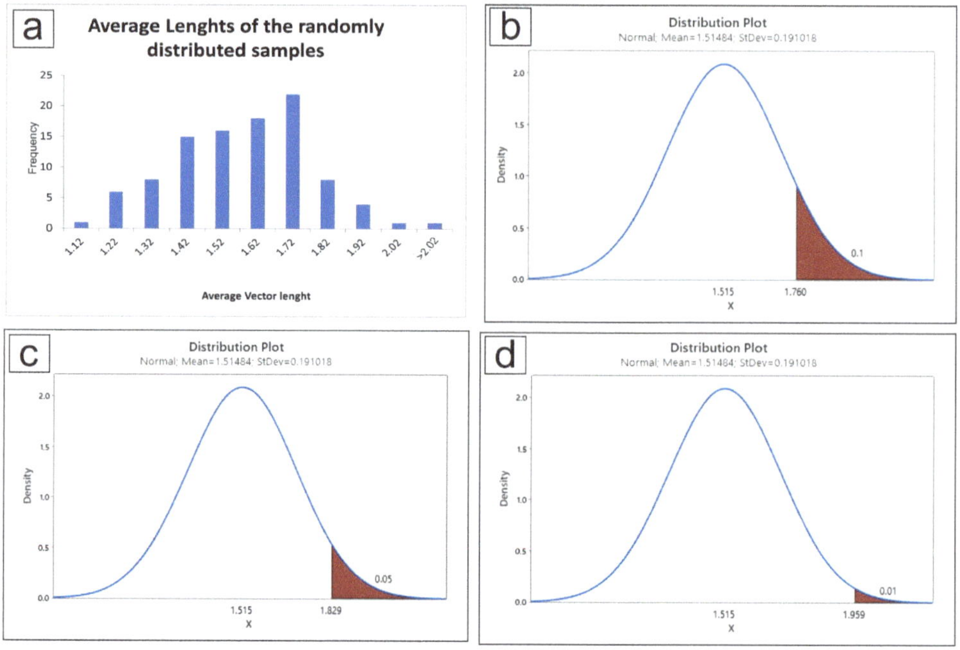

Figure 7. (**a**) Histogram of the average vector length of the 100 randomly distributed datasets. (**b**) Estimation of vector length L90 for a significance level of 0.1. (**c**) Estimation of vector length L95 for a significance level of 0.05 and (**d**) Estimation of vector length L99 for a significance level of 0.01. The red part of the plots corresponds to the critical area for the significance tests.

The analysis of grain-size trends has revealed a complex sediment transport network (Figure 8), highlighting the different processes that control sediment distribution in the marine environment. In the eastern part of the study area, sediment transport is mainly towards the north or northeast. Transport vectors are oriented towards the west or east near the coastal area. In the easternmost area, close to Alexandroupoli, the NNE trend is sustained for a long distance. Strong N-NE transport patterns have been identified along Thasos Island's eastern and northern coastline. Thasos's western and southwestern coasts show W-SW trends similar to the sediment transport vectors estimated for the open marine area westwards of Thasos. In the periphery of the semicircular depression defined by the Athos Peninsula and Thasos Island, the vector pattern is radial, with sediment transport directed towards the deepest central part of the depression. Between Thasos and

Samothraki, especially near the margins of the North Aegean Plateau, sediment transport vectors are less significant, in terms of vector length.

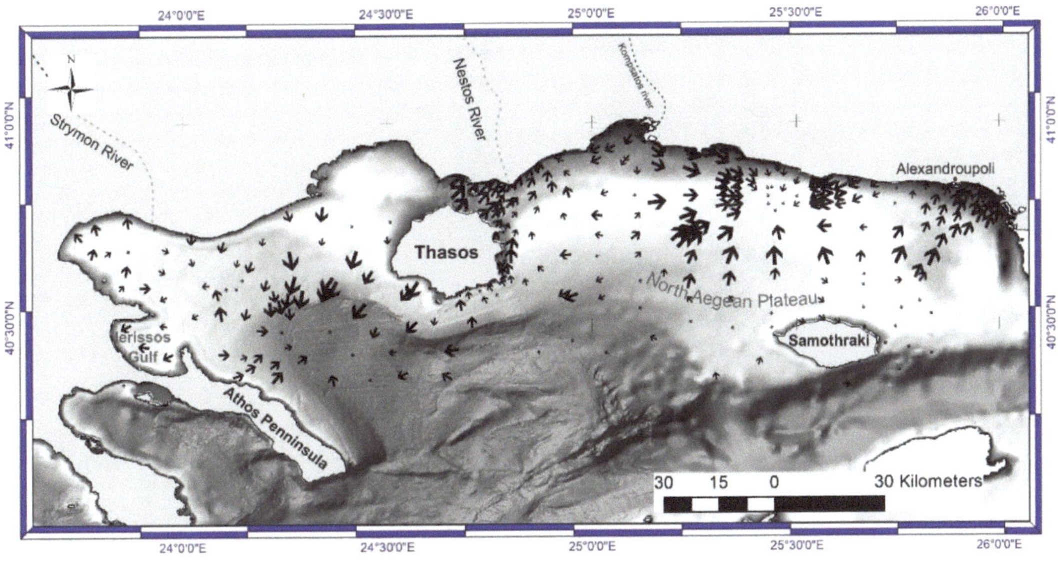

Figure 8. Net sediment transport pattern across the study area.

The direction of sediment transport is not always the same as the direction of river flow, especially in areas near river deltas where sediment inflow is higher. For instance, in the Strymon River delta, the vectors indicating sediment transport have a weak tendency towards the south-southeast. On the other hand, in the Nestos River delta, sediment transport vectors are oriented towards the north-northeast. Finally, in the case of the Kompsatos River, net transport vectors align with the direction of sediment inflow.

4. Discussion

4.1. North Aegean Water Circulation Pattern

Various factors, such as sediment sources, hydrodynamic conditions, and topography features, influence the distribution of grain-size parameters in an area. In the Aegean region, the circulation pattern is driven by a thermohaline circulation close to the surface that involves the outflow of low-salinity water from the Black Sea (salinity ranging from 22.50 to 27.50 psu and an influx of 5.000 to 15.000 m^3/s [38]) and the inflow of freshwater from rivers that flow into the Aegean [39] (Figures 9 and 10). Additionally, the Aegean region receives saline water from the eastern Mediterranean. In the northern Aegean, the river runoff controls the sediment distribution in close sub-basins such as Thermaikos Gulf [40–42] which are not exposed neither to the open water's circulation of the Aegean nor to the influx of Black Sea water masses. Studies conducted in the area [38] support the idea that the water masses from the Black Sea that enter the northern Aegean region dominate the combined effect of all freshwater influx (annual river water influx according to [38] has been estimated to 50 m^3/s for Nestos and 100 m^3/s for Strymonas and Evros respectively). This influx explains the lower salinities in northern Aegean [38,43–45]. Seasonal variations occur in the water circulation pattern. However, the general context is that the circulation in the Aegean Sea is controlled by the buoyancy input from the Black Sea to a lesser degree in North Aegean (except Thermaikos Gulf) by the river discharge, and the effects of winds in the region. The overall tendency of the buoyancy force is to create a cyclonic circulation around the Aegean. Another significant process is the transportation of dense waters formed on the shallow parts during the cold winter months that slowly descend toward the deep basins. According to [43], dense waters originating from the north Aegean shelf areas can generally remain in the adjacent deep basins for long periods. Due to their thickness, separating sills exceeding the bottom layer's depth impede flow.

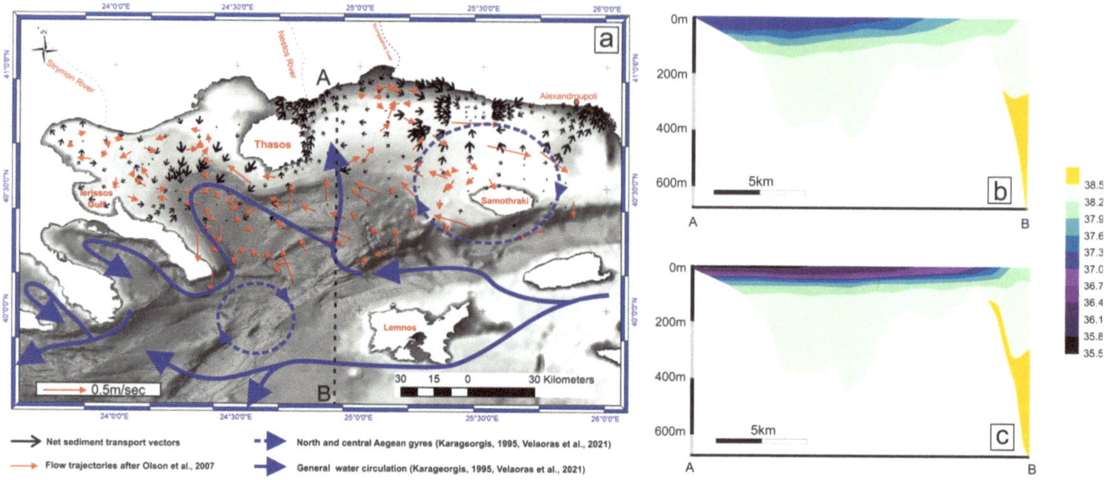

Figure 9. (**a**) Generalized water circulation pattern where the flow circulation vectors of the various research are also shown [39,46,47]. (**b**) Vertical salinity distribution across section AB for winter (modified from [38]). (**c**) Vertical salinity distribution across section AB for summer (modified from [38]).

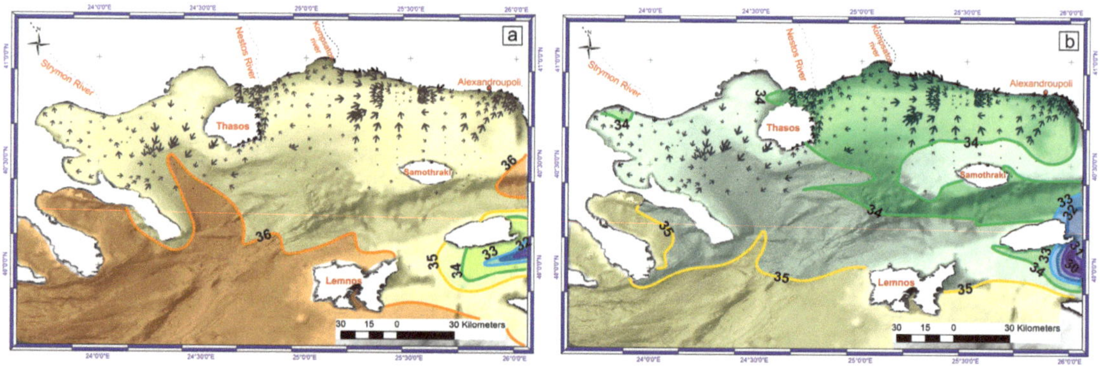

Figure 10. (**a**) Surface salinity during winter. (**b**) Surface salinity during summer (modified from [38]).

4.2. Water Circulation Pattern and Sediment Transport

Sediment transport patterns can provide insights into the complexity of water circulation and the influence of basin topography. The area located northwards of Samothraki Island, as defined by the direction of transport vectors, is predominantly influenced by the north Aegean gyre. The transport vectors exhibit a consistent N-NE orientation, which is aligned with the flow trend of the gyre in this specific area. The impact of the gyre is also reflected in the spatial distribution of sediment components and statistical parameters. The data suggest that sediments in the gyre area are predominantly sand, and skewness values are symmetrical or coarse skewed, indicating that fine fractions are transported northwards, following reworking. Near the coast, the impact of the gyre diminishes, and longshore drift becomes the dominant factor, as indicated by the transport vector trending parallel to the coast. A more complex and difficult-to-assess pattern regarding sediment transport is persistent in the area defined as NE of Samothraki to the Alexandroupoli coast. The dominant trend towards the northeast suggests that masses of low salinity, such as those dispersed in the area (Figure 10a,b), combined with the north Aegean gyre, result in a composite transport vector that forces fine-grained sediment towards the coast. This process is reflected in the increased clay component in the specific area (Figure 4).

In the studied region, sediment transport vectors exhibit bidirectional shifting towards the northwestern and northeastern directions between Samothraki and Thasos islands. A sector with less significant sediment transport vectors is identified where the BSW and North Aegean gyre are clearly separated and follow northwestern and northeastern directions, respectively. Sediment transport vectors become more intense in areas where the plume of low-salinity water expands on the shelf regions, and the depth decreases. Along the eastern coast of Thasos, BSW circulation is consistent with sediment transport vectors. On the northern coast of Samothraki, there is a complex situation where sediment transport vectors are moving towards the northeast despite the low-salinity plume expanding towards the west. This situation can be attributed to the combined effect of Nestos River's outflow and the Black Sea Water (BSW) fluxes, resulting in a configuration that shifts the transport vectors towards the northeast.

The region between the Thasos and Athos Peninsula displays a radial pattern in the areas near the shelf break. The fine-grained sediment from the shelf (mainly silts) is re-deposited at the deeper parts of the morphological depression, where water circulation is reduced, and stagnant conditions are prevalent. This regime is also supported by the small length of the net transport vectors. This supports [43]'s hypothesis that dense waters formed in shallow areas during the cold winter months are transported to the deep sea, ultimately remaining in adjacent deep basins for prolonged periods due to the difficulty in flowing over separating sills that exceed the thickness of the bottom layers. Consequently, sediment transport vectors in the area are insignificant. At the open shelf to the coast region, sediment transport patterns are generally consistent with the flow trajectories proposed by [39]. The significance of net transport vectors may not be as prominent as those affected by BSW or North Aegean gyre. However, assessing the controlling factors that impact sediment dispersal is still essential. Close to the coast, the outflow of the Strymon River leads to a group of vectors trending towards the south, despite the fact that the length of the vectors indicates a weak trend of sediment reworking by this process. In Ierissos Gulf, the transport vectors steadily trend toward the W, consistent with the prevailing wind direction [48]. This results in a washout of the suspended fine-grained material, accumulating in the coastline, as indicated by the mean phi values in the specific area (Figure 5a). Lastly, the net transport vectors in the area defined by the northern coast of the Athos Peninsula (northwards of Ierissos Gulf) reflect the outflow of two local-scale river streams, resulting in N or NE directions.

5. Conclusions

Multiple factors, such as the sources of sediment, hydrodynamic conditions, and topographical features, influence the distribution of grain-size parameters in a particular area. The complex net sediment transport patterns in the North Aegean basin have been identified by analyzing statistical parameters like mean size, sorting, and skewness. These patterns reveal the contribution of various hydrodynamic factors and their impact on the area.

The sediment distribution in the central and eastern parts of the study area is primarily influenced by two factors, namely, the north Aegean gyre and the Black Sea water mass (BSW), which enters the Aegean space through the Marmara Sea. The BSW, characterized by low-salinity waters, generates a plume that spreads towards the W-NW. As this plume reaches lower depths at the shelf areas, it triggers a reworking of basin floor sediments, reflected in the measured net transport vectors. The north Aegean gyre also favors sediment reworking, as net transport vectors are generally oriented with the direction of the gyre.

In the shallow water coastal areas, the dominant process is the longshore drift, which is indicated by vectors that trend parallel to the coast.

The net transport vectors are insignificant in the deeper parts of the basin beyond the shelf edge, indicating the formation of stagnant conditions. This is due to the gradual movement of cold, dense waters downslope, which become trapped in the deeper parts and remain there for long periods. This process helps to maintain the initial textural characteristics of the accumulated sediments on the basin floor, preventing further reworking.

In the western part of the study area, two processes dominate the sediment distribution pattern. Near the shelf edge, the circulation of bottom seawater (BSW) plays a significant role, while in the shelf–coastal areas, the outflow of rivers is more important. In small gulfs such as Ierissos, the direction of prevailing winds also contributes to sediment distribution.

Regarding the significance of sediment distribution patterns, the area where transport vectors suggest the intense reworking of sediments could potentially favor the enrichment of sediment in placers, considering that minerals are not so mobile. Such an area has been identified northwestern of Samothraki, where sand component is dominant implying the wash out of finer material. The composition of placers in the specific area is under research. Sediment transport is also important in terms of the existing and planned infrastructures. The eastern part of the study area accommodates underwater cable systems for telecommunications and electric energy transport, while a major cable project is planned for the connection of the northeastern Aegean islands with the National System of Electricity Transport. Finally, the identification of the seafloor sediment dynamics in this area is crucial given that in the easternmost part of the Samothraki–Alexandroupoli shelf is a priority area for the potential installation of anchored offshore windfarms, and thus desktop studies for underwater foundations and electricity transport are in progress. Sediment transport vectors can help identify areas where cables or the foundation of the aforementioned infrastructures can be exposed or even damaged by intense sediment reworking.

Author Contributions: Conceptualization, I.V. and I.Z.; methodology, I.V. and I.Z.; validation, I.V. and I.Z.; data curation, I.V. and I.Z.; writing—original draft preparation, I.V. and I.Z.; writing—review and editing, I.V. and I.Z.; visualization, I.V. and I.Z. All authors have read and agreed to the published version of the manuscript.

Funding: This research was funded by NSRF 2014–2020, grant number 5030010 ("Geological Mapping of Greece for the support of innovation and entrepreneurship—GEOINFRA"), and European Commission contract No. SI2.658129 ("European Marine Observation and Data Network—Lot 2 Geology"). The APC was funded by European Commission contract No. SI2.658129 ("European Marine Observation and Data Network—Lot 2 Geology").

Institutional Review Board Statement: Not applicable.

Informed Consent Statement: Not applicable.

Data Availability Statement: Data is contained within the article.

Acknowledgments: Used data were collected in the frame of the Research and Development program of the European Communities (E.E.E., 1984–1989) and the NSRF/Operational Programme Competitiveness & Entrepreneurship (2013–2015), project No. 351008 ("YPOTHER: Marine geology and mineral research study along the coastline between Chalkidiki and Kavala, and the caldera of Santorini").

Conflicts of Interest: The authors declare no conflicts of interest.

Appendix A

Table A1. Grain-size parameters.

ID	x	y	Mean	Sort	Skew
1	564,117	4,495,734	2.12	0.21	1.29
2	565,142	4,499,443	2.51	0.41	0.1
3	564,877	4,502,987	0.47	0.45	−0.25
4	664,621	4,521,654	3.5	0.48	0.06
5	561,735	4,513,753	2.51	0.52	−0.02
6	662,132	4,522,987	3.5	0.55	0.03
7	572,623	4,526,958	1.3	0.59	−0.11
8	564,133	4,507,607	3.06	0.61	−0.11
9	561,072	4,494,629	2.75	0.62	0.2
10	571,002	4,525,247	3.17	0.63	−0.15
11	604,855	4,531,510	3.42	0.65	−0.06
12	622,224	4,528,077	−0.85	0.68	0.61
13	568,684	4,522,757	2.65	0.68	0.03
14	584,505	4,531,557	0.14	0.72	−0.08
15	632,223	4,485,682	−0.05	0.77	−1.09
16	627,886	4,484,681	−0.39	0.81	0.19
17	538,723.7	4,468,109	5.83	0.82	0.38
18	530,532.2	4,481,180	6.44	0.86	0.2
19	620,525	4,518,866	7.08	0.89	−0.23
20	520,544.2	4,481,179	6.47	0.97	0.2
21	653,457	4,523,264	3.98	0.99	0.32
22	543,395.6	4,475,578	6.34	1.11	0.26
23	515,674.6	4,474,244	6.34	1.12	0.04
24	536,425.5	4,486,993	6.92	1.12	−0.52
25	539,524.2	4,480,960	5.95	1.13	0.43
26	576,686	4,530,084	2.84	1.15	−0.61
27	599,227	4,497,661	7.16	1.15	−0.16
28	643,503	4,480,647	−0.04	1.16	0.43
29	516,552.9	4,481,735	6.22	1.17	−0.11
30	621,283	4,521,277	7.3	1.18	0.05
31	620,447	4,521,096	7.31	1.18	0.03
32	521,826.1	4,477,227	6.46	1.18	0.15
33	580,491	4,479,024	6.82	1.19	0.09
34	524,872.9	4,473,734	6.18	1.19	0.21
35	507,686.6	4,486,452	6.54	1.19	0.12
36	579,912	4,491,104	5.86	1.25	−0.39
37	614,096	4,525,320	7.23	1.26	−0.12
38	545,347.7	4,469,495	5.98	1.27	0.31
39	650,533	4,477,987	1.67	1.28	0.19

Table A1. *Cont.*

ID	x	y	Mean	Sort	Skew
40	565,173	4,522,571	3.04	1.28	0.26
41	504,645.3	4,493,972	7	1.28	−0.18
42	573,619	4,475,900	7.21	1.29	0
43	617,571	4,521,085	7.22	1.31	−0.11
44	531,091.4	4,468,284	6.05	1.32	0.32
45	597,344	4,479,903	6.98	1.34	−0.2
46	668,092	4,518,492	3.95	1.35	0.54
47	657,933	4,522,124	3.97	1.36	0.57
48	660,789	4,520,088	4.38	1.37	0.47
49	491,560.1	4,477,278	7.18	1.37	0.01
50	664,646	4,514,250	7.51	1.38	−0.09
51	492,158.7	4,493,472	7.16	1.38	−0.07
52	512,381.4	4,495,458	6.82	1.4	0.18
53	481,810.7	4,509,859	7.33	1.41	0.11
54	522,251.5	4,500,630	6.28	1.42	0.24
55	656,900	4,515,623	8.08	1.43	−0.21
56	487,873.2	4,482,525	7	1.44	−0.19
57	657,604	4,521,038	4.29	1.45	0.56
58	631,609	4,472,824	1.22	1.46	0.38
59	626,452	4,518,931	6.86	1.46	−0.37
60	635,417	4,484,814	0.92	1.48	0.45
61	591,374	4,534,106	1.05	1.49	0.17
62	558,846	4,493,686	4.46	1.49	0.42
63	554,411.7	4,469,214	6.58	1.49	0.29
64	535,054.8	4,475,005	6.42	1.5	0.24
65	542,773.5	4,511,316	2.24	1.51	0.03
66	587,523	4,475,865	5.86	1.54	−0.28
67	522,080	4,486,185	6.88	1.55	−0.04
68	663,687	4,516,233	7.15	1.57	−0.06
69	598,300	4,532,038	0.04	1.58	0.03
70	586,382	4,490,871	7.45	1.58	−0.03
71	668,183	4,514,483	7.52	1.58	−0.04
72	566,111	4,496,214	3.08	1.59	0.35
73	478,119.4	4,507,424	7.23	1.59	−0.05
74	594,319	4,495,593	6.25	1.6	−0.16
75	585,973	4,495,801	6.93	1.6	−0.31
76	496,246.8	4,503,711	6.83	1.61	−0.17
77	495,485.8	4,497,867	7.03	1.62	−0.21
78	575,958	4,521,287	7.62	1.63	−0.15
79	647,569	4,483,503	7.38	1.67	−0.11
80	508,892	4,493,121	6.68	1.67	0.24

Table A1. Cont.

ID	x	y	Mean	Sort	Skew
81	648,567	4,471,734	6.87	1.68	0.22
82	608,251	4,531,252	0.31	1.69	0.2
83	630,399	4,521,338	6.32	1.69	−0.67
84	501,636.3	4,500,436	6.99	1.69	−0.11
85	666,081	4,516,162	7.56	1.7	−0.34
86	487,496.9	4,466,948	4.91	1.7	0.35
87	516,467.1	4,506,008	6.62	1.7	0.01
88	632,470	4,523,336	6.45	1.71	−0.06
89	508,583.4	4,475,158	6.08	1.71	−0.1
90	650,488	4,519,531	7.79	1.72	−0.29
91	648,645	4,488,614	7.86	1.72	−0.33
92	612,766	4,518,982	8.31	1.72	−0.68
93	513,830	4,499,073	6.92	1.72	−0.04
94	571,450	4,522,476	7.06	1.73	−0.05
95	480,993.8	4,502,872	7.03	1.73	−0.16
96	666,463	4,512,746	7.56	1.74	−0.41
97	660,582	4,516,689	7.67	1.74	−0.27
98	594,415	4,488,039	8.24	1.74	−0.66
99	511,084.8	4,492,952	6.18	1.74	−0.11
100	641,097	4,482,959	0.93	1.75	0.42
101	563,372	4,494,956	3.45	1.75	0.47
102	565,626	4,497,906	3.78	1.75	0.34
103	495,900.1	4,482,883	2.15	1.75	0.25
104	632,567	4,516,899	2.83	1.78	0.28
105	614,182	4,518,784	6.83	1.78	−0.31
106	524,392	4,490,001	1.78	1.78	0.37
107	494,219.2	4,509,572	5.87	1.78	0.28
108	614,211	4,516,588	6.14	1.79	−0.42
109	519,773.5	4,469,054	5.27	1.79	−0.33
110	496,666.2	4,474,488	7.06	1.8	−0.08
111	519,143	4,472,329	6.55	1.81	0.2
112	547,067.3	4,482,398	1.52	1.82	0.29
113	535,175.7	4,500,982	2.32	1.83	0.23
114	512,642.3	4,474,267	6.5	1.83	0.03
115	661,975	4,519,435	4.66	1.84	0.63
116	667,413	4,517,396	7.16	1.84	−0.04
117	511,540.7	4,469,297	5.56	1.84	−0.23
118	647,640	4,509,570	2.6	1.86	0.27
119	578,494	4,491,862	6.66	1.86	−0.46
120	574,990	4,524,299	7.13	1.86	0.13
121	630,052	4,522,196	7.83	1.87	−0.37

Table A1. *Cont.*

ID	x	y	Mean	Sort	Skew
122	556,389	4,523,023	6.03	1.89	−0.16
123	623,600	4,521,188	6.79	1.89	0.14
124	630,593	4,516,862	2.78	1.9	0.21
125	664,328	4,519,024	6.48	1.9	0.32
126	610,969	4,529,442	7.29	1.9	−0.1
127	528,753.8	4,506,074	1.39	1.9	0.3
128	611,368	4,516,528	6.69	1.91	−0.52
129	570,588	4,495,021	7.29	1.91	−0.22
130	655,432	4,518,522	7.33	1.91	−0.04
131	562,258	4,521,312	7.03	1.92	0.09
132	541,984.6	4,506,608	2.05	1.92	0.05
133	635,420	4,516,931	2.85	1.93	0.35
134	563,884	4,522,714	5.14	1.93	0.46
135	655,327	4,476,332	1.94	1.94	0.34
136	488,431	4,511,368	7.35	1.94	0.07
137	655,058	4,502,781	1.69	1.95	0.17
138	561,205	4,492,780	6.75	1.95	−0.32
139	577,404	4,528,550	3.64	1.96	0.66
140	506,877.7	4,491,552	6.36	1.96	−0.06
141	565,310	4,520,260	6.21	2.01	0.41
142	564,144	4,519,787	6.86	2.01	0.06
143	568,149	4,517,201	6.89	2.01	0.06
144	666,286	4,520,610	7.36	2.01	−0.23
145	665,721	4,519,672	6.76	2.02	0.25
146	487,116.7	4,493,485	6.74	2.02	−0.35
147	610,429	4,469,538	0.94	2.03	0.14
148	611,256	4,522,984	7.42	2.05	−0.01
149	604,820	4,525,805	2.42	2.08	−0.05
150	622,653	4,509,270	2.46	2.08	0.28
151	647,168	4,521,439	7.21	2.08	−0.12
152	508,000.7	4,504,417	6.46	2.08	−0.27
153	495,954.8	4,490,603	6.77	2.1	−0.27
154	514,094.7	4,471,404	5.37	2.11	−0.39
155	563,198	4,493,413	6.66	2.12	−0.28
156	592,967	4,519,015	3.16	2.13	0.25
157	558,864	4,491,374	4.03	2.13	0.68
158	562,284	4,490,167	6.13	2.14	−0.47
159	640,251	4,473,889	3.55	2.17	0.54
160	659,649	4,518,612	6.34	2.17	0.35
161	555,381.1	4,476,142	5.65	2.17	−0.25
162	632,126	4,510,975	3.2	2.2	0.48

Table A1. *Cont.*

ID	x	y	Mean	Sort	Skew
163	613,319	4,490,460	6.41	2.2	−0.39
164	626,330	4,523,199	7.01	2.2	−0.06
165	517,027.2	4,494,024	5.41	2.2	−0.27
166	518,255.9	4,487,260	3.95	2.2	0.51
167	653,480	4,510,892	1.95	2.21	0.33
168	653,369	4,521,874	5.97	2.22	0.57
169	521,592.5	4,489,595	5.52	2.23	−0.2
170	573,213	4,526,501	6.53	2.26	0.13
171	645,558	4,478,932	0.77	2.27	0.55
172	648,115	4,503,411	1.92	2.27	0.14
173	585,313	4,502,114	5.61	2.3	−0.2
174	566,119	4,521,655	5.86	2.32	0.44
175	653,055	4,497,342	1.61	2.33	0.35
176	557,240	4,488,739	6.92	2.33	−0.27
177	652,048	4,499,290	1.51	2.34	0.92
178	659,594	4,510,281	2.53	2.35	0.33
179	522,659.1	4,493,944	3.93	2.36	0.19
180	632,512	4,519,078	4.53	2.39	0.29
181	611,622	4,532,691	5.91	2.39	0.19
182	530,854.5	4,492,842	3.33	2.39	0.45
183	623,743	4,516,775	4.12	2.41	0.56
184	604,936	4,484,479	4.65	2.41	0.42
185	573,729	4,498,731	5.56	2.43	−0.24
186	591,669	4,538,582	6.54	2.43	0.17
187	511,161.2	4,500,122	5.34	2.44	−0.36
188	567,463	4,502,548	2.81	2.45	0.66
189	579,136	4,519,779	3.63	2.45	0.49
190	642,988	4,519,506	5.52	2.45	−0.26
191	622,254	4,497,850	7.11	2.45	−0.42
192	638,909	4,518,349	3.83	2.46	0.56
193	620,555	4,516,660	5.32	2.46	−0.3
194	620,429	4,525,387	1.71	2.48	−0.1
195	635,385	4,519,172	4.27	2.49	0.31
196	556,978	4,492,129	4.95	2.49	0.61
197	630,509	4,519,024	5.57	2.5	−0.34
198	591,201	4,529,171	6.93	2.5	−0.17
199	637,608	4,467,992	7.56	2.5	0.25
200	604,362	4,508,527	3.27	2.51	0.52
201	638,245	4,522,501	6.37	2.56	−0.39
202	502,374.4	4,485,617	4.79	2.56	−0.21
203	545,862.2	4,491,199	4.04	2.58	0.2

Table A1. *Cont.*

ID	x	y	Mean	Sort	Skew
204	604,921	4,510,478	3.4	2.59	0.48
205	626,339	4,523,198	6.2	2.59	−0.43
206	520,131.9	4,487,833	3.55	2.59	0.53
207	528,672.6	4,513,176	3.5	2.6	0.53
208	621,423	4,491,206	6	2.62	0.17
209	512,939.9	4,485,970	3.58	2.62	0.6
210	626,446	4,516,814	3.97	2.63	0.62
211	565,216	4,517,792	5.83	2.63	−0.21
212	641,092	4,483,531	1.31	2.64	0.48
213	547,375.6	4,492,928	5.3	2.65	0.56
214	534,941.5	4,510,079	5.39	2.66	−0.02
215	501,634.8	4,506,610	5.87	2.66	0.5
216	633,234	4,522,204	5.3	2.67	0.05
217	551,867.5	4,484,249	3.35	2.68	0.22
218	614,114	4,523,035	4.39	2.69	0.16
219	614,113	4,523,018	5.4	2.7	−0.22
220	602,250	4,508,497	3.04	2.72	0.54
221	529,344.6	4,491,880	3.65	2.72	0.5
222	656,838	4,507,446	1.68	2.73	0.28
223	621,974	4,514,811	4.57	2.73	0.42
224	597,796	4,517,072	4.11	2.75	0.51
225	488,571.2	4,503,505	3.31	2.75	0.6
226	586,635	4,519,554	3.64	2.76	0.41
227	585,064	4,513,522	3.86	2.76	0.35
228	621,909	4,483,040	5.46	2.76	0.41
229	529,860.2	4,494,108	4.95	2.76	−0.37
230	518,943.6	4,490,862	3.85	2.76	0.53
231	652,603	4,490,698	3.64	2.77	0.57
232	564,741	4,491,576	5.83	2.8	−0.51
233	567,407	4,521,513	5.73	2.81	0.14
234	567,372	4,516,785	3.19	2.87	0.51
235	585,715	4,507,979	3.87	2.87	0.53
236	565,888	4,494,824	4.21	2.87	0.62
237	630,250	4,517,573	4.47	2.87	0.65
238	578,732	4,502,659	4.21	2.88	0.22
239	603,449	4,506,664	2.8	2.9	0.44
240	640,464	4,510,512	3.75	2.9	0.55
241	646,896	4,517,422	3.35	2.91	0.39
242	659,987	4,513,777	2.95	2.92	0.73
243	594,648	4,533,994	4.02	2.92	−0.02
244	521,812	4,512,913	3.49	2.93	0.41

Table A1. *Cont.*

ID	x	y	Mean	Sort	Skew
245	604,173	4,513,613	3.63	2.94	0.51
246	612,755	4,496,621	3.75	2.94	0.52
247	613,552	4,482,908	4.03	2.98	0.33
248	647,904	4,496,312	2.21	2.99	0.49
249	611,303	4,522,959	3.24	2.99	0.44
250	619,349	4,473,746	4.3	2.99	0.2
251	614,931	4,530,428	2.22	3	0.51
252	640,821	4,497,870	3.03	3	0.44
253	559,746	4,484,134	3.18	3	0.64
254	632,495	4,521,390	7.32	3	−0.03
255	593,507	4,513,315	3.65	3.02	0.13
256	605,668	4,507,465	3.68	3.02	0.21
257	621,730	4,486,738	4.05	3.03	0.41
258	621,939	4,502,781	4.69	3.03	0.56
259	653,374	4,516,012	3.48	3.06	0.5
260	567,524	4,495,917	5.84	3.08	−0.22
261	662,166	4,510,707	3.89	3.09	0.44
262	614,143	4,521,038	2.81	3.12	0.28
263	571,767	4,518,778	3.74	3.12	0.61
264	613,663	4,514,060	5.96	3.13	−0.24
265	559,636	4,521,105	3.41	3.15	0.49
266	598,017	4,526,945	2.36	3.16	0.52
267	567,672	4,505,479	3.46	3.16	0.37
268	571,757	4,507,677	4.9	3.19	0.41
269	630,426	4,523,290	5.41	3.19	0.1
270	616,905	4,522,819	5.95	3.19	−0.39
271	578,964	4,524,866	3.74	3.21	0.7
272	647,840	4,522,994	2.53	3.22	0.56
273	587,542	4,531,902	3.57	3.22	0.59
274	578,506	4,512,987	4.43	3.22	0.06
275	604,918	4,502,365	2.38	3.23	0.42
276	656,495	4,512,529	3.59	3.24	0.37
277	607,033	4,510,416	4.22	3.24	0.09
278	640,020	4,502,944	3.67	3.29	0.42
279	638,097	4,524,193	3.2	3.3	−0.15
280	635,352	4,521,380	5.96	3.3	−0.29
281	564,879	4,516,093	3.92	3.32	0.42
282	538,150.4	4,496,083	3.42	3.32	0.35
283	593,457	4,508,072	3.69	3.33	0.51
284	632,019	4,490,614	3.26	3.34	0.24
285	566,909	4,498,842	3.6	3.34	0.44

Table A1. *Cont.*

ID	x	y	Mean	Sort	Skew
286	611,984	4,508,792	3.77	3.34	0.67
287	488,069.3	4,498,286	4.13	3.35	0.53
288	611,214	4,519,012	2.83	3.37	0.65
289	565,246	4,514,399	3.32	3.38	0.42
290	564,313	4,513,775	2.63	3.39	0.59
291	640,879	4,488,617	3.75	3.39	0.36
292	617,501	4,525,368	2.82	3.4	0.39
293	557,261	4,516,246	6.09	3.4	−0.35
294	630,490	4,523,901	2.32	3.42	0.55
295	572,626	4,514,777	5.44	3.42	−0.19
296	640,756	4,495,092	3.61	3.49	0.4
297	604,800	4,519,018	3.56	3.53	0.44
298	563,477	4,515,618	4.09	3.57	0.31
299	562,048	4,518,381	3.55	3.58	0.46
300	611,863	4,524,520	3.04	3.62	0.38
301	593,760	4,502,678	3.4	3.62	0.51
302	604,735	4,490,334	3.74	3.62	0.29
303	604,659	4,495,731	3.45	3.63	0.49
304	632,621	4,502,964	3.71	3.63	0.44
305	584,773	4,528,631	3.75	3.63	0.44
306	655,149	4,487,373	6.44	3.63	0.1
307	484,153.2	4,502,856	4.27	3.64	0.61
308	590,532	4,528,047	7.89	3.67	0.21
309	597,731	4,530,796	4.8	3.7	−0.02
310	604,871	4,530,432	3.28	3.71	0.46
311	603,649	4,509,133	2.88	3.73	0.52
312	622,366	4,526,538	4.56	3.75	0.22
313	564,689	4,511,003	4.81	3.76	0.04
314	568,067	4,513,500	3.17	3.77	0.47
315	613,013	4,502,793	4.79	3.77	0.06
316	623,499	4,525,449	2.87	3.78	0.38
317	566,690	4,510,249	3.96	3.83	0.36
318	556,672	4,519,694	4.77	3.9	−0.07
319	568,349	4,508,414	3.3	3.93	0.39
320	593,469	4,525,652	2.92	3.94	0.45
321	642,416	4,522,487	3.8	4.05	0.08
322	632,510	4,496,022	5.02	4.11	−0.25
323	611,297	4,521,007	5	5.48	0.72

References

1. Atkins, R.J. Sediment Suspension by Waves. In *Encyclopedia of Coastal Science*; Schwartz, M.L., Ed.; Springer: Dordrecht, The Netherlands, 2005; pp. 850–853.
2. Bagnold, R.J. Mechanics of marine sedimentation. *Sea* **1963**, *3*, 507–528.
3. Middleton, G.V.; Hampton, M.A. *Part I. Sediment Gravity Flows: Mechanics of Flow and Deposition*; SEPM: Tulsa, OK, USA, 1973.
4. Cieśla, M.; Gruca-Rokosz, R.; Bartoszek, L. Significance of organic matter in the process of aggregation of suspended sediments in retention reservoirs. *Sci. Total Environ.* **2022**, *815*, 152850. [CrossRef]
5. Lee, B.J.; Kim, J.; Hur, J.; Choi, I.H.; Toorman, E.A.; Fettweis, M.; Choi, J.W. Seasonal Dynamics of Organic Matter Composition and Its Effects on Suspended Sediment Flocculation in River Water. *Water Resour. Res.* **2019**, *55*, 6968–6985. [CrossRef]
6. Pettijohn, F.; Potter, P.E.; Siever, R. Transport, Deposition, and Deformation of Sand. In *Sand and Sandstone*; Springer: New York, NY, USA, 1972; pp. 327–382.
7. McLaren, P. An interpretation of trends in grain size measurements. *J. Sediment. Petrol.* **1981**, *51*, 611–624.
8. McLaren, P.; Bowles, D. The effects of sediment transport on grain-size distribution. *J. Sediment. Petrol.* **1985**, *55*, 457–470.
9. Gao, S.; Collins, M. Net sediment transport patterns inferred from grain-size trends, based upon definition of "transport vectors". *Sediment. Geol.* **1992**, *81*, 47–60. [CrossRef]
10. Zananiri, I.; Mitropoulos, D.; Zimianitis, V.; Ioakim, C.; Papadopoulos, V.; Efthimiou, G. Marine geology data accessibility in the European Framework: The I.G.M.E. participation in the GEO-SEAS project. *Bull. Geol. Soc. Greece* **2016**, *47*, 1590–1597. [CrossRef]
11. Folk, R.L. The distinction between grain size and mineral composition in sedimentary rock nomenclature. *J. Geol.* **1954**, *62*, 344–359. [CrossRef]
12. Mitropoulos, D.; Angelopoulos, I.; Perissoratis, C.; Zimianitis, E. *Surficial Sediment Map of the Bottom of the Aegean Sea in Scale 1:200,000, North Sporades Sheet*; IGME Publications: Athens, Greece, 1994.
13. Perissoratis, C.; Mitropoulos, D.; Angelopoulos, I. *Surficial Sediment Map of the Bottom of the Aegean Sea in Scale 1:200,000, Ierissos-Kavala Sheet*; IGME Publications: Athens, Greece, 1988.
14. Zananiri, I.; Vakalas, I. Geostatistical mapping of marine surficial sediment types in the Northern Aegean Sea using indicator kriging. *Geo-Mar. Lett.* **2019**, *39*, 363–376. [CrossRef]
15. Sakellariou, D.; Tsampouraki-Kraounaki, K. Plio-quaternary extension and strike-slip tectonics in the Aegean. In *Transform Plate Boundaries and Fracture Zones*; Duarte, J., Ed.; Elsevier: Amsterdam, The Netherlands, 2019; Chapter 14.
16. Giamali, C.; Koskeridou, E.; Antonarakou, A.; Ioakim, C.; Kontakiotis, G.; Karageorgis, A.P.; Roussakis, G.; Karakitsios, V. Multiproxy ecosystem response of abrupt Holocene climatic changes in the northeastern Mediterranean sedimentary archive and hydrological regime. *Quat. Res.* **2019**, *92*, 665–685. [CrossRef]
17. Panagiotopoulos, I.; Kapsimalis, V.; Ioakim, C.; Karageorgis, A.; Rousakis, G.; Morfis, I.; Hatiris, G.; Anagnostou, C.; Koukoulis, A.; Papatrechas, C.; et al. High-resolution geomorphological mapping of the shallow continental shelf west of the Kavala Bay, North Aegean. *Bull. Geol. Soc. Greece* **2016**, *50*, 448–457. [CrossRef]
18. EMODnet. EMODnet Digital Bathymetry (DTM). 2016. Available online: https://sextant.ifremer.fr/record/c7b53704-999d-4721-b1a3-04ec60c87238/ (accessed on 2 February 2018). [CrossRef]
19. Karageorgis, A.P.; Zananiri, I.; Kanellopoulos, T.D.; Ioakim, C.; Vakalas, I.; Kaberi, H.; Botsou, F.; Anagnostou, C. *Seabed Sedimentology and Elemental Geochemistry of the Aegean Sea*; Springer: Berlin/Heidelberg, Germany, 2023; pp. 1–33.
20. Karageorgis, A.P.; Anagnostou, C. Particulate matter spatial-temporal distribution and associated surface sediment properties: Thermaikos gulf and Sporades Basin, NW Aegean Sea. *Cont. Shelf Res.* **2001**, *21*, 2141–2153. [CrossRef]
21. Karageorgis, A.P.; Anagnostou, C.L.; Kaberi, H. Geochemistry and mineralogy of the NW Aegean Sea surface sediments: Implications for river runoff and anthropogenic impact. *Appl. Geochem.* **2005**, *20*, 69–88. [CrossRef]
22. Lykousis, V.; Chronis, G. Mechanisms of sediment transport and deposition: Sediment sequences and accumulation during the Holocene on the Thermaikos plateau, the continental slope, and basin (Sporadhes basin), northwestern Aegean sea, Greece. *Mar. Geol.* **1989**, *87*, 15–26. [CrossRef]
23. Perissoratis, C.; Angelopoulos, I.; Mitropoulos, D. Exploring the Offshore Area of N.E. Greece for Placer Deposits: Geologic Framework and Preliminary Results. In *Marine Minerals*; Springer: Dordrecht, The Netherlands, 1987.
24. Zananiri, I.; Vakalas, I.; Mitropoulos, D.; Zimianitis, V. *The Application of GIS Methods and Techniques for the Elaboration of Surficial Sediment Maps of the Bottom of the Aegean Sea: Ierissos–Kavala Sheet*; IGME Report; IGME Publications: Athens, Greece, 2016.
25. Perissoratis, C.; Mitropoulos, D.; Angelopoulos, I. *Surficial Sediment Map of the Bottom of the Aegean Sea in Scale 1:200,000, Thassos-Samothraki Sheet*; IGME Publications: Athens, Greece, 1986.
26. Ioakim, C.; Zananiri, I.; Zimianitis, V.; Efthymiou, G.; Vakalas, I.; Giamali, C.; Valaouris, A.; Drosopoulou, E.; Gidoni, E.; Economou, G.; et al. *Final Comprehensive Report of Marine Geology and Sedimentology Research in the Aegean Sea: Work Carried out in the Frame of YPOTHER Project*; IGME Publications: Athens, Greece, 2016; p. 339.
27. Fotiadis, A.; Zananiri, I. *Harmonized Digital Lithology and Geochronology Map of Greece, in Scale 1:1,000,000, in Compliance with OneGeology Europe*; IGME Publications: Athens, Greece, 2016; p. 21.
28. Folk, R.L. *Petrology of Sedimentary Rocks*; Hemphill Publishing Company: Sutton, UK, 1980.
29. Folk, R.L.; Ward, W.C. Brazos River bar [Texas]; a study in the significance of grain size parameters. *J. Sediment. Res.* **1957**, *27*, 3–26. [CrossRef]

30. Gao, S.; Collins, M. Analysis of grain size trends, for defining sediment transport pathways in marine environments. *J. Coast. Res.* **1994**, *10*, 70–78.
31. Balsinha, M.; Fernandes, C.; Oliveira, A.; Rodrigues, A.; Taborda, R. Sediment transport patterns on the Estremadura Spur continental shelf: Insights from grain-size trend analysis. *J. Sea Res.* **2014**, *93*, 28–32. [CrossRef]
32. Liang, J.; Liu, J.; Xu, G.; Chen, B. Grain-size characteristics and net transport patterns of surficial sediments in the Zhejiang nearshore area, East China Sea. *Oceanologia* **2020**, *62*, 12–22. [CrossRef]
33. Yamashita, S.; Naruse, H.; Nakajo, T. Reconstruction of sediment-transport pathways on a modern microtidal coast by a new grain-size trend analysis method. *Prog. Earth Planet. Sci.* **2018**, *5*, 1–18. [CrossRef]
34. Zhang, W.; Zheng, J.; Xiaomei, J.; Hoitink, A.; Van Der Vegt, M.; Zhu, Y. Surficial sediment distribution and the associated net sediment transport pattern retain--> in the Pearl River Estuary, South China. *Cont. Shelf Res.* **2013**, *61*, 41–51. [CrossRef]
35. Choi, T.-J.; Park, J.-Y. *MATLAB Version of the GSTA Model*. 2018. Available online: https://www.mathworks.com/matlabcentral/fileexchange/72698-matlab-version-of-the-gsta-model (accessed on 7 February 2024).
36. Gao, S. A FORTRAN program for grain-size trend analysis to define net sediment transport pathways. *Comput. Geosci.* **1996**, *22*, 449–452. [CrossRef]
37. Davis, J.C.; Sampson, R.J. *Statistics and Data Analysis in Geology*; Wiley: New York, NY, USA, 1986; Volume 646.
38. Kourafalou, V.H.; Barbopoulos, K. High-resolution simulations on the North Aegean Sea seasonal circulation. *Ann. Geophys.* **2003**, *21*, 251–265. [CrossRef]
39. Olson, D.B.; Kourafalou, V.H.; Johns, W.E.; Samuels, G.; Veneziani, M. Aegean Surface Circulation from a Satellite-Tracked Drifter Array. *J. Phys. Oceanogr.* **2007**, *37*, 1898–1917. [CrossRef]
40. Kontoyiannis, H.; Kourafalou, V.H.; Papadopoulos, V. The seasonal characteristics of the hydrology and circulation in the northwest Aegean Sea (eastern Mediterranean): Observations and modeling. *J. Geophys. Res. Oceans* **2003**, *108*, 3302. [CrossRef]
41. Kourafalou, V.H.; Savvidis, Y.G.; Koutitas, C.G.; Krestenitis, Y.N. Modeling studies on the processes that influence matter transfer on the Gulf of Thermaikos (NW Aegean Sea). *Cont. Shelf Res.* **2004**, *24*, 203–222. [CrossRef]
42. Poulos, S.E.; Drakopoulos, P.G.; Collins, M.B. Seasonal variability in sea surface oceanographic conditions in the Aegean Sea (eastern Mediterranean): An overview. *J. Mar. Syst.* **1997**, *13*, 225–244. [CrossRef]
43. Zervakis, V.; Georgopoulos, D.; Drakopoulos, P.G. The role of the North Aegean in triggering the recent eastern Mediterranean climate changes. *J. Geophys. Res. Oceans* **2000**, *105*, 26103–26116. [CrossRef]
44. Zodiatis, G. Advection of the Black Sea water in the north Aegean Sea. *Glob. Atmos. Ocean Syst.* **1994**, *2*, 41–60.
45. Zodiatis, G.; Balopoulos, E. Structure and characteristics of fronts in the North Aegean Sea.Boll. *Oceanol. Teor. Appl.* **1993**, *11*, 113–124.
46. Karageorgis, A. *Integration of Remote Sensing, In Situ Observations and Numerical Modelling, Applied to the North-East Atlantic*; Ecole des Mines de Paris: Paris, France, 1995; p. 30.
47. Velaoras, D.; Zervakis, V.; Theocharis, A. The Physical Characteristics and Dynamics of the Aegean Water Masses. In *The Handbook of Environmental Chemistry*; Springer: Berlin/Heidelberg, Germany, 2021.
48. Vagenas, C.; Anagnostopoulou, C.; Tolika, K. Climatic Study of the Marine Surface Wind Field over the Greek Seas with the Use of a High Resolution RCM Focusing on Extreme Winds. *Climate* **2017**, *5*, 29. [CrossRef]

MDPI AG
Grosspeteranlage 5
4052 Basel
Switzerland
Tel.: +41 61 683 77 34

Journal of Marine Science and Engineering Editorial Office
E-mail: jmse@mdpi.com
www.mdpi.com/journal/jmse

MDPI

Academic Open
Access Publishing

mdpi.com

ISBN 978-3-7258-2558-5

www.ingramcontent.com/pod-product-compliance
Lightning Source LLC
Chambersburg PA
CBHW042012260926
51850CB00020B/3178
* 9 7 8 3 7 2 5 8 2 5 5 7 8 *